To the Student

1. The Audio Tutorial Approach

You are about to embark on a new adventure in learning arithmetic, namely the audio-tutorial method. This method utilizes a book designed to correlate with audio tapes. You will read through a unit and at the same time listen to an explanation of the problems and theory. Also, throughout each unit are study exercises for you to work in order to periodically check your progress.

This technique for the teaching of arithmetic was developed, tested, and used at Fullerton Junior College, Fullerton, California. It has been received with widespread student approval since its inception. In general, students have indicated that this technique greatly improved study habits, resulting in a better understanding of mathematics with a commensurate higher grade.

In the study of mathematics, most students are accustomed to the traditional lecture-textbook method where they read a certain section in the text, go to class for a lecture on the material, then are left to try to work the exercises. The frustrations of this type of situation are many. First, most students have difficulty reading a mathematics textbook; second, once the classroom lecture is over, the chalk board is erased and the explanatory lecture is lost forever; and third, many students have great difficulty working the homework assignment five or six hours after the lecture.

The audio-tutorial method of studying mathematics is an attempt to remedy the defects of the traditional lecture-textbook method. In the audio-tutorial approach, the lecture and other explanations are put on audio tape. The chalkboard illustrations are put in the text. Therefore, each student has a permanent record of the material generally presented in the class. The advantages of this system are numerous. The student may go through the lecture at his own rate and any part of the lecture may be reviewed as often as desired by simply reversing the recorder and turning back in the text.

2. Format of the Text

There are 38 units contained in the text. Each unit has been divided into the following parts:

(a) Frames
(b) Study Exercises
(c) Review Exercises
(d) Solutions to Review Exercises
(e) Supplementary Problems
(f) Solutions to Study Exercises

(a) Frames. Each unit consists of 15 to 35 frames. Each frame has a circled "frame number" located at the lower right hand corner. The frames are to be studied as you listen to the commentary on the audio tape. Remember to watch the circled "frame numbers," as this is how to keep the tape synchronized with the text.

(b) Study Exercises. Some of the frames are devoted to study exercises. When you come to a study exercise, turn off the recorder. The exercises should be worked and your answers checked before you turn the recorder on and proceed to the next frame. If you do poorly on a study exercise, repeat that part of the unit.

(c) Review Exercises. When you come to the end of the frames, you should turn off the recorder and do the review exercises. Check your answers with the ones immediately following. The review exercises are extremely important, in that they will be a measure of your understanding of the entire unit. If you don't do well on the review exercises, go back and review the entire unit.

(d) Solutions to Review Exercises. The solutions to the review exercises immediately follow the review exercises. They are, for the most part, detailed solutions. Be sure to compare your work carefully with the details of these solutions. Many of your mistakes can be understood and corrected in this section.

(e) **Supplementary Problems.** Immediately following solutions to review exercises is a section of supplementary problems. There are no answers furnished for these problems in the text. Your instructor may wish to use these for homework problems or for quizzes. However, you are encouraged to try as many of these as you wish. Remember, in mathematics, practice is the key to success.

(f) **Solutions to Study Exercises.** At the end of each unit you will find the detailed solutions to the study exercises which were located in the frames. Before you begin each unit with the recorder, you will want to locate and mark this section for ready reference. Remember, as you proceed through the unit you will be asked to work study exercises and to check your results with this section.

3. Study Techniques

(a) The recorder may be stopped at any time to give more time for analysis of a frame.
(b) The recorder may be reversed to review a particular portion of a unit.
(c) Cumulative practice tests with answers are spaced periodically throughout the text. These practice tests cover several units and should be worked prior to each examination since they contain questions similar to those you will encounter on the examination itself.
(d) Before an examination, review all of the units involved, examine the review exercises and work the practice test.
(e) Before the final examination, you should read over each frame and rework the review exercises and practice tests.
(f) Mathematics pyramids. That is, it builds on itself. Don't get behind. In fact, it is a good idea to try and stay a unit ahead.
(g) Many students have indicated that they had tremendous success by doing a "constant review"—that is, constantly going back and re-doing earlier units.

The authors wish to express their appreciation to the students and the Mathematics Department of Fullerton Junior College. Without their cooperation, an undertaking of this magnitude would have been impossible.

January 1971

Robert G. Moon
Arthur H. Konrad
Gus Klentos
Joseph Newmyer

BASIC ARITHMETIC

ROBERT G. MOON
ARTHUR H. KONRAD
GUS KLENTOS
JOSEPH NEWMYER

Fullerton Junior College

CHARLES E. MERRILL PUBLISHING COMPANY

A Bell & Howell Company

Columbus, Ohio

International Standard Book Number: 0-675-09230-2

Library of Congress Catalog Card Number: 78-144087

9 10—76 75

PRINTED IN THE UNITED STATES OF AMERICA

Table of Contents

unit

1

Whole Numbers and the Place Value System

Objectives:

By the end of this unit you should be able to:

1. Show how place value and multiples of ten are used to name whole numbers.
2. Read any numerals representing whole numbers.
3. Write any numerals representing whole numbers.
4. Round off whole numbers to a specified degree of accuracy. (1)

Whole numbers may be used to count objects in a collection.

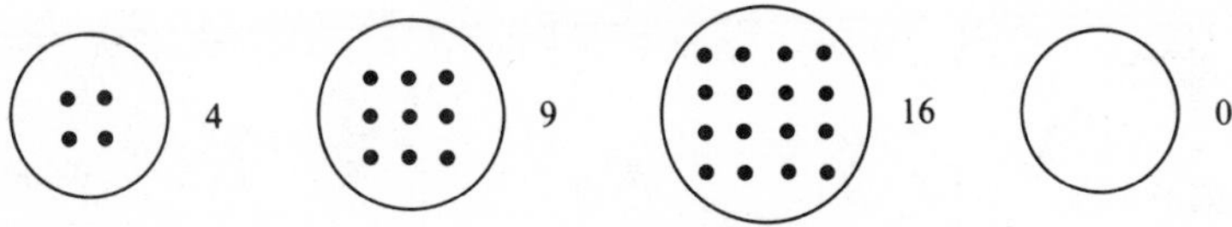

The whole numbers are represented by the following symbols called *numerals:* 0, 1, 2, 3, 4, 5, 6, 7, 8, 9, 10, 11, 12, 13, 14, 15, ... (2)

Since primitive man counted on his fingers, we have a *base ten* or *decimal* system.

The symbol for ten is *10*, and means *one group of ten plus no units.*

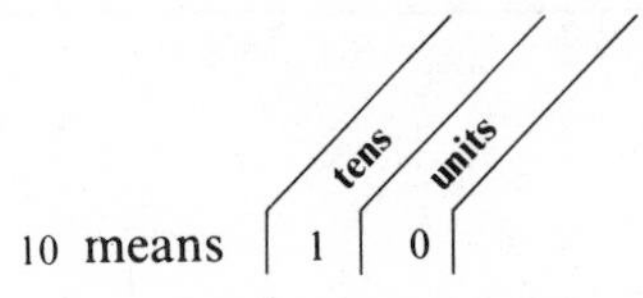

(3)

Each numeral representing a whole number may be composed of any of the following *basic digits*: 0, 1, 2, 3, 4, 5, 6, 7, 8, 9.

Examples:

1. 23 **2.** 456 **3.** 9,178 **4.** 631,205,479 (4)

Place Value

When the basic digits are placed together to form symbols for whole numbers, each digit has a *place value.* (5)

Example 1. 23
23 means *2 tens plus 3 units.*

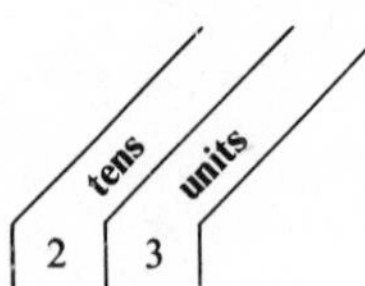

The place value of the digit 2 is *tens.*
The place value of the digit 3 is *units.*

Example 2. 456
456 means *4 hundreds plus 5 tens plus 6 units.*

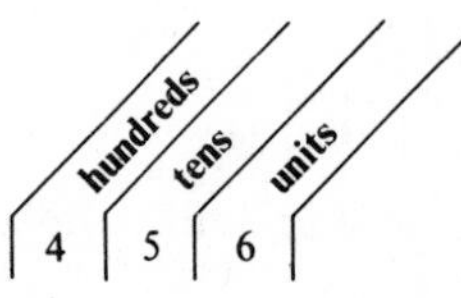

The place value of the digit 4 is *hundreds.*
The place value of the digit 5 is *tens.*
The place value of the digit 6 is *units.*

Example 3. 9178
9178 means *nine thousands plus one hundred plus seven tens plus 8 units.*

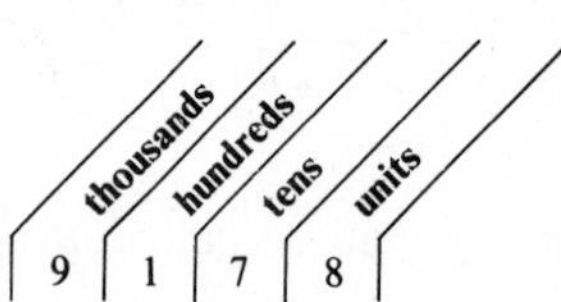

The place value of the digit 9 is *thousands.*
The place value of the digit 1 is *hundreds.*
The place value of the digit 7 is *tens.*
The place value of the digit 8 is *units.*

(8)

Example 4. 631,205,479
This numeral means:

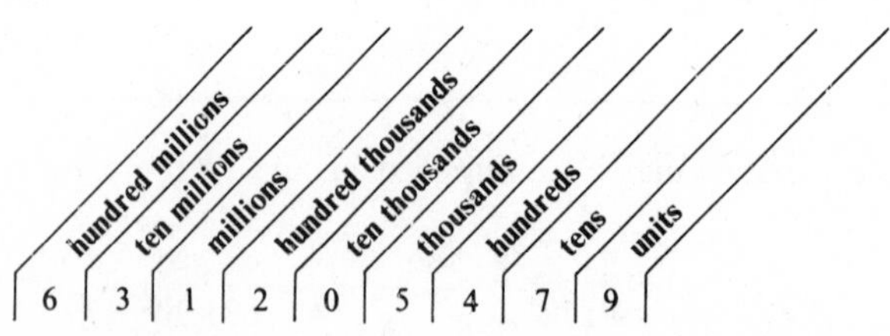

(9)

Place Value Chart

This chart should be memorized!

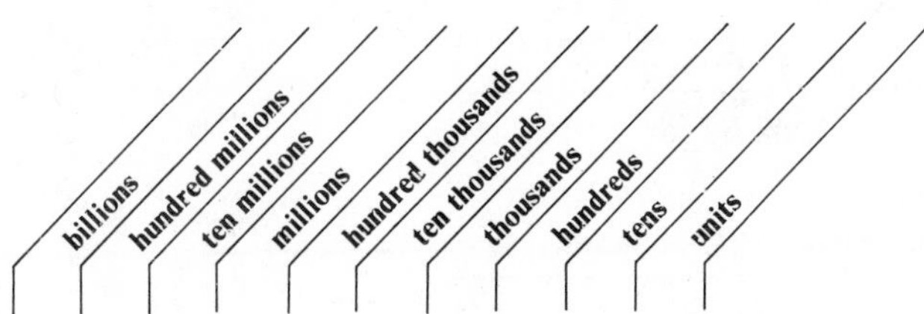

Each place is ten times the one to its immediate right.

Study Exercise One

1. 8,975 means eight_________ plus_________ hundreds plus seven_________ plus _________ units.
2. Use the previous chart, if necessary, to determine the place value of each digit in the following numeral: 5,062,791,438.
3. Write a numeral having 2 in the tens' place, 8 in the thousands' place, 3 in the units' place, 0 in the hundreds' place, 7 in the ten thousands' place, 2 in the millions' place, and 9 in the hundred thousands' place.

Reading Numerals Representing Whole Numbers

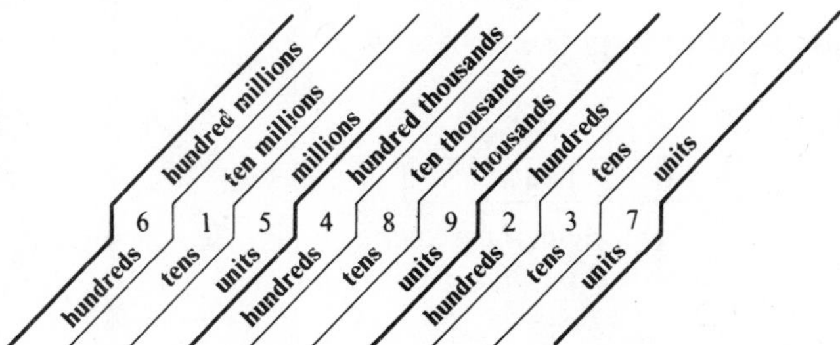

Six hundred fifteen million, four hundred eighty-nine thousand, two hundred thirty-seven.

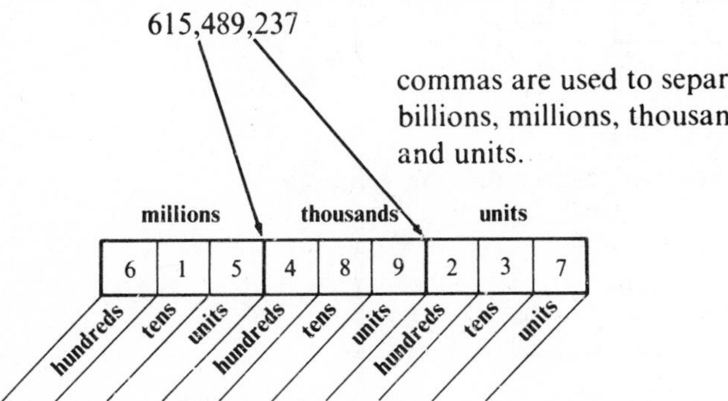

commas are used to separate billions, millions, thousands, and units.

Six hundred fifteen million, four hundred eighty-nine thousand, two hundred thirty-seven.

More Examples

1. 121,648

thousands			units		
1	2	1	6	4	8

One hundred twenty-one thousand, six hundred forty-eight.

2. 37,502

thousands			units		
	3	7	5	0	2

Thirty-seven thousand, five hundred two.

3. 5,201,310,007

billions			millions			thousands			units		
		5	2	0	1	3	1	0	0	0	7

Five billion, two hundred one million, three hundred ten thousand, seven. (14)

Study Exercise Two

Write in words the names for the following numerals.

1. 512 **2.** 1,602 **3.** 35,011

4. 615,901,002 **5.** 2,007,500,400 **6.** 5,465,932,213 (15)

Writing Numerals Representing Whole Numbers

Example 1: Two thousand, three hundred forty-five.

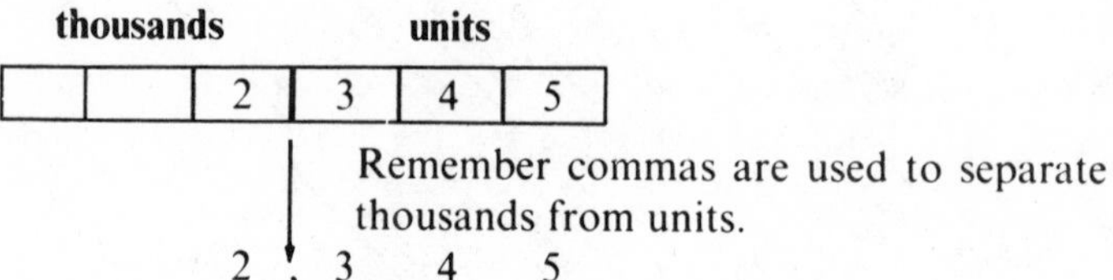

Example 2: Five hundred twenty-six thousand, one hundred sixty-three.

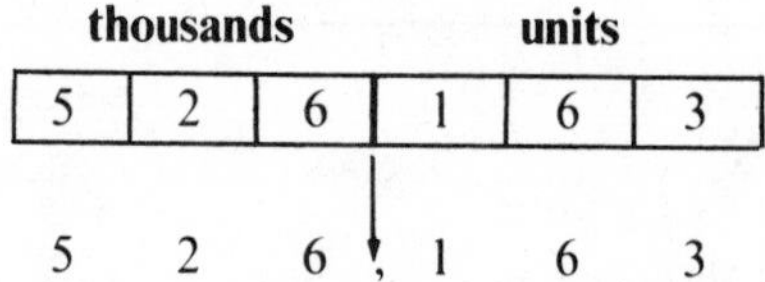

Example 3: Thirty-one million, four hundred eighty-two thousand, nine hundred twenty-eight.

millions			thousands			units		
	3	1	4	8	2	9	2	8

3 1 , 4 8 2 , 9 2 8

(Frame 16, contd.)

Example 4: Three hundred million.

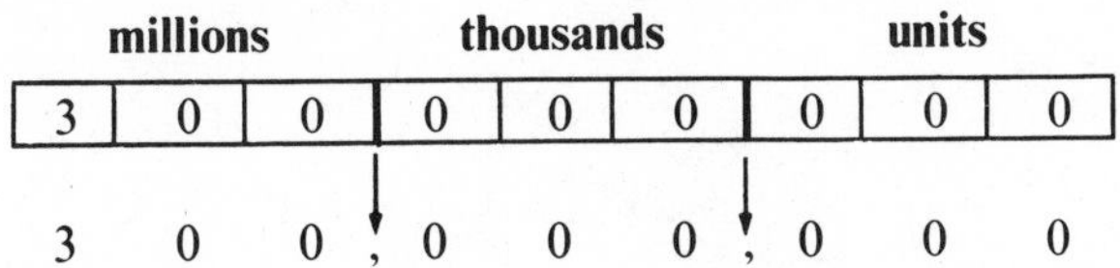

millions
thousands
units
3 0 0 0 0 0 0 0 0
3 0 0 , 0 0 0 , 0 0 0

Example 5: Nine billion, four million, six hundred five thousand, fourteen.

billions			millions			thousands			units		
		9	0	0	4	6	0	5	0	1	4

9 , 0 0 4 , 6 0 5 , 0 1 4 (16)

Study Exercise Three

A. Turn on the recorder. These exercises will be given orally.

B. Use the ten basic digits to write numerals for the following whole numbers.

8. Three hundred four.
9. Four thousand, nine hundred eleven.
10. Forty thousand, six hundred two.
11. Eight hundred fifty-five thousand, one.
12. Twenty-five million.
13. Thirty-two million, one thousand, nine hundred ten.
14. Three billion, four hundred twelve million, six hundred twenty-two thousand, three hundred ninety-five. (17)

Rounding Off Whole Numbers

When great accuracy is not required, whole numbers may be rounded off to a given place value by replacing certain digits with zeros.

Example 1: Rounding 184,628 to the nearest thousand, we have 185,000.

Example 2: Rounding 184,256 to the nearest thousand, we have 184,000.

Procedure for Rounding Off

Step One: Locate the place value to which you are rounding off and draw a box around this digit.

Step Two: Examine the digit which is to the immediate right of the box.

a) If this digit is 5, 6, 7, 8, or 9, then *increase by one the digit in the box, and replace all digits to the right with zeros.* (If the digit in the box is nine, it can be raised one by writing a zero and increasing by one the digit on its left.)

b) If this digit is 0, 1, 2, 3, or 4, then *leave the digit in the box as is and replace all digits to the right with zeros.* (18)

Examples:

1. Round off 63,295 to the nearest hundred.
   Step (1) 6 3, [2] 9 5
   Step (2) 6 3, 3 0 0
2. Round off 67,973 to the nearest hundred.
   Step (1) 6 7, [9] 7 3
   Step (2) 6 8, 0 0 0
3. Round off 63,235 to the nearest hundred.
   Step (1) 6 3, [2] 3 5
   Step (2) 6 3, 2 0 0
4. Round off 2,586,913 to the nearest one hundred thousand.
   Step (1) 2, [5] 8 6, 9 1 3
   Step (2) 2, 6 0 0, 0 0 0
5. Round off 5,214,692 to the nearest million.
   Step (1) [5] , 2 1 4, 6 9 2
   Step (2) 5, 0 0 0, 0 0 0

Study Exercise Four

A. Round off each of the following whole numbers to the indicated place value.

1. 628 to the nearest ten.
2. 628 to the nearest hundred.
3. 628 to the nearest thousand.
4. 3,864,950 to the nearest million.
5. 3,864,950 to the nearest ten thousand.
6. 3,864,950 to the nearest hundred.

B. In the following paragraph, round all whole numbers to the nearest hundred thousand.

The assets of Valley Savings and Loan are $75,672,452 for the current fiscal year. Last year the assets were $61,971,412. The increase in assets has been $13,701,040.

REVIEW EXERCISES

A. Give the place value for each digit in the following numerals.
1. 6,201 **2.** 5,906,871,234

B. Write in words the name for each of the following numerals.
3. 16 **4.** 496 **5.** 3,002
6. 46,950 **7.** 100,000 **8.** 3,000,000,000
9. 3,000,000 **10.** 9,070,021,004 **11.** 2,673,332

C. Write the numeral for each of the following whole numbers.
12. Twenty-two
13. Five hundred, two
14. Sixteen thousand, nine hundred twelve
15. Fifty thousand, four hundred
16. Twenty million
17. Two billion
18. Two hundred sixty-five million, four hundred twenty-one thousand, two hundred eighty-one

D. Round off 4,619,745 to the nearest:
19. Million **20.** Hundred thousand **21.** Thousand
22. Hundred **23.** Ten

E. Fill in the blanks.
24. 21,034 means two________ plus________ thousands plus________ hundreds plus three ________ plus ________ units. (21)

SOLUTIONS TO REVIEW EXERCISES

1.

Digit	Place Value
6	thousands
2	hundreds
0	tens
1	units

2.

Digit	Place Value
5	billions
9	hundred millions
0	ten millions
6	millions
8	hundred thousands
7	ten thousands
1	thousands
2	hundreds
3	tens
4	units

3. Sixteen
4. Four hundred ninety-six
5. Three thousand, two
6. Forty-six thousand, nine hundred fifty
7. One hundred thousand
8. Three billion
9. Three million
10. Nine billion, seventy million, twenty-one thousand, four

SOLUTIONS TO REVIEW EXERCISES, CONTD.

(Frame 22, contd.)

11. Two million, six hundred seventy-three thousand, three hundred thirty-two.

12. 22 **13.** 502 **14.** 16,912

15. 50,400 **16.** 20,000,000 **17.** 2,000,000,000

18. 265,421,281 **19.** 5,000,000 **20.** 4,600,000

21. 4,620,000 **22.** 4,619,700 **23.** 4,619,750

24. 21,034 means two ten thousands plus one thousands plus zero hundreds plus three tens plus four units. (22)

SUPPLEMENTARY PROBLEMS

A. Give the place value for each digit which is underlined:

1. $\underline{3}\,0,\,2\,\underline{0}\,8$

2. $\underline{5},\,0\,\underline{6}\,3,\,\underline{1}\,4\,\underline{9}$

B. Write, in words, names for the following numerals:

3. 56 **4.** 506 **5.** 5,006

6. 20,316 **7.** 200,316 **8.** 4,953,629

9. 95,000,000 **10.** 9,500,000,000 **11.** 2,673,498,212

C. Write numerals for the following whole numbers:

12. Three thousand, four hundred twenty-nine.

13. Thirty thousand, four hundred twenty-nine.

14. Eight hundred fifty-six million, two hundred fifteen thousand, one hundred forty-one.

15. Three billion, four million, two thousand, five.

16. Thirteen million.

17. One billion.

18. Five hundred thousand.

D. Round off the whole number, 7,936,429 to the nearest:

19. Million **20.** Ten thousand **21.** Hundred

22. Ten

E. Round off each whole number in the following paragraph to the nearest million.

23. The 1960 census showed that the United States population was 179,323,175 people. The 1970 census showed there were 204,821,693 people. This represents an increase of 25,498,518 people.

24. A corporation showed yearly receipts of $67,692,456. Its yearly expenditures were $35,449,991. This gives a gross profit of $32,242,465.

SOLUTIONS TO STUDY EXERCISES

Study Exercise One (Frame 11)

1. 8,975 means eight thousands plus nine hundreds plus seven tens plus five units.

2.

Digit	Place Value
5	billions
0	hundred millions
6	ten millions
2	millions
7	hundred thousands
9	ten thousands
1	thousands
4	hundreds
3	tens
8	units

3. 2,978,023

11A

Study Exercise Two (Frame 15)

A.
1. Five hundred twelve.
2. One thousand, six hundred two.
3. Thirty-five thousand, eleven.
4. Six hundred fifteen million, nine hundred one thousand, two.
5. Two billion, 7 million, five hundred thousand, four hundred.
6. Five billion, four hundred sixty-five million, nine hundred, thirty-two thousand, two hundred thirteen.

15A

Study Exercise Three (Frame 17)

B. **8.** 304 **9.** 4,911 **10.** 40,602

11. 855,001 **12.** 25,000,000 **13.** 32,001,910

14. 3,412,622,395

17A

Study Exercise Four (Frame 20)

A.

1. *Step (1)* 6[2]8
Step (2) 6 3 0

2. *Step (1)* [6]2 8
Step (2) 6 0 0

3. *Step (1)* [0]6 2 8
Step (2) 1 0 0 0

4. *Step (1)* [3], 8 6 4, 9 5 0
Step (2) 4 , 0 0 0, 0 0 0

5. *Step (1)* 3, 8[6]4, 9 5 0
Step (2) 3, 8 6 0, 0 0 0

6. *Step (1)* 3, 8 6 4,[9]5 0
Step (2) 3, 8 6 5, 0 0 0

B. $75,700,000; $62,000,000; $13,700,000.

20A

unit

2

Addition of Whole Numbers

Objectives:

By the end of this unit you should:

1. be able to use three methods of addition.
2. be able to identify the terms *addend* and *sum* in an addition problem.
3. know what is meant by addition being commutative and associative.
4. be able to add whole numbers using the concept of carrying.

①

Methods of Addition

1. *Combining collections of objects.*

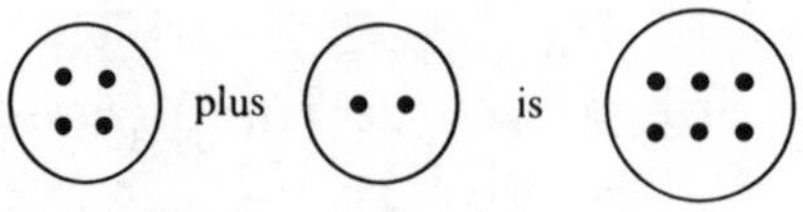

(• • • •) plus (• •) is (• • • • • •)

4 + 2 = 6

2. *Counting*

1,2,3,4 5,6

4 + 2 = 6

3. *Number Line*

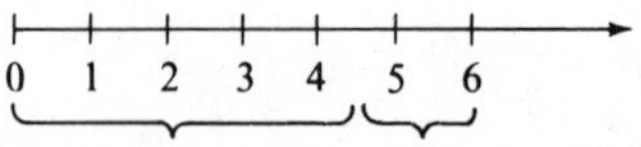

0 1 2 3 4 5 6

4 + 2 = 6

②

Horizontal Form	Vertical Form
4 + 2 = 6	4 ← addends
addends ↑ ↑ sum ↑	+2 ←
	6 ← sum

The numbers which are added are called *addends*.
The answer or total is called the *sum*.

③

Study Exercise One

1. Use the three methods of addition to show that $5 + 4 = 9$
2. Given this addition in horizontal form, $5 + 3 = 8$, which numbers are called addends? Which number is called the sum?
3. Given this addition in vertical form,

$$\begin{array}{r} 6 \\ +3 \\ \hline 9 \end{array}$$

which numbers are called addends? Which number is called the sum?

Horizontal Form	Vertical Form	
$5 + 4 = 9$	$\begin{array}{r} 5 \\ +4 \\ \hline 9 \end{array}$	$\begin{array}{r} 4 \\ +5 \\ \hline 9 \end{array}$
$4 + 5 = 9$		

The addends may be interchanged but the sum remains the same.
Addition is *commutative.*

Addition Using Three Addends

Example 1:

Horizontal Form

$$\underbrace{2 + 3} + 4$$

$$\boxed{5} + 4$$

$$9$$

Vertical Form

$$\begin{array}{r} \left.\begin{array}{r} 2 \\ 3 \end{array}\right\} \boxed{5} \\ 4 \\ \hline 9 \end{array}$$

Example 2:

Horizontal Form

$$2 + \underbrace{3 + 4}$$

$$2 + \boxed{7}$$

$$9$$

Vertical Form

$$\begin{array}{r} 2 \\ \left.\begin{array}{r} 3 \\ 4 \end{array}\right\} \boxed{7} \\ \hline 9 \end{array}$$

Addends may be added in any order and the sum remains the same.
Addition is *associative.*

Checking Addition Problems

The commutative check:

Horizontal Form	Vertical Form	
$6 + 1 = 7$	$\begin{array}{r} 6 \\ +1 \\ \hline 7 \end{array}$	$\begin{array}{r} 1 \\ +6 \\ \hline 7 \end{array}$
$1 + 6 = 7$		

7

The associative check:

Horizontal Form

$$\underbrace{1 + 4} + 2 \qquad 1 + \underbrace{4 + 2}$$

$$\boxed{5} + 2 \qquad 1 + \boxed{6}$$

$$7 \qquad 7$$

Vertical Form

$$\begin{array}{r} \left.\begin{array}{r} 1 \\ 4 \end{array}\right\} \boxed{5} \\ 2 \\ \hline 7 \end{array} \qquad \begin{array}{r} 1 \\ \left.\begin{array}{r} 4 \\ 2 \end{array}\right\} \boxed{6} \\ \hline 7 \end{array}$$

8

Study Exercise Two

Check the following sums by using the commutative or associative check. (Fill in the boxes.)

1. $3 + 2 = 5$
   $\square + 3 = 5$

2. $\begin{array}{r} 3 \\ +2 \\ \hline 5 \end{array}$ $\qquad$ $\begin{array}{r} 2 \\ +\square \\ \hline 5 \end{array}$

3. $\underbrace{1 + 3} + 2$ $\qquad$ $1 + \underbrace{3 + 2}$
   $\square + 2$ $\qquad$ $1 + \square$
   6 $\qquad$ 6

4. $\underbrace{4 + 0} + 5$ $\qquad$ $4 + \underbrace{0 + 5}$
   $\square + 5$ $\qquad$ $4 + \square$
   9 $\qquad$ 9

5. $\begin{array}{r} \left.\begin{array}{r} 1 \\ 6 \end{array}\right\} \square \\ 2 \\ \hline 9 \end{array}$ $\qquad$ $\begin{array}{r} 1 \\ \left.\begin{array}{r} 6 \\ 2 \end{array}\right\} \square \\ \hline 9 \end{array}$

6. $\begin{array}{r} \left.\begin{array}{r} 4 \\ 1 \end{array}\right\} \square \\ 3 \\ \hline 8 \end{array}$ $\qquad$ $\begin{array}{r} 4 \\ \left.\begin{array}{r} 1 \\ 3 \end{array}\right\} \square \\ \hline 8 \end{array}$

(9)

Review of Place Value Notation

Example 1: Change 38 to place value notation.

Solution:

line (a) 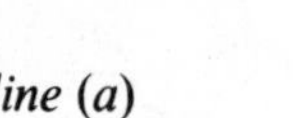

tens | units

3 | 8

line (b) 3 tens + 8 units

Example 2: Change 58,423 to place value notation.

Solution:

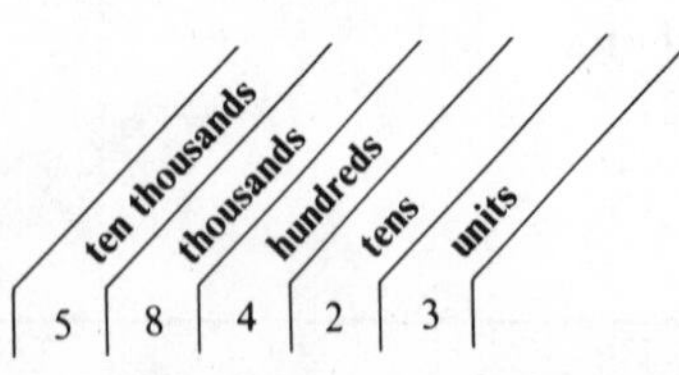

line (a)

line (b) 5 ten-thousands + 8 thousands + 4 hundreds + 2 tens + 3 units

Study Exercise Three

Change each of the following to place value notation.

1. 15 $\qquad$ 2. 129 $\qquad$ 3. 4,923 $\qquad$ 4. 23,050

(11)

Addition Using Place Value Notation

Example 1: 32 + 15

Solution: Write in vertical form, change to place value notation and add the corresponding columns.

Step (1) Write in vertical form

$$\begin{array}{r} 32 \\ +15 \\ \hline \end{array}$$

Step (2) Change to place value notation

$$\begin{array}{r} 32 \\ +15 \\ \hline \end{array} \quad \text{means} \quad \begin{array}{l} 3 \text{ tens} + 2 \text{ units} \\ 1 \text{ ten} + 5 \text{ units} \\ \hline \end{array}$$

Step (3) Add the corresponding columns

$$\begin{array}{l} 3 \text{ tens} + 2 \text{ units} \\ 1 \text{ ten} + 5 \text{ units} \\ \hline 4 \text{ tens} + 7 \text{ units} \end{array}$$

Step (4) 4 tens + 7 units is 47

Example 2: 23 + 35 + 41

Solution:

Step (1) Write in vertical form

$$\begin{array}{r} 23 \\ 35 \\ 41 \\ \hline \end{array}$$

Step (2) Change to place value notation

$$\begin{array}{r} 23 \\ 35 \\ 41 \\ \hline \end{array} \quad \text{means} \quad \begin{array}{l} 2 \text{ tens} + 3 \text{ units} \\ 3 \text{ tens} + 5 \text{ units} \\ 4 \text{ tens} + 1 \text{ unit} \\ \hline \end{array}$$

Step (3) Add the corresponding columns

$$\begin{array}{l} 2 \text{ tens} + 3 \text{ units} \\ 3 \text{ tens} + 5 \text{ units} \\ 4 \text{ tens} + 1 \text{ unit} \\ \hline 9 \text{ tens} + 9 \text{ units} \end{array}$$

Step (4) 9 tens + 9 units is 99

Example 3: 4,205 + 361 + 2,131

Solution:

Step (1) Write in vertical form

$$\begin{array}{r} 4{,}205 \\ 361 \\ 2{,}131 \\ \hline \end{array}$$

Step (2) Change to place value notation

$$\begin{array}{r} 4{,}205 \\ 361 \\ 2{,}131 \\ \hline \end{array} \quad \text{means} \quad \begin{array}{r} 4 \text{ thousands} + 2 \text{ hundreds} + 0 \text{ tens} + 5 \text{ units} \\ 3 \text{ hundreds} + 6 \text{ tens} + 1 \text{ unit} \\ 2 \text{ thousands} + 1 \text{ hundred} + 3 \text{ tens} + 1 \text{ unit} \\ \hline \end{array}$$

(Frame 14, contd.)

Step (*3*) Add the corresponding columns

4 thousands + 2 hundreds + 0 tens + 5 units
3 hundreds + 6 tens + 1 unit
2 thousands + 1 hundred + 3 tens + 1 unit
―――――――――――――――――――――
6 thousands + 6 hundreds + 9 tens + 7 units

Step (*4*) 6 thousands + 6 hundreds + 9 tens + 7 units is 6,697

(14)

A Shorter Method

Write in vertical form and add the corresponding place value columns.

Example 1: 32 + 15 (Compare with Frame 12)

Step (*1*)

```
 32
+15
———
↓↓
```

Step (*2*)

```
 32
+15
———
 47
```

Example 2: 23 + 35 + 41 (Compare with Frame 13)

Step (*1*)

```
23
35
41
——
↓↓
```

Step (*2*)

```
23
35
41
——
99
```

(16)

Example 3: 4,205 + 361 + 2,131 (Compare with Frame 14)

Step (*1*)

```
4,205
  361
2,131
—————
↓ ↓↓↓
```

Step (*2*)

```
4,205
  361
2,131
—————
6,697
```

(17)

Checking Addition Problems

Since addition is commutative and associative, the sum may be checked by adding the corresponding place value columns from bottom to top.

Example:

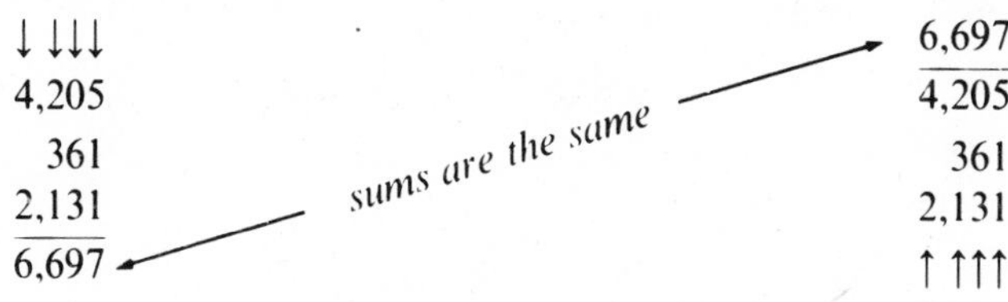

↓ ↓↓↓	6,697
4,205	4,205
361	361
2,131	2,131
6,697	↑ ↑↑↑

(18)

Study Exercise Four

Find the sum for each problem by first, converting to place value notation, and second, using the shortcut. Check answers by adding from bottom to top.

Example: 62 + 35

Solution. 6 tens + 2 units
3 tens + 5 units
9 tens + 7 units

↓↓	97
62	62
+35	+35
97	↑↑

1. 17 + 21

2. 54 + 23 + 20

3. 131 + 2,644 + 5,013

4. 53,074 + 125 + 2,700 + 31,100

(19)

Addition Involving Carrying

Example: 65 + 28

line (a) 65
28
↓

line (b) Since 5 + 8 = 13 we carry the "1" to the tens' column. Leave 3 in the units' column.

line (c)
1
65
28
3

line (d)
1
65
28
93

More Examples

Example 1: 34 + 18

Solution:
1 ← space for carrying
34
18
52

Example 2: 342 + 279

Solution:
11 ← space for carrying
342
279
621

(Frame 21, contd.)

Example 3: 56,214 + 39,412 + 15,609

```
           121 1 ←——space for carrying
Solution:  56,214
           39,412
           15,609
          -------
          111,235
```

Example 4: 29 + 6 + 1,024 + 15 + 638

```
             13 ←——space for carrying
Solution:    29
              6
          1,024
             15
            638
          -----
          1,712
```

(21)

Study Exercise Five

Find the sum for each of the following using the shortcut.
Check your answers by adding from bottom to top.

1. 78 + 16
2. 78 + 44
3. 296 + 459
4. 79,468 + 15,291 + 36,982 + 98,512
5. 12 + 6,915 + 298 + 3,001 + 79 + 907

REVIEW EXERCISES

1. Use the three methods of addition to show that $4 + 3 = 7$.
2. Given the following additions, name the addends and sum.
   (a) $29 + 48 = 77$
   (b)
$$\begin{array}{r} 1{,}203 \\ +6{,}908 \\ \hline 8{,}111 \end{array}$$
3. Addition is said to be____________ because the addends may be interchanged but the sum remains the same.
4. Addition is said to be____________ because the addends may be added in any order.
5. Change 29,058 into place value notation.
6. Add the following by changing to place value notation.
$$602 + 10{,}111 + 3{,}272$$

Perform the following additions and check your answers.

7. $9 + 8 + 2 + 1 + 0 + 5$
8. $19 + 23 + 56 + 7 + 14$
9. $2{,}619 + 586 + 4{,}012 + 89$
10.
$$\begin{array}{r} 51{,}612 \\ 92{,}915 \\ 21{,}713 \\ \hline \end{array}$$
11.
$$\begin{array}{r} 15 \\ 10{,}026 \\ 5{,}948 \\ 36{,}712 \\ 4{,}893 \\ \hline \end{array}$$
12.
$$\begin{array}{r} 2{,}698{,}512 \\ 5{,}207{,}060 \\ 318{,}945 \\ \hline \end{array}$$

SOLUTIONS TO REVIEW EXERCISES

1. (a)

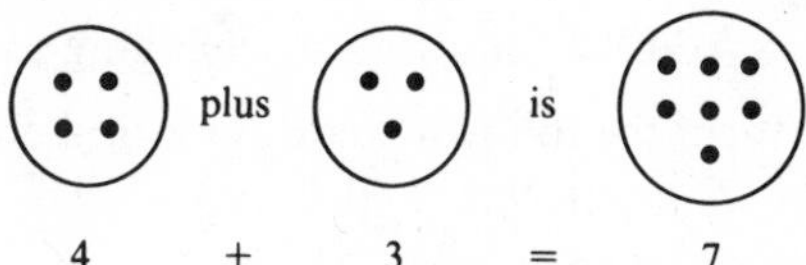

(b) 1, 2, 3, 4   5, 6, 7

4 + 3 = 7

(c)

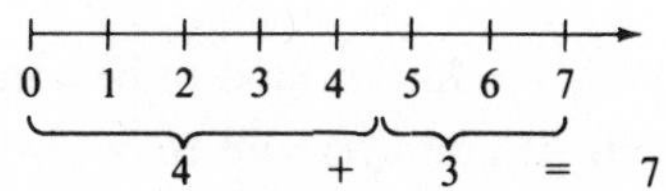

SOLUTIONS TO REVIEW EXERCISES, CONTD.

(Frame 24, contd.)

2. (a) addends are 29 and 48; sum is 77.
(b) addends are 1,203 and 6,908; sum is 8,111.

3. commutative

4. associative

5. 2 ten-thousands + 9 thousands + 0 hundreds + 5 tens + 8 units

6.

		6 hundreds +	0 tens +	2 units
1 ten-thousand +	0 thousands +	1 hundred +	1 ten +	1 unit
	3 thousands +	2 hundreds +	7 tens +	2 units
1 ten-thousand +	3 thousands +	9 hundreds +	8 tens +	5 units or 13, 985

7.
9
8
2
1
0
5
———
25

8.
2
19
23
56
7
14
———
119

9.
1 22
2,619
586
4,012
89
———
7,306

10.
2 1
51,612
92,915
21,713
———
166,240

11.
12 12
15
10,026
5,948
36,712
4,893
———
57,594

12.
1 121 1
2,698,512
5,207,060
318,945
———
8,224,517

SUPPLEMENTARY PROBLEMS

1. Use the three methods of addition to show that, 2 + 3 = 5.

2. Add the following by changing to place value notation. 2,605 + 3,171 + 212

Perform the following additions and check your answers.

3. 2 + 6 + 9 + 1 + 0 + 3

4. 25 + 65 + 183

5. 5,809 + 3,095 + 12,649

6. 53,941 + 68,011 + 92,349 + 129,814

7.
129
20,692
3,413
12,971
15,604
———

8.
1,295,642
5,468,711
2,905,829
6,002,193
———

9. Why is addition said to be commutative?

10. Why is addition said to be associative?

SOLUTIONS TO STUDY EXERCISES

Study Exercise One (Frame 4)

1. (a)

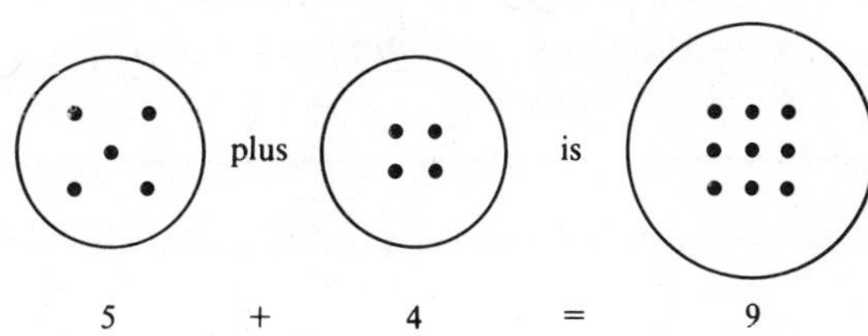

5 + 4 = 9

(b) 1, 2, 3, 4, 5   6, 7, 8, 9

(c) 5 + 4 = 9

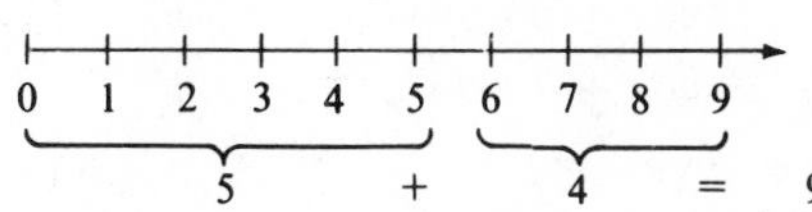

2. 5 and 3 are the addends. 8 is the sum.

3. 6 and 3 are the addends. 9 is the sum.

4A

Study Exercise Two (Frame 9)

1. 3 + 2 = 5

[2] + 3 = 5

2. $\begin{array}{r} 3 \\ +2 \\ \hline 5 \end{array}$ $\begin{array}{r} 2 \\ +\boxed{3} \\ \hline 5 \end{array}$

3. $\underbrace{1 + 3} + 2$ $1 + \underbrace{3 + 2}$

[4] + 2   1 + [5]

6   6

4. $\underbrace{4 + 0} + 5$ $4 + \underbrace{0 + 5}$

[4] + 5   4 + [5]

9   9

5. $\left.\begin{array}{r} 1 \\ 6 \end{array}\right\} \boxed{7}$, 2, sum 9   1, $\left.\begin{array}{r} 6 \\ 2 \end{array}\right\} \boxed{8}$, sum 9

6. $\left.\begin{array}{r} 4 \\ 1 \end{array}\right\} \boxed{5}$, 3, sum 8   4, $\left.\begin{array}{r} 1 \\ 3 \end{array}\right\} \boxed{4}$, sum 8

9A

SOLUTIONS TO STUDY EXERCISES, CONTD.

Study Exercise Three (Frame 11)

1. 1 ten + 5 units
2. 1 hundred + 2 tens + 9 units
3. 4 thousands + 9 hundreds + 2 tens + 3 units
4. 2 ten-thousands + 3 thousands + 0 hundreds + 5 tens + 0 units

11A

Study Exercise Four (Frame 19)

1.

		38
1 ten + 7 units	17	17
2 tens + 1 unit	+21	+21
3 tens + 8 units	38	

2.

		97
5 tens + 4 units	54	54
2 tens + 3 units	23	23
2 tens + 0 units	20	20
9 tens + 7 units	97	

3.

		7,788
1 hundred + 3 tens + 1 unit	131	131
2 thousands + 6 hundreds + 4 tens + 4 units	2,644	2,644
5 thousands + 0 hundreds + 1 ten + 3 units	5,013	5,013
7 thousands + 7 hundreds + 8 tens + 8 units	7,788	

4.

		86,999
5 ten-thousands + 3 thousands + 0 hundreds + 7 tens + 4 units	53,074	53,074
1 hundred + 2 tens + 5 units	125	125
2 thousands + 7 hundreds + 0 tens + 0 units	2,700	2,700
3 ten-thousands + 1 thousand + 1 hundred + 0 tens + 0 units	31,100	31,100
8 ten-thousands + 6 thousands + 9 hundreds + 9 tens + 9 units	86,999	

19A

Study Exercise Five (Frame 22)

1.
```
 1
 78
 16
 --
 94
```

2.
```
  1
 78
 44
---
122
```

3.
```
 11
296
459
---
755
```

4.
```
  32 21
 79,468
 15,291
 36,982
 98,512
-------
230,253
```

5.
```
 2 23
    12
 6,915
   298
 3,001
    79
   907
------
11,212
```

22A

unit

3

Subtraction of Whole Numbers

Objectives:

By the end of this unit you should be able to:

1. Identify the terms *minuend, subtrahend,* and *difference* in a subtraction problem.
2. Use addition as a check for subtraction.
3. Perform subtraction using the concept of borrowing.

(1)

Subtraction

Horizontal Form

$$\underset{\text{minuend}}{9} - \underset{\text{subtrahend}}{6} = \underset{\text{difference}}{3}$$

Vertical Form

$$\begin{array}{rl} 9 & \leftarrow \textit{minuend} \\ -6 & \leftarrow \textit{subtrahend} \\ \hline 3 & \leftarrow \textit{difference} \end{array}$$

(2)

Check for Subtraction

The difference plus the subtrahend equals the minuend.

	The Problem	*The Check*
1.	$\begin{array}{r} 8 \\ -5 \\ \hline 3 \end{array}$	$\begin{array}{r} 3 \\ +5 \\ \hline 8 \end{array}$
2.	$8 - 5 = 3$	$3 + 5 = 8$
3.	$\begin{array}{r} 7 \\ -0 \\ \hline 7 \end{array}$	$\begin{array}{r} 7 \\ +0 \\ \hline 7 \end{array}$
4.	$7 - 0 = 7$	$7 + 0 = 7$
5.	$\begin{array}{r} 4 \\ -3 \\ \hline 1 \end{array}$	$\begin{array}{r} 1 \\ +3 \\ \hline 4 \end{array}$
6.	$4 - 3 = 1$	$1 + 3 = 4$

(3)

Study Exercise One

Subtract the following and check your answers.

1. 9 − 5

2. $\begin{array}{r} 8 \\ -2 \\ \hline \end{array}$

3. 10 − 6

4. $\begin{array}{r} 8 \\ -1 \\ \hline \end{array}$

5. 3 − 0

6. $\begin{array}{r} 10 \\ -\ 1 \\ \hline \end{array}$

7. 3 − 2

8. $\begin{array}{r} 5 \\ -2 \\ \hline \end{array}$

Subtraction Using Place Value Notation

Example: 96 − 32

Solution: Write in vertical form, change to place value notation and subtract in the corresponding columns.

Step (1) Write in vertical form.

$$\begin{array}{r} 96 \\ -32 \\ \hline \end{array}$$

Step (2) Change to place value notation.

$$\begin{array}{r} 96 \\ -32 \\ \hline \end{array} \quad \text{means} \quad \begin{array}{r} 9 \text{ tens} + 6 \text{ units} \\ 3 \text{ tens} + 2 \text{ units} \\ \hline \end{array}$$

Step (3) Subtract in the corresponding columns.

$$\begin{array}{r} 9 \text{ tens} + 6 \text{ units} \\ 3 \text{ tens} + 2 \text{ units} \\ \hline 6 \text{ tens} + 4 \text{ units} \end{array}$$

Step (4) 6 tens + 4 units is 64

Step (5) Check the answer.

$$\begin{array}{r} 64 \\ +32 \\ \hline 96 \end{array}$$

Another Example: 6,765 − 523

Step (1) Write in vertical form.

$$\begin{array}{r} 6{,}765 \\ -\ \ 523 \\ \hline \end{array}$$

Step (2) Change to place value notation.

$$\begin{array}{r} 6{,}765 \\ -\ \ 523 \\ \hline \end{array} \quad \text{means} \quad \begin{array}{r} 6 \text{ thousands} + 7 \text{ hundreds} + 6 \text{ tens} + 5 \text{ units} \\ 5 \text{ hundreds} + 2 \text{ tens} + 3 \text{ units} \\ \hline \end{array}$$

Step (3) Subtract in the corresponding columns.

$$\begin{array}{r} 6 \text{ thousands} + 7 \text{ hundreds} + 6 \text{ tens} + 5 \text{ units} \\ 5 \text{ hundreds} + 2 \text{ tens} + 3 \text{ units} \\ \hline 6 \text{ thousands} + 2 \text{ hundreds} + 4 \text{ tens} + 2 \text{ units} \end{array}$$

(Frame 6, contd.)

Step (4) 6 thousands + 2 hundreds + 4 tens + 2 units is 6,242

Step (5) Check the answer.

$$\begin{array}{r} 6{,}242 \\ +\ 523 \\ \hline 6{,}765 \end{array}$$

(6)

A Shorter Method

Example: 96 − 32

Solution: Write in vertical form and subtract the digits in the corresponding place value columns.

Step (1) Write in vertical form.

$$\begin{array}{r} 96 \\ -32 \\ \hline \end{array}$$

Step (2) Subtract the digits in the corresponding place value columns.

$$\begin{array}{r} 96 \\ -32 \\ \hline 64 \end{array}$$

Step (3) Check the answer.

$$\begin{array}{r} 64 \\ +32 \\ \hline 96 \end{array}$$

(7)

Another Example: 6,765 − 523

Step (1) Write in vertical form.

$$\begin{array}{r} 6{,}765 \\ -\ 523 \\ \hline \end{array}$$

Step (2) Subtract the digits in the corresponding place value columns.

$$\begin{array}{r} 6{,}765 \\ -\ 523 \\ \hline 6{,}242 \end{array}$$

Step (3) Check the answer.

$$\begin{array}{r} 6{,}242 \\ +\ 523 \\ \hline 6{,}765 \end{array}$$

(8)

Study Exercise Two

A. Subtract the following by writing each numeral in place value notation. Check your answers by adding the difference to the subtrahend.

1. 58 − 25 **2.** 6,982 − 4,711 **3.** 28,976 − 17,005

B. Subtract the following using the shorter method. Check your answers by adding the difference to the subtrahend.

4. 74 − 21 **5.** 8,978 − 651

6. 58,976 − 31,760 **7.** 565,938 − 22,705

(9)

Subtraction Using Borrowing

Example: 84 − 46

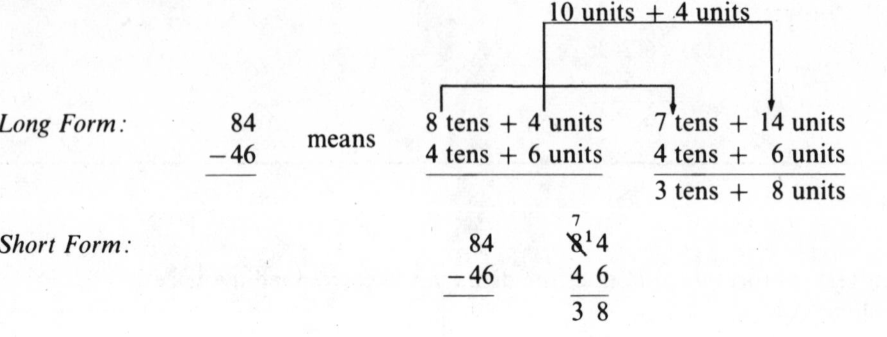

⑩

In a similar fashion we may borrow from any column since our place value system is such that each place value is 10 times the value on its right.

Example: 632 − 285

```
              2            1
                          5 2
Solution:   6 3 ¹2       6 3 ¹2
           −2 8  5      −2 8  5
           ---------    ---------
            ? ?  7       3 4  7   (answer)

             11
Check:      347
           +285
           ----
            632
```

⑪

More Examples

Example 1: 4347 − 1478

```
                                  1          1 1
               3                 2 3        3 2 3
Solution:   4 3 4 ¹7          4 3 4 ¹7     4 3 4 ¹7
           −1 4 7  8         −1 4 7  8    −1 4 7  8
           ----------        ----------   ----------
            ? ? ?  9          ? ? 6  9     2 8 6  9   (answer)

             111
Check:      2869
           +1478
           -----
            4347
```

Example 2: 900 − 265

```
             8             8 9
Solution:    9 ¹0 0        9 ¹0 ¹0
            −2  6 5       −2  6  5
            --------      ---------
             ?  ? ?        6  3  5   (answer)

             11
Check:      635
           +265
           ----
            900
```

(Frame 12, contd.)

Example 3: 26,050 − 9,768

Solution:

```
    4             5  4              5 9 4          1 5 9 4
 26,05¹0        26,¹05¹0          26,¹0 5¹0      2 6,¹0 5¹0
 −9,76 8        −9, 76 8          −9, 7 6 8      −  9, 7 6 8
 ?? ?? 2        ??  ?? 2          ??  2 8 2      1 6, 2 8 2 (answer)
```

Check:

```
 11 11
 16,282
 +9,768
 26,050
```

(12)

Study Exercise Three

Subtract the following and check your answers.

1. 517 − 279
2. 4,681 − 3,022
3. 1,684 − 255
4. 4,200 − 1,892
5. 3,000 − 1,784
6. 48,000 − 19,736
7. 40,300 − 21,856

(13)

Uses of Subtraction

Subtraction is used when we wish to find:

1. the difference between two quantities,
2. how much greater the larger quantity is than the smaller, or
3. how much is left when a quantity is taken away.

(14)

Example 1: The odometer on Harry's car registered 21,491 miles before he left on a trip. Upon his return the odometer registered 22,078 miles. How many miles did he travel on the trip?

Solution: Find the difference of 22,078 and 21,491.

```
   1               1 9
 22,¹078         22,¹0¹78
−21, 491        −21, 4 91
       7             5 87
```

The trip was 587 miles.

Example 2: Jim's annual salary is $9,750. Bob's annual salary is $12,225. How much greater is Bob's salary than Jim's?

Solution:

```
  1  1
 12, 2¹25
 −9, 7 50
  2, 4 75
```

Bob's salary is $2,475 greater than Jim's.

(Frame 15, contd.)

Example 3: Mr. Jones had 1,500 feet of wire and used 378 feet. How many feet did he have left?

Solution:

$$\begin{array}{r} \overset{4}{\cancel{5}}\,\overset{9}{\cancel{0}}\,{}^{1}0 \\ 1,\,5\;0\;0 \\ -\;3\;7\;8 \\ \hline 1,1\;2\;2 \end{array}$$

Mr. Jones had 1,122 feet of wire left. 

Study Exercise Four

1. Ted averaged 221 pins in his bowling league. George averaged 204 pins. Ted's average is how much greater than George's?
2. A contractor agrees to construct a building for \$32,075. His materials and labor cost him \$28,980. How much is his profit?
3. Dick bought a new tire guaranteed for 40,000 miles. He drove 26,750 miles and the tire was worn out. According to the guarantee, how many more miles should the tire have lasted?
4. On January 1 a salesman's odometer registered 13,642 miles. On February 1st the odometer registered 16,571 miles. How many miles did the salesman drive for the month of January? 

REVIEW EXERCISES

1. In the following subtraction problems, name the minuend, subtrahend and difference.
   (a) $9 - 7 = 2$
   (b)
   $$\begin{array}{r} 486 \\ -297 \\ \hline 189 \end{array}$$
2. Explain how addition may be used to check a subtraction problem.
3. Perform the following subtraction by writing in place value notation.
   5,691 − 2,380

Perform the following subtractions and check your answers.

4. 38 − 21
5. 4,695 − 2,374
6. 2,829 − 706
7. 71 − 58
8. 800 − 138
9. 3,865 − 2,984
10. 305,060 − 68,459
11. $\begin{array}{r} 2{,}915{,}698 \\ -879{,}409 \\ \hline \end{array}$
12. $\begin{array}{r} 2{,}000{,}000 \\ -70{,}007 \\ \hline \end{array}$

Solve the following problems by using subtraction.

13. A used car shows 38,924 miles on the odometer. The car is under warranty by the manufacturer for 50,000 miles. For how many more miles is the car under warranty?
14. Bob's annual salary in his present job is $11,250. He is offered a new job paying an annual salary of $13,825. How much of a salary increase would Bob obtain if he accepted the new job?
15. An electrician had a 600-foot roll of electrical wire. In the process of wiring a room he used 237 feet. How much wire was left on the roll?
16. Jim decided to trade his car in on a later model costing $1,395. The dealer allowed Jim a $500 "trade-in" allowance. Exclusive of tax and license, how much extra would Jim have to pay for the later model?

SOLUTIONS TO REVIEW EXERCISES

1. (a) minuend 9; subtrahend 7; difference 2.
   (b) minuend 486; subtrahend 297; difference 189.
2. The difference plus the subtrahend must equal the minuend.
3. 5 thousands + 6 hundreds + 9 tens + 1 unit
   2 thousands + 3 hundreds + 8 tens + 0 units
   ―――――――――――――――――――――――――――――
   3 thousands + 3 hundreds + 1 ten + 1 unit or 3,311
4. 17
5. 2,321
6. 2,123
7. 13
8. 662
9. 881
10. 236,601
11. 2,036,289
12. 1,929,993
13. 11,076 miles left under warranty.
14. $2,575 increase in annual salary.
15. 363 feet left on the roll
16. $895 extra to trade cars.

18

SUPPLEMENTARY PROBLEMS

1. Study the following problems:

$$\begin{array}{r} 834 \\ +\ 58 \\ \hline 892 \end{array} \qquad \begin{array}{r} 562 \\ -\ 86 \\ \hline 476 \end{array}$$

Which of the above numbers is:
(a) a sum
(b) an addend
(c) a difference
(d) a subtrahend
(e) a minuend

2. Perform the following subtraction by writing each number in place value notation. 5,963 − 402

Perform the following additions and check your answers.

3. 529 + 238
4. 678 + 2,947
5. 1,008 + 3,092
6. 461 + 38 + 692
7. 83,394 + 705 + 192,647 + 315
8. 43,081 + 2,009 + 346 + 35,976

Perform the following subtractions and check your answers.

9. 5,869 − 3,143
10. 7,639 − 3,528
11. 12,484 − 6,597
12. 60,502 − 28,767
13. 7,000 − 2,513
14. 73,008 − 49,679
15. 3,205,612 − 1,951,946

Solve the following problems using addition and/or subtraction.

16. Disneyland had the following attendance for each day of the week: Sunday, 31,362; Monday, 28,795; Tuesday, 23,647; Wednesday, 26,004; Thursday, 19,793; Friday, 20,782; Saturday, 36,271.
What was the total attendance for the week?
17. Before Bill left on a trip the odometer of his automobile read 48,652 miles. At the end of the trip the odometer reading was 50,028 miles. How far did Bill travel on his trip?
18. Subtract 695 from the sum of 219 and 1,142.
19. Fred had $110 and then he spent $25 for theatre tickets, $7 for a corsage and $6 for gasoline. How much money does he have left?
20. A motion picture theatre has 500 seats. 386 tickets were sold for a performance. How many seats were vacant?
21. A contractor agreed to build a house for $37,225. His costs were as follows: labor, $18,120; materials, $16,495; insurance, $115. How much was his profit?

SOLUTIONS TO STUDY EXERCISES

Study Exercise One (Frame 4)

	Answer	Check		Answer	Check
1.	4	$4 + 5 = 9$	**2.**	6	$6 + 2 = 8$
3.	4	$4 + 6 = 10$	**4.**	7	$7 + 1 = 8$
5.	3	$3 + 0 = 3$	**6.**	9	$9 + 1 = 10$
7.	1	$1 + 2 = 3$	**8.**	3	$3 + 2 = 5$

4A

Study Exercise Two (Frame 9)

A. **1.**

5 tens + 8 units	33
2 tens + 5 units	+25
3 tens + 3 units	58

2.

6 thousands + 9 hundreds + 8 tens + 2 units	2,271
4 thousands + 7 hundreds + 1 ten + 1 unit	+4,711
2 thousands + 2 hundreds + 7 tens + 1 unit	6,982

3.

2 ten-thousands + 8 thousands + 9 hundreds + 7 tens + 6 units	11,971
1 ten-thousand + 7 thousands + 0 hundreds + 0 tens + 5 units	17,005
1 ten-thousand + 1 thousand + 9 hundreds + 7 tens + 1 unit	28,976

B.

	Problem	Check		Problem	Check
4.	74	53	**5.**	8,978	8,327
	−21	+21		− 651	+ 651
	53	74		8,327	8,978
6.	58,976	27,216	**7.**	565,938	543,233
	−31,760	+31,760		− 22,705	+ 22,705
	27,216	58,976		543,233	565,938

Study Exercise Three (Frame 13)

	Problem	Check		Problem	Check
1.	5̸ 1̸ ¹7 (borrowed: 4 ¹0)	238	**2.**	4,6 8̸ ¹1 (borrowed: 7)	1,659
	−2 7 9	+279		−3,02 2	+3,022
	2 3 8	517		1,65 9	4,681

SOLUTIONS TO STUDY EXERCISES, CONTD.

Study Exercise Three (Frame 13, contd.)

	Problem	Check		Problem	Check
3.	7 1,68¹4 − 25 5 1,42 9	1 1,429 + 255 1,684	4.	1 3 19 4, 20¹0 −1, 89 2 2, 30 8	1 11 2,308 +1,892 4,200
5.	2 9 9 3,¹0¹0¹0 −1, 7 8 4 1, 2 1 6	1 11 1,216 +1,784 3,000	6.	1 3 7 99 4 8,00¹0 −1 9,73 6 2 8,26 4	11 11 28,264 +19,736 48,000
7.	1 39 29 40, 30¹0 −21, 85 6 18, 44 4	11 11 18,444 +21,856 40,300			

13A

Study Exercise Four (Frame 16)

1. 1
22¹1
−20 4
1 7 — Ted's average is 17 pins greater than George's.

2. 1
2 1 9
3 2,0¹75
−2 8,9 80
3,0 95 — The contractor's profit is $3,095.

3. 39 9
40,0¹00
−26,7 50
13,2 50 — The tire should have lasted 13,250 miles longer.

4. 5 6
16,¹57¹1
−13, 64 2
2, 92 9 — The salesman drove 2,929 miles for the month of January.

16A

unit

4

Multiplication of Whole Numbers

Objectives:

By the end of this unit you should be able to:

1. tell how multiplication is related to addition.
2. identify the terms *multiplicand*, *multiplier*, *factors* and *product* in a multiplication problem.
3. multiply by multiples of ten.
4. use a tabular form to show how place value is used in multiplication.
5. multiply two numbers using the concept of carrying.

①

Multiplication means *repeated addition* and is indicated by the multiplication sign, ×.

Line (*a*) 2×4 means $2 + 2 + 2 + 2$ or 8.

Line (*b*) 2×4 means $4 + 4$ or 8.

②

More Examples

1. 3×2 means $3 + 3$ or 6.
   3×2 means $2 + 2 + 2$ or 6.
2. 4×3 means $4 + 4 + 4$ or 12.
   4×3 means $3 + 3 + 3 + 3$ or 12.
3. 5×2 means $5 + 5$ or 10.
   5×2 means $2 + 2 + 2 + 2 + 2$ or 10.

③

Study Exercise One

Find the value of each multiplication two ways, using addition.

1. 2×3 **2.** 3×5 **3.** 2×6 **4.** 4×5 **5.** 6×3

④

Horizontal Form

$3 \times 2 = 6$

3: *multiplicand*; 2: *multiplier*; 6: *product*

Vertical Form

$$\begin{array}{r} 3 \leftarrow \textit{multiplicand} \\ \times 2 \leftarrow \textit{multiplier} \\ \hline 6 \leftarrow \textit{product} \end{array}$$

⑤

The multiplicand and multiplier are also said to be *factors*.

Horizontal Form	**Vertical Form**
3 × 2 = 6 (3 and 2: *factors*; 6: *product*)	$\begin{array}{r} 3 \\ \times 2 \\ \hline 6 \end{array}$ (3 and 2: *factors*; 6: *product*)

⑥

Multiplication is *commutative* because the factors may be interchanged but the product remains the same.

Examples:

1. 3 × 2 = 6   2 × 3 = 6

2. $\begin{array}{r} 4 \\ \times 5 \\ \hline 20 \end{array}$ $\begin{array}{r} 5 \\ \times 4 \\ \hline 20 \end{array}$

⑦

Study Exercise Two

A. In the following multiplication problems find the multiplicand, multiplier and product.

1. 6 × 2 = 12

2. $\begin{array}{r} 3 \\ \times 6 \\ \hline 18 \end{array}$

3. 5 × 3 = 15

4. $\begin{array}{r} 6 \\ \times 2 \\ \hline 12 \end{array}$

B. In the following multiplication problems find the factors and product.

5. 5 × 2 = 10

6. $\begin{array}{r} 3 \\ \times 7 \\ \hline 21 \end{array}$

7. 8 × 2 = 16

8. $\begin{array}{r} 9 \\ \times 3 \\ \hline 27 \end{array}$

C. Rewrite each of the following multiplication problems to show that multiplication is commutative.

9. 4 × 6 = 24

10. $\begin{array}{r} 6 \\ \times 3 \\ \hline 18 \end{array}$

11. 2 × 9 = 18

12. $\begin{array}{r} 3 \\ \times 7 \\ \hline 21 \end{array}$

⑧

Multiplication by Zero

1. 0 × 3 or 3 × 0 means 0 + 0 + 0 or 0.

2. 2 × 0 or 0 × 2 means 0 + 0 or 0.

3. 8 × 0 = 0.

4. 0 × 8 = 0.

If any number is multiplied by 0, the product is 0.

⑨

Multiplication by One

1. 2 × 1 or 1 × 2 means 1 + 1 or 2.
2. 1 × 4 or 4 × 1 means 1 + 1 + 1 + 1 or 4.
3. 7 × 1 = 7
4. 1 × 7 = 7

If any number is multiplied by 1, the product is the original number.

(10)

Study Exercise Three

Find the product for each of the following.

1. 6 × 0	**2.** 0 × 6	**3.** 8 × 1
4. 1 × 8	**5.** 32 × 1	**6.** 32 × 0
7. 5 × 0	**8.** 1 × 2	**9.** 12 × 0
10. 169 × 1	**11.** 169 × 0	**12.** 1 × 1
13. 0 × 0		

(11)

Multiplication by Ten

1. 2 × 10 means 10 + 10 or 20.
2. 3 × 10 means 10 + 10 + 10 or 30.
3. 6 × 10 = 60.
4. 9 × 10 = 90.
5. 10 × 10 = 100.
6. 100 × 10 = 1,000.

(12)

When multiplying by 10 the product may be obtained by attaching a zero to the right of the other factor.

Examples:

1. 6 × 10 = 60

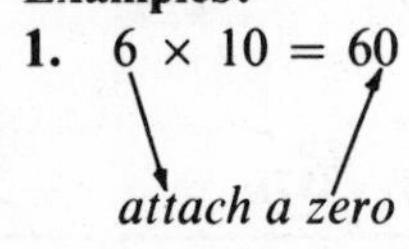

2. 10 × 10 = 100

3. 132 → *attach a zero*
× 10
1,320

(13)

Multiplication by One Hundred

1. 3 × 100 means 100 + 100 + 100 or 300.
2. 2 × 100 means 100 + 100 or 200.
3. 10 × 100 = 1,000.
4. 100 × 100 = 10,000.
5. 1,000 × 100 = 100,000.

(14)

When multiplying by 100, the product may be obtained by attaching *two zeros* to the right of the other factor.

Examples:

1. 7 × 100 = 700

attach two zeros

2. 1,000 × 100 = 100,000

attach two zeros

3.

```
  1,026 ← attach two zeros
  × 100
102,600
```

⑮

Study Exercise Four

A. Multiply each of the following by 10.

1. 2	**2.** 5	**3.** 10
4. 10,000	**5.** 635	**6.** 630
7. 1,295	**8.** 100,000	

B. Multiply each of the following by 100.

9. 6	**10.** 9	**11.** 100
12. 1,000	**13.** 763	**14.** 760
15. 2,642	**16.** 100,000	

⑯

Multiples of Ten

1	unit
10	ten
100	one hundred
1,000	one thousand
10,000	ten thousand
100,000	one hundred thousand
1,000,000	one million

⑰

When multiplying by a multiple of ten attach the number of zeros that follow the one.

Examples:

1. 63 × 1,000 = 63,000

attach three zeros

2.

```
  4,397      attach two zeros
  × 100
439,700
```

(Frame 18, contd.)

3. 925 × 10,000 = 9,250,000

attach four zeros

4. 100 × 100,000 = 10,000,000

attach five zeros

5. 93 × 1 = 93

attach no zeros

(18)

Study Exercise Five

A. Multiply each of the following by 1,000.

1. 8 **2.** 230 **3.** 10 **4.** 100

B. Multiply each of the following by 100,000.

5. 6 **6.** 50 **7.** 10 **8.** 1,000

C. How many zeros would be attached to the other factor if you multiplied by:

9. one hundred **10.** ten **11.** one unit
12. ten thousand **13.** one thousand **14.** ten million

(19)

Multiplication Involving Place Values

1.	units × units = units	1 × 1 = 1
2.	units × tens = tens	1 × 10 = 10
3.	units × hundreds = hundreds	1 × 100 = 100
4.	tens × units = tens	10 × 1 = 10
5.	tens × tens = hundreds	10 × 10 = 100
6.	tens × hundreds = thousands	10 × 100 = 1,000
7.	hundreds × units = hundreds	100 × 1 = 100
8.	hundreds × tens = thousands	100 × 10 = 1,000
9.	hundreds × hundreds = ten-thousands	100 × 100 = 10,000

(20)

Multiplication of 21 × 4 Long Form

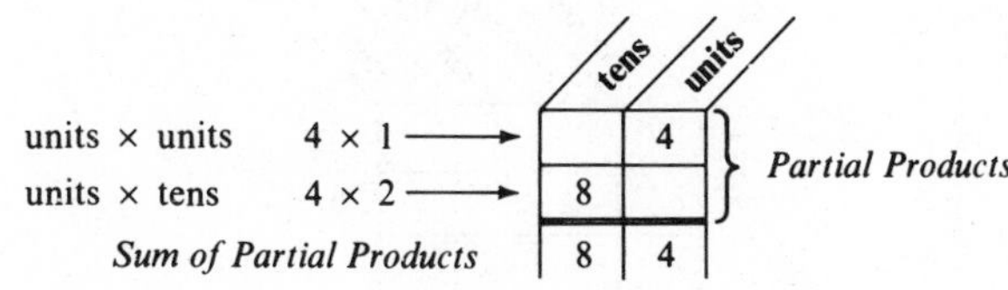

(21)

Multiplication of 21 × 4 Short Form

$$\begin{array}{r} 21 \\ \times 4 \\ \hline 84 \end{array}$$

(22)

Multiplication of 102 × 34 Long Form

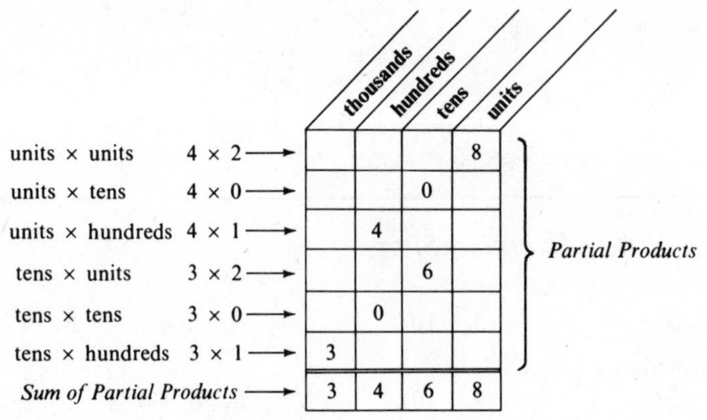

thousands
hundreds
tens
units
units × units 4 × 2
units × tens 4 × 0
units × hundreds 4 × 1
tens × units 3 × 2
tens × tens 3 × 0
tens × hundreds 3 × 1
Sum of Partial Products
8
0
4
6
0
3
3 4 6 8
Partial Products

(23)

Multiplication of 102 × 34 Short Form

	102	
	× 34	
Line (a) ⟶	408	*Partial Products*
Line (b) ⟶	306	
Line (c) ⟶	3,468	*Sum of Partial Products*

Each partial product begins in the same place value position as the multiplier digit. (24)

Study Exercise Six

A. Fill in the following table for the long form multiplication of 121 × 32.

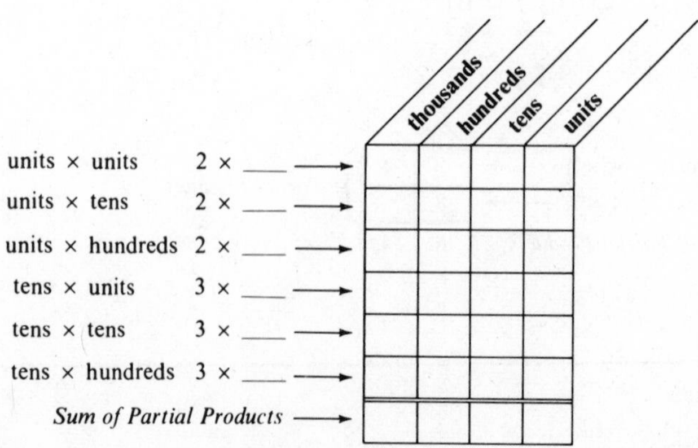

(Frame 25, contd.)

B. Multiply the following using the short form.

1.
```
 121
×32
```
2.
```
 112
×22
```
3.
```
 201
×42
```

4. 311 × 21 **5.** 412 × 12 **6.** 1,012 × 123

Difficulties with Zero in Multiplication

Example 1:
```
   1,231        1,231
   × 102        × 102   ┐
   -----        -----   │   insert one zero
    2462         2462   │   and start next
   0000         12310 ←─┘   partial product.
  1231         -------
 -------       125,562
 125,562
```

Example 2:
```
    12,131         12,131
   × 2,001        × 2,001   ┐
   -------        -------   │   insert two zeros
    12 131         12 131   │   and start next
   000 00       24 262 00 ←─┘   partial product.
  0 000 0       ----------
 24 262         24,274,131
 ----------
 24,274,131
```

Example 3:
```
     23,112          23,112
   × 3,000         × 3,000 ─┐   insert three
   -------       ----------  │   zeros.
    00 000       69,336,000 ←┘
   000 00
  0 000 0
 69 336
 ----------
 69,336,000
```

26

Study Exercise Seven

Multiply each of the following by the short form in two ways: (a) keep all zeros in partial products; (b) delete as many zeros for partial products as possible.

1.
```
 2,131
 × 201
```
2.
```
  2,131
× 2,001
```

3. 21,311 × 1,020 **4.** 12,112 × 4,000

Multiplication in Short Form with Carrying

Example 1: 457 × 8
```
  4 5 ←──── carrying space for units' multiplier.
  457
   ×8
 -----
 3,656
```

(Frame 28, contd.)

Example 2: 457 × 28

```
 11 ← carrying space for tens' digit.
 45 ← carrying space for units' digit.
 457
× 28
-----
 3 656
 9 14
-----
12,796
```

(28)

More Examples

1. 3,529 × 56

```
   2 14
   3 15
  3,529
   × 56
 ------
 21 174
176 45
 ------
197,624
```

2. 826 × 294

```
      1
     25
     12
    826
  × 294
 ------
  3 304
 74 34
165 2
 ------
242,844
```

3. 50,695 × 3,050

```
     2 21
    13 42
    50,695
   × 3,050
 ----------
 2 534 750
152 085 0
 ----------
154,619,750
```

(29)

Checking Answers to Multiplication Problems

Multiplication is commutative. Therefore, the factors may be interchanged but the product must remain the same.

Example: 492 × 67

Problem	*Check*
	2
5 1	6
6 1	1
492	67
× 67	× 492
3 444	134
29 52	6 03
32,964	26 8
	32,964

Study Exercise Eight

Multiply the following and check your answers by interchanging the factors.

1. 23 × 7 **2.** 296 × 78 **3.** 304 × 203

4. 2,396 × 259 **5.** 50,406 × 2,080

Written Problems Using Multiplication

Example 1: A company employs 237 people at an average salary of \$938 a month. How much is the company's total monthly payroll?

Solution: Multiply 938 by 237.

```
    1
   12
   25
   938
 × 237
 6 566
 28 14
187 6
222,306
```

The total monthly payroll is \$222,306.

Example 2: On the average each person will use 60 gallons of water per day. How many gallons of water will a city use per day if its population is 600,000?

Solution: Multiply 600,000 by 60.

```
  600,000
     × 60
36,000,000
```

The city will use 36,000,000 gallons of water per day.

Study Exercise Nine

Solve the following written problems.

1. Ed's salary is $1,246 per month. How much is his salary for one year?
2. A hydroelectric plant can generate 675,000 kilowatts per day. How many kilowatts of electrical energy can it generate in a year? 

REVIEW EXERCISES

A. Find the value of each multiplication two ways, using addition.

1. 3 × 4 **2.** 2 × 5

B. Fill in the blanks.

3. In the multiplication problem 5 × 8 = 40, 5 is called the ___________, 8 is called the ___________, and 40 is called the ___________. 5 and 8 may also be called ___________.

4. Multiplication is said to be ___________ because the factors may be interchanged but the product remains the same.

C. Find the product for each of the following.

5. 56 × 0 **6.** 56 × 1 **7.** 75 × 10
8. 75 × 100 **9.** 75 × 1,000 **10.** 75 × 10,000
11. 0 × 1 **12.** 3,958 × 1,000 **13.** 1,000 × 1,000

14. Fill in the table for the long form multiplication of 213 × 23.

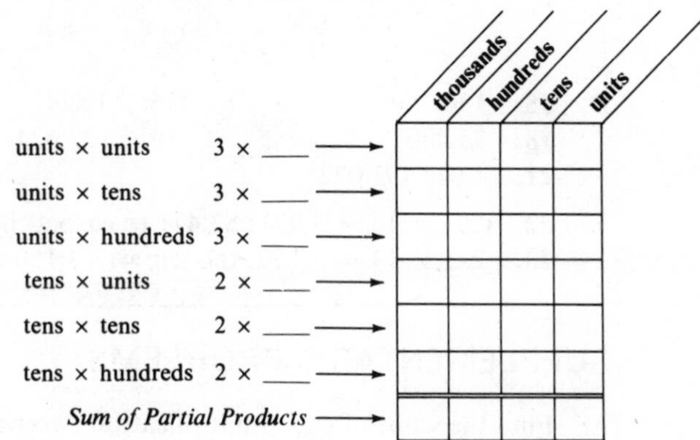

D. Multiply the following using the short form and check your answers.

15. 52 × 3 **16.** 201 × 24 **17.** 318 × 30
18. 318 × 300 **19.** 2,968 × 43 **20.** 50,208 × 2,005
21. 318,659 × 3,258

E. Solve the following written problems.

22. If a certain carpet sells for $13 a square yard installed, how much will it cost, less tax, to carpet a house containing 267 square yards?

23. If an automobile averages 14 miles per gallon of gas, how many miles were driven on a trip, if 225 gallons of gas were used? (34)

SOLUTIONS TO REVIEW EXERCISES

A. **1.** 3 + 3 + 3 + 3 = 12 or 4 + 4 + 4 = 12.
2. 2 + 2 + 2 + 2 + 2 = 10 or 5 + 5 = 10.

B. **3.** multiplicand, multiplier, product. factors.
4. commutative.

C. **5.** 0 **6.** 56 **7.** 750
8. 7,500 **9.** 75,000 **10.** 750,000
11. 0 **12.** 3,958,000 **13.** 1,000,000

SOLUTIONS TO REVIEW EXERCISES, CONTD.

(Frame 35, contd.)

14.

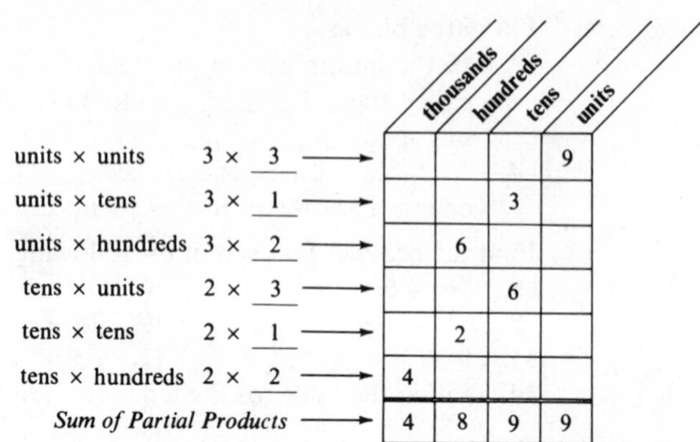

thousands
hundreds
tens
units
units × units 3 × 3
units × tens 3 × 1
units × hundreds 3 × 2
tens × units 2 × 3
tens × tens 2 × 1
tens × hundreds 2 × 2
Sum of Partial Products
9
3
6
6
2
4
4 8 9 9

D. **15.** 156 **16.** 4,824 **17.** 9,540
18. 95,400 **19.** 127,624 **20.** 100,667,040
21. 1,038,191,022

E. **22.** 267 × 13 = 3,471; $3,471 to carpet the house.
23. 225 × 14 = 3,150; the trip was 3,150 miles. 35

SUPPLEMENTARY PROBLEMS

A. Find the value of each multiplication two ways, using addition.
1. 5 × 6 **2.** 4 × 2

B. Fill in the blanks.
3. In the multiplication problem 6 × 9 = 54, ______ is called the multiplicand, ______ is called the multiplier and ______ is called the product. ______ and ______ are also called factors.
4. Repeated addition is called __________.
5. 4 × 8 = 32 and 8 × 4 = 32 shows that multiplication is __________.

C. Find the product for each of the following.
6. 5 × 1 **7.** 0 × 5 **8.** 1 × 5
9. 5 × 0 **10.** 2,695 × 0 **11.** 2,695 × 1
12. 2,695 × 10 **13.** 2,695 × 100 **14.** 4,695 × 10,000
15. 100 × 10 **16.** 10 × 10 **17.** 10,000 × 10,000
18. Using frame 23 as a guide, make a table showing the long form multiplication of 2,113 × 122.

D. Multiply the following using the short form and check your answers.
19. 32 × 9 **20.** 3,076 × 38 **21.** 2,529 × 205
22. 40,105 × 3,050 **23.** 695 × 400 **24.** 23,960 × 8,671

E. Solve the following written problems.
25. Jim earns $563 per month. How much will he earn in one year?
26. If building costs run $23 a square foot, how much would it cost to build a home containing 2,250 square feet?

SOLUTIONS TO STUDY EXERCISES

Study Exercise One (Frame 4)

1. $2 + 2 + 2 = 6$ or $3 + 3 = 6$.
2. $3 + 3 + 3 + 3 + 3 = 15$ or $5 + 5 + 5 = 15$.
3. $2 + 2 + 2 + 2 + 2 + 2 = 12$ or $6 + 6 = 12$.
4. $4 + 4 + 4 + 4 + 4 = 20$ or $5 + 5 + 5 + 5 = 20$.
5. $6 + 6 + 6 = 18$ or $3 + 3 + 3 + 3 + 3 + 3 = 18$.

4A

Study Exercise Two (Frame 8)

A.

	Multiplicand	*Multiplier*	*Product*
1.	6	2	12
2.	3	6	18
3.	5	3	15
4.	6	2	12

B.

	Factors	*Product*
5.	5 and 2	10
6.	3 and 7	21
7.	8 and 2	16
8.	9 and 3	27

C. **9.** $6 \times 4 = 24$

10. $\begin{array}{r} 3 \\ \times 6 \\ \hline 18 \end{array}$

11. $9 \times 2 = 18$

12. $\begin{array}{r} 7 \\ \times 3 \\ \hline 21 \end{array}$

8A

Study Exercise Three (Frame 11)

1. 0 **2.** 0 **3.** 8
4. 8 **5.** 32 **6.** 0
7. 0 **8.** 2 **9.** 0
10. 169 **11.** 0 **12.** 1
13. 0

11A

Study Exercise Four (Frame 16)

A. **1.** $2 \times 10 = 20$ **2.** $5 \times 10 = 50$
3. $10 \times 10 = 100$ **4.** $10{,}000 \times 10 = 100{,}000$
5. $635 \times 10 = 6{,}350$ **6.** $630 \times 10 = 6{,}300$
7. $1{,}295 \times 10 = 12{,}950$ **8.** $100{,}000 \times 10 = 1{,}000{,}000$

B. **9.** $6 \times 100 = 600$ **10.** $9 \times 100 = 900$
11. $100 \times 100 = 10{,}000$ **12.** $1{,}000 \times 100 = 100{,}000$
13. $763 \times 100 = 76{,}300$ **14.** $760 \times 100 = 76{,}000$
15. $2{,}642 \times 100 = 264{,}200$ **16.** $100{,}000 \times 100 = 10{,}000{,}000$

SOLUTIONS TO STUDY EXERCISES, CONTD.

Study Exercise Five (Frame 19)

A. **1.** 8 × 1,000 = 8,000 **2.** 230 × 1,000 = 230,000
3. 10 × 1,000 = 10,000 **4.** 100 × 1,000 = 100,000

B. **5.** 6 × 100,000 = 600,000 **6.** 50 × 100,000 = 5,000,000
7. 10 × 100,000 = 1,000,000 **8.** 1,000 × 100,000 = 100,000,000

C. **9.** two **10.** one
11. none **12.** four
13. three **14.** seven

19A

Study Exercise Six (Frame 25)

A.

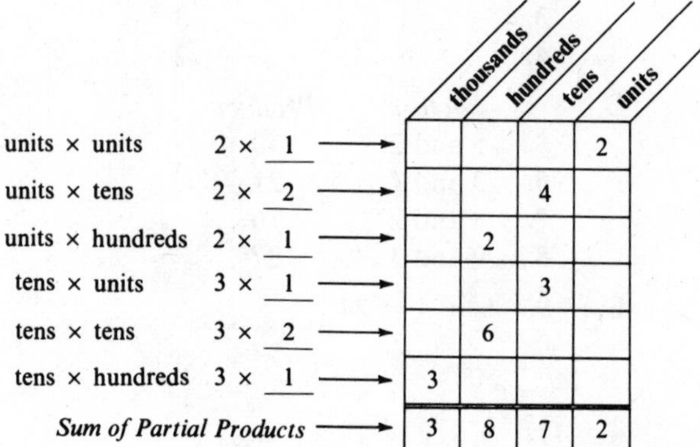

thousands
hundreds
tens
units
units × units 2 × 1
units × tens 2 × 2
units × hundreds 2 × 1
tens × units 3 × 1
tens × tens 3 × 2
tens × hundreds 3 × 1
Sum of Partial Products
2
4
2
3
6
3
3 8 7 2

B.

1.
```
  121
 × 32
  242
 3 63
3,872
```

2.
```
  112
 × 22
  224
 2 24
2,464
```

3.
```
  201
 × 42
  402
 8 04
8,442
```

4.
```
  311
 × 21
  311
 6 22
6,531
```

5.
```
  412
 × 12
  824
 4 12
4,944
```

6.
```
  1,012
  × 123
  3 036
  20 24
  101 2
124,476
```

25A

Study Exercise Seven (Frame 27)

1.
```
  2,131
  × 201
  2 131
  00 00
  426 2
428,331
```

```
  2,131
  × 201
  2 131
 426 20
428,331
```

SOLUTIONS TO STUDY EXERCISES, CONTD.

Study Exercise Seven (Frame 27, contd.)

2.

```
    2,131             2,131
  × 2 001           × 2 001
  -------           -------
    2 131             2 131
   00 00          4 262 00
  000 0           ---------
 4 262            4,264,131
---------
4,264,131
```

3.

```
    21,311             21,311
   × 1,020            × 1,020
   -------            -------
    00 000            426 220
   426 22           21 311 0
  0 000 0          ----------
 21 311            21,737,220
----------
21,737,220
```

4.

```
    12,112             12,112
   × 4,000            × 4,000
   -------         ----------
    00 000         48,448,000
   000 00
  0 000 0
 48 448
----------
48,448,000
```

27A

Study Exercise Eight (Frame 31)

1.

```
  2
 23          7
× 7       × 23
---       ----
161         21
           14
          ----
           161
```

2.

```
                   1
                   7
  64               4
  74
   296             78
  × 78          × 296
 -----          -----
 2 368            468
20 72            7 02
------          15 6
23,088          ------
                23,088
```

3.

```
    1               1
   304             203
 × 203           × 304
 -----           -----
   912             812
 60 80           60 90
------          ------
61,712          61,712
```

SOLUTIONS TO STUDY EXERCISES, CONTD.

Study Exercise Eight (Frame 31, contd.)

4.

```
    11
   1 43
   3 85
   2,396
   ×259
  21 564
 119 80
 479 2
 620,564
```

```
     11
     12
     58
     35
    259
  × 2 396
   1 554
  23 31
  77 7
 518
 620,564
```

5.

```
      1
    3  4
   50,406
  × 2,080
 4 032 480
100 812 0
104,844,480
```

```
        4
        3
        4
    2,080
  × 50,406
    12 480
   832 00
104 000
104,844,480
```

31A

Study Exercise Nine (Frame 33)

1.

```
   1
 1,246
  × 12
 2 492
12 46
14,952
```

Ed's salary is $14,952 per year.

2.

```
      21
      43
      32
   675,000
     × 365
 3 375 000
40 500 00
202 500 0
246,375,000
```

246,375,000 kilowatts may be generated in a year.

33A

unit

5

Division of Whole Numbers—Part One

Objectives:

By the end of this unit you should be able to:

1. give three symbols that are used to represent division.
2. identify the terms *dividend, divisor, quotient* and *remainder* in a division problem.
3. check division by using multiplication and addition.
4. show that division by zero is impossible.
5. perform long division when the divisor is a single digit. (1)

Three Symbols for Division

$$6 \div 3 \qquad 3\overline{)6} \qquad \frac{6}{3}$$

Each of these is read, 6 *divided by* 3. (2)

---

6 divided by 3 asks the question: How many three's are in six?

$$6 \div 3 = 2 \qquad 3\overline{)6}^{\,2} \qquad \frac{6}{3} = 2$$

Each of these is correct. There are 2 three's in 6, because $3 \times 2 = 6$. (3)

Parts of a Division Problem

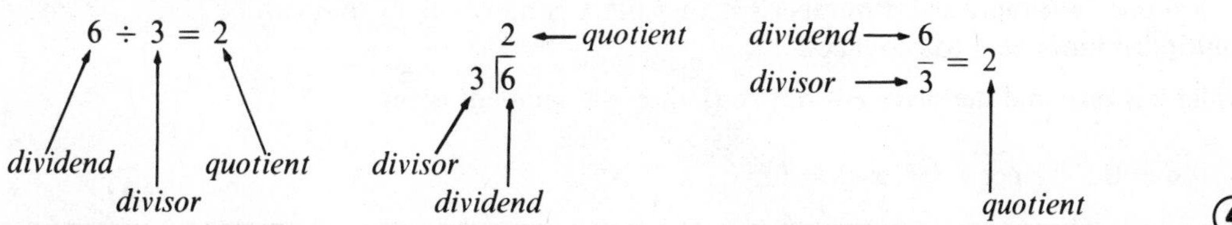

(4)

The Division Check

1. $6 \div 3 = 2$ since $3 \times 2 = 6$.

2. $3\overline{)6}^{\,2}$ since $3 \times 2 = 6$.

3. $\frac{6}{3} = 2$ since $3 \times 2 = 6$.

In a division problem the *divisor times the quotient must equal the dividend.* (5)

Examples:

1. $8 \div 2 = 4$ since $2 \times 4 = 8$.
2. $\begin{array}{r} 5 \\ 2\overline{)10} \end{array}$ since $2 \times 5 = 10$.
3. $3 \div 3 = 1$ since $3 \times 1 = 3$.
4. $\begin{array}{r} 4 \\ 3\overline{)12} \end{array}$ since $3 \times 4 = 12$.
5. $\frac{15}{3} = 5$ since $3 \times 5 = 15$.

(6)

Study Exercise One

A. For each division problem, give the dividend, divisor, and quotient.

1. $6 \div 2 = 3$ **2.** $16 \div 2 = 8$

3. $\begin{array}{r} 8 \\ 3\overline{)24} \end{array}$ **4.** $\begin{array}{r} 3 \\ 5\overline{)15} \end{array}$

5. $\frac{21}{3} = 7$

B. Find the quotient for each of the following and check the answer using multiplication.

6. $15 \div 5$ **7.** $8\overline{)16}$

8. $\frac{27}{9}$ **9.** $3\overline{)21}$

10. $42 \div 7$ **11.** $6 \div 1$

12. $6\overline{)24}$ **13.** $\frac{18}{2}$

14. $30 \div 5$ **15.** $7\overline{)35}$

(7)

Properties of Division

1. Any number divided by 1 produces the original number.

Example: $6 \div 1 = 6$ since $1 \times 6 = 6$.

2. The divisor may *never* be 0.

Example: $5 \div 0$ does not represent a number because there is no quotient that can be multiplied times zero to produce 5.

3. If the dividend is zero and the divisor is not zero, then the quotient is zero.

Example: $0 \div 6 = 0$ since $6 \times 0 = 0$.

(8)

Examples:

1. $7 \div 1 = 7$ **2.** $243 \div 1 = 243$

3. $0 \div 1 = 0$ **4.** $0 \div 8 = 0$

5. $8 \div 0 = ?$ impossible **6.** $0 \div 0 = ?$ impossible

(9)

Study Exercise Two

Find the quotient for each of the following. If the division cannot be performed, write "impossible." Check each answer.

1. $12 \div 1$ **2.** $6 \div 0$ **3.** $0 \div 6$

4. $957 \div 1$ **5.** $0 \div 2{,}963$

The Long Division Form

Example 1: $8 \div 2$

Line (a)

$$2\overline{)8}$$ with an empty box □ above ← *2 times "what number" is 8?*

Line (b)

$$\begin{array}{r} \boxed{4} \\ 2\overline{)8} \\ 8 \end{array}$$ ← *2 times* $\boxed{4}$ *is 8.*

Line (c)

$$\begin{array}{r} 4 \\ 2\overline{)8} \\ 8 \\ \hline 0 \end{array}$$

8 ← *Subtract*

0 ← *Remainder* (amount left over)

There are 4 two's in 8 and zero left over. Thus, 8 is *evenly divisible* by 2 because the remainder is zero.

(11)

Example 2: $9 \div 2$

Line (a)

$$2\overline{)9}$$ with an empty box □ above ← *2 times "what number" is 9?*

Line (b)

$$\begin{array}{r} \boxed{4} \\ 2\overline{)9} \\ 8 \end{array}$$ ← *2 times* $\boxed{4}$ *is 8.*

Line (c)

$$\begin{array}{r} 4 \\ 2\overline{)9} \\ 8 \\ \hline 1 \end{array}$$

8 ← *Subtract*

1 ← *Remainder* (must be less than 2)

There are 4 two's in 9 and a remainder of 1 left over. Therefore, $9 \div 2 = 4$ R1.

(12)

Checking Division Involving Remainders

Multiply the quotient times the divisor and add the remainder. This result must equal the dividend.

Problem: $9 \div 2 = 4$ R1.

Check: $4 \times 2 = 8$ and $8 + 1 = 9$

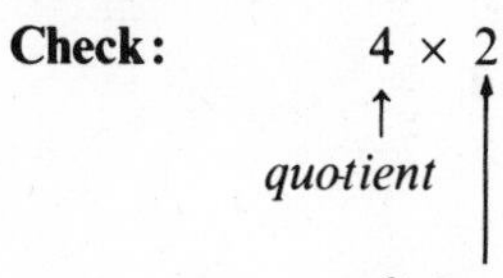

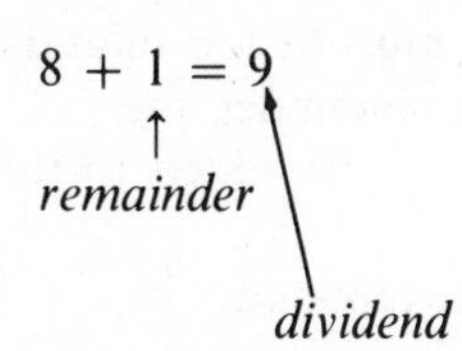

dividend

(13)

More Examples

1. $7 \div 2$

Solution: $\begin{array}{r} 3 \\ 2\overline{)7} \\ \underline{6} \\ 1 \end{array}$ Therefore, $7 \div 2 = 3$ R1.

Check: $3 \times 2 = 6$ and $6 + 1 = 7$.

2. $23 \div 4$

Solution: $\begin{array}{r} 5 \\ 4\overline{)23} \\ \underline{20} \\ 3 \end{array}$ Therefore, $23 \div 4 = 5$ R3.

Check: $5 \times 4 = 20$ and $20 + 3 = 23$.

3. $41 \div 9$

Solution: $\begin{array}{r} 4 \\ 9\overline{)41} \\ \underline{36} \\ 5 \end{array}$ Therefore, $41 \div 9 = 4$ R5.

Check: $4 \times 9 = 36$ and $36 + 5 = 41$.

4. $98 \div 10$

Solution: $\begin{array}{r} 9 \\ 10\overline{)98} \\ \underline{90} \\ 8 \end{array}$ Therefore, $98 \div 10 = 9$ R8.

Check: $9 \times 10 = 90$ and $90 + 8 = 98$.

Study Exercise Three

A. For each of the following, indicate the dividend, divisor, quotient, and remainder.

1. $\begin{array}{r} 2 \\ 5\overline{)14} \\ \underline{10} \\ 4 \end{array}$

2. $14 \div 5 = 2$ R4

3. $\begin{array}{r} 8 \\ 10\overline{)86} \\ \underline{80} \\ 6 \end{array}$

4. $86 \div 10 = 8$ R6

B. Fill in the blanks.

5. In a division problem the remainder must be less than the ____________.

6. To check a division problem having a remainder, the ____________ is multiplied times the divisor and the ____________ is added to this product. This result must equal the ____________.

7. 15 is evenly divisible by 3 because the remainder is ____________.

C. Perform the following divisions and check your answers.

8. $9 \div 6$

9. $27 \div 4$

10. $\frac{13}{3}$

11. $10\overline{)56}$

12. $9\overline{)72}$

(15)

More Long Division

Example 1: 95 ÷ 4

Line (a)

```
  □ ←—— First digit of quotient goes above the "9".
4|9 5
```

4 is smaller than 9.

Line (b)

```
   2
4|95
  8 ←—— 4 × 2
  1 ←—— Subtract
```

Line (c)

```
   2
4|95
  8↓
  15←—— Bring down the "5".
```

Line (d)

```
  23
4|95
  8
  15
  12←—— 4 × 3
   3←—— Remainder
```

Line (e) 95 ÷ 4 = 23 R3. (16)

Example 2: 137 ÷ 6

Line (a)

```
   □ ←—— First digit of quotient goes above the "3".
6|1 3 7
```

6 is smaller than 13.

Line (b)

```
   2
6|137
  12 ←—— 6 × 2
   1 ←—— Subtract
```

Line (c)

```
   2
6|137
  12↓
   17←—— Bring down the "7".
```

Line (d)

```
   22
6|137
  12
   17
   12←—— 6 × 2
    5←—— Remainder
```

Line (e) 137 ÷ 6 = 22 R5. (17)

Example 3: $823 \div 7$

$\square$ ← *First digit of quotient goes above the "8".*

Line (a) $7\overline{)\,8\;2\;3}$

7 is smaller than 8.

Line (b)
```
   1
7 |823
   7   ← 7 × 1
   1   ← Subtract
```

Line (c)
```
   1
7 |823
   7↓
   12  ← Bring down the "2".
```

Line (d)
```
   11
7 |823
   7
   12
    7  ← 7 × 1
    5  ← Subtract
```

Line (e)
```
   11
7 |823
   7
   12
    7
    53 ← Bring down the "3".
```

Line (f)
```
   117
7 |823
   7
   12
    7
    53
    49 ← 7 × 7
     4 ← Remainder
```

Line (g) $823 \div 7 = 117$ R4. (18)

Example 4: 1805 ÷ 3

```
              □  ←——First digit of quotient goes above the "8".
Line (a)   3 |1 8 0 5
              
3 is smaller than 18.

              6
Line (b)   3 |1805
             18    ←——3 × 6
             —
              0    ←——Subtract

              6
Line (c)   3 |1805
             18↓
             —
             00    ←——Bring down the "0".

             60
Line (d)   3 |1805
             18
             —
             00
              0    ←——3 × 0
             —
              0    ←——Subtract

             60
Line (e)   3 |1805
             18  |
             —   |
             00  |
              0  ↓
             —
             05    ←——Bring down the "5".

             601
Line (f)   3 |1805
             18
             —
             00
              0
             —
             05
              3    ←——3 × 1
             —
              2    ←——Remainder
```

Line (g) 1805 ÷ 3 = 601 R2.

(19)

Study Exercise Four

Perform the following by long division and check your answers.

1. 87 ÷ 3 **2.** 239 ÷ 8 **3.** 6 |755

4. 7 |2106 **5.** $\frac{307}{3}$

REVIEW EXERCISES

A. Indicate the dividend, divisor, quotient and remainder.

1.
$$\begin{array}{r} 35 \\ 6\,\overline{)215} \\ \underline{18} \\ 35 \\ \underline{30} \\ 5 \end{array}$$

2. (Fill in the blanks.) To check the preceding problem: 6 × ____ = 210 and 210 + ____ = 215.

B. Find the quotient and check your answer. If the division cannot be performed, write "impossible."

3. $16 \div 4$ **4.** $\frac{25}{5}$ **5.** $21 \div 7$

6. $0 \div 7$ **7.** $7 \div 0$ **8.** $12 \div 1$

9. $72 \div 8$ **10.** $6\,\overline{)42}$ **11.** $3\,\overline{)12}$

12. $\frac{15}{1}$ **13.** $\frac{0}{9}$ **14.** $\frac{9}{0}$

15. $35 \div 7$

C. Perform the following long divisions and check your answers.

16. $5\,\overline{)43}$ **17.** $7\,\overline{)61}$ **18.** $2\,\overline{)34}$

19. $3\,\overline{)63}$ **20.** $9\,\overline{)126}$ **21.** $7\,\overline{)341}$

22. $262 \div 8$ **23.** $2{,}256 \div 3$ **24.** $1{,}805 \div 9$

25. $1{,}322 \div 5$ **26.** $\frac{3516}{6}$ 21

SOLUTIONS TO REVIEW EXERCISES

A. **1.** dividend is 215; divisor is 6; quotient is 35; remainder is 5.
2. 6 × <u>35</u> = 210 and 210 + <u>5</u> = 215.

B.

	Answer	Check
3.	4	4 × 4 = 16
4.	5	5 × 5 = 25
5.	3	7 × 3 = 21
6.	0	7 × 0 = 0
7.	impossible	
8.	12	1 × 12 = 12
9.	9	8 × 9 = 72
10.	7	6 × 7 = 42
11.	4	3 × 4 = 12
12.	15	1 × 15 = 15
13.	0	9 × 0 = 0
14.	impossible	
15.	5	7 × 5 = 35

SOLUTIONS TO REVIEW EXERCISES, CONTD.

(Frame 22, contd.)

C.

	Answer	Check
16.	8 R3	$5 \times 8 = 40$ and $40 + 3 = 43$
17.	8 R5	$7 \times 8 = 56$ and $56 + 5 = 61$
18.	17 or 17 R0	$2 \times 17 = 34$ and $34 + 0 = 34$
19.	21 or 21 R0	$3 \times 21 = 63$ and $63 + 0 = 63$
20.	14 or 14 R0	$9 \times 14 = 126$ and $126 + 0 = 126$
21.	48 R5	$7 \times 48 = 336$ and $336 + 5 = 341$
22.	32 R6	$8 \times 32 = 256$ and $256 + 6 = 262$
23.	752 or 752 R0	$3 \times 752 = 2{,}256$ and $2{,}256 + 0 = 2{,}256$
24.	200 R5	$9 \times 200 = 1{,}800$ and $1{,}800 + 5 = 1{,}805$
25.	264 R2	$5 \times 264 = 1{,}320$ and $1{,}320 + 2 = 1{,}322$
26.	586 or 586 R0	$6 \times 586 = 3{,}516$ and $3{,}516 + 0 = 3{,}516$

(22)

SUPPLEMENTARY PROBLEMS

A. Indicate the dividend, divisor, quotient and remainder (if any).

1. $8 \div 2 = 4$

2. $\frac{20}{4} = 5$

3. $$6\overline{)18}\quad\text{quotient } 3$$

4.
$$\begin{array}{r} 539 \\ 3\overline{)1618} \\ \underline{15} \\ 11 \\ \underline{9} \\ 28 \\ \underline{27} \\ 1 \end{array}$$

B. Fill in the blanks.

5. To check a division problem, the quotient is multiplied times the ___________ and the ___________ is added to this product.

6. A dividend is said to be evenly divisible by the divisor if the remainder is ___________.

C. Find the following quotients and check your answers. If the division cannot be performed, write "impossible."

7. $\frac{28}{7}$ **8.** $\frac{30}{6}$ **9.** $49 \div 7$

10. $0 \div 4$ **11.** $0 \div 0$ **12.** $4 \div 0$

13. $8\overline{)56}$ **14.** $6\overline{)18}$ **15.** $\frac{9}{1}$

16. $\frac{9}{9}$ **17.** $10 \div 2$ **18.** $81 \div 9$

D. Perform the following long divisions and check your answers.

19. $3\overline{)13}$ **20.** $5\overline{)17}$ **21.** $7\overline{)21}$

22. $132 \div 6$ **23.** $258 \div 3$ **24.** $2{,}110 \div 3$

25. $453 \div 5$ **26.** $621 \div 7$ **27.** $\frac{295}{9}$

SUPPLEMENTARY PROBLEMS, CONTD.

28. $\dfrac{1{,}234}{5}$ **29.** $\dfrac{2{,}691}{4}$ **30.** $\dfrac{9{,}607}{7}$

31. $4{,}251 \div 2$ **32.** $6{,}987 \div 8$ **33.** $21{,}456 \div 3$

34. $93{,}071 \div 9$ **35.** $41{,}251 \div 4$ **36.** $\dfrac{9{,}076}{4}$

37. $180 \div 3$ **38.** $90 \div 2$ **39.** $100 \div 4$

40. $\dfrac{10{,}905}{9}$

SOLUTIONS TO STUDY EXERCISES

Study Exercise One (Frame 7)

A.

	Dividend	Divisor	Quotient
1.	6	2	3
2.	16	2	8
3.	24	3	8
4.	15	5	3
5.	21	3	7

B.

	Quotient	Check
6.	3	$5 \times 3 = 15$
7.	2	$8 \times 2 = 16$
8.	3	$9 \times 3 = 27$
9.	7	$3 \times 7 = 21$
10.	6	$7 \times 6 = 42$
11.	6	$1 \times 6 = 6$
12.	4	$6 \times 4 = 24$
13.	9	$2 \times 9 = 18$
14.	6	$5 \times 6 = 30$
15.	5	$7 \times 5 = 35$

7A

Study Exercise Two (Frame 10)

	Quotient	Check
1.	12	$1 \times 12 = 12$
2.	impossible	
3.	0	$6 \times 0 = 0$
4.	957	$1 \times 957 = 957$
5.	0	$2,963 \times 0 = 0$

10A

Study Exercise Three (Frame 15)

A.

	Dividend	Divisor	Quotient	Remainder
1.	14	5	2	4
2.	14	5	2	4
3.	86	10	8	6
4.	86	10	8	6

B. **5.** divisor
6. quotient, remainder, dividend
7. zero

SOLUTIONS TO STUDY EXERCISES, CONTD.

Study Exercise Three (Frame 15, contd.)

C.

	Answer	Check
8.	1 R3	$1 \times 6 = 6, \quad 6 + 3 = 9$
9.	6 R3	$6 \times 4 = 24, \quad 24 + 3 = 27$
10.	4 R1	$4 \times 3 = 12, \quad 12 + 1 = 13$
11.	5 R6	$5 \times 10 = 50, \quad 50 + 6 = 56$
12.	8 or 8 R0	$8 \times 9 = 72, \quad 72 + 0 = 72$

15A

Study Exercise Four (Frame 20)

	Problem	Answer	Check
1.	29 3 ⟌87 6 27 27 0	29 R0	$29 \times 3 = 87$ and $87 + 0 = 87$
2.	29 8 ⟌239 16 79 72 7	29 R7	$29 \times 8 = 232$ and $232 + 7 = 239$
3.	125 6 ⟌755 6 15 12 35 30 5	125 R5	$125 \times 6 = 750$ and $750 + 5 = 755$
4.	300 7 ⟌2106 21 00 0 06 0 6	300 R6	$300 \times 7 = 2{,}100$ and $2{,}100 + 6 = 2{,}106$
5.	102 3 ⟌307 3 00 0 07 6 1	102 R1	$102 \times 3 = 306$ and $306 + 1 = 307$

20A

unit

6

Division of Whole Numbers—Part Two

Objectives:

By the end of this unit you should know:

1. the terms *trial divisor*, *trial dividend* and *trial quotient*.
2. how to perform long division where the divisor contains two or more digits.
3. how to use division in solving word problems.
4. how to use the divisibility tests for 2, 3, 5, and 10. (1)

In the last unit we learned to perform long division where the divisor was a single digit. In this unit we will learn to perform long division where the divisor contains two or more digits. To do this we must examine the concepts of trial divisor, trial dividend, and trial quotient. (2)

Trial Divisor—Trial Dividend—Trial Quotient

Example 1: 78,469 ÷ 52

	Trial Divisor	Trial Dividend	Trial Quotient
Solution: $52\overline{)78,469}$ (quotient digit [1])	5~~2~~ → 5	7~~8~~ → 7	7 ÷ 5 → 1

52 is smaller than 78.

Example 2: 178,469 ÷ 52

	Trial Divisor	Trial Dividend	Trial Quotient
Solution: $52\overline{)178,469}$ (quotient digit [3])	5~~2~~ → 5	17~~8~~ → 17	17 ÷ 5 → 3

52 is smaller than 178.

Example 3: 68,405 ÷ 223

	Trial Divisor	Trial Dividend	Trial Quotient
$223\overline{)68,405}$ (quotient digit [3])	2~~23~~ → 2	6~~84~~ → 6	6 ÷ 2 → 3

223 is smaller than 684.

(Frame 3, contd.)

Example 4: 16,840 ÷ 223

Solution:

```
            [8] <-------------------------------------------
223 |16,840    22̸3̸ → 2     168̸4̸ → 16     16 ÷ 2 → 8
```

223 is smaller than 1,684.

(3)

Checking the Trial Quotient

Example 1: 78,469 ÷ 52

Solution:

```
         1
52 |78,469
    52   <------  52 × 1
    26   <------  Difference is less than 52.
```

Example 2: 178,469 ÷ 52

Solution:

```
         3
52 |178,469
    156  <------  52 × 3
     22  <------  Difference is less than 52.
```

Example 3: 68,405 ÷ 223

Solution:

```
           3
223 |68,405
     66 9  <------ 223 × 3
      1 5  <------ Difference is less than 223.
```

Example 4: 16,840 ÷ 223

Solution:

```
            8        Trial quotient is too large.
223 |16,840
     17 84  <------ 223 × 8
            <------ No difference can be obtained.
```

(4)

Adjusting the Trial Quotient

```
            8
223 |16,840
     17 84

                   If too large adjust by making smaller.

            7
223 |16,840
     15 61   <------  223 × 7
      1 23   <------  Difference is less than 223.
```

(5)

More Examples

Example 1: 4,268 ÷ 25

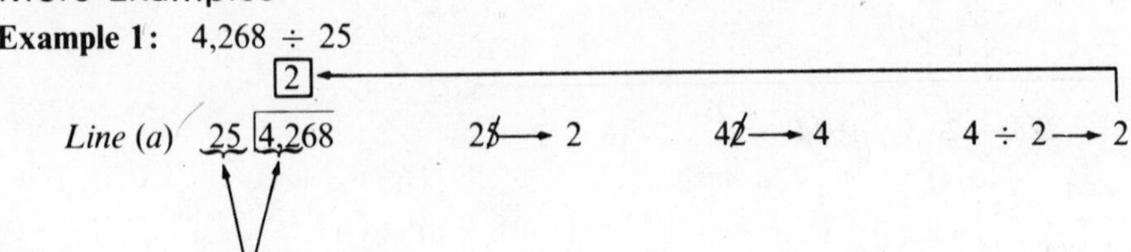

25 is smaller than 42.

Line (b) 25)4,268 — quotient 2 ← Trial quotient is too large.
50 ← 25 × 2
← No difference can be obtained.

Line (c) 25)4,268 — quotient 1 ← Adjust by making smaller.
25 ← 25 × 1
17 ← Difference is less than 25.

⑥

Example 2: 142,130 ÷ 362

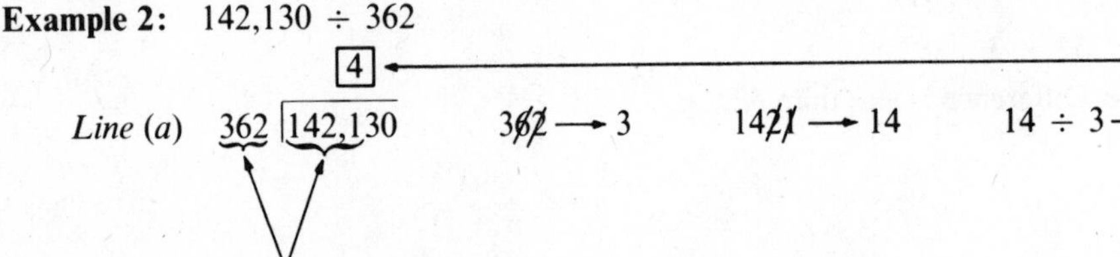

362 is smaller than 1,421.

Line (b) 362)142,130 — quotient 4 ← Trial quotient is too large.
144 8 ← 362 × 4
← No difference can be obtained.

Line (c) 362)142,130 — quotient 3 ← Adjust by making smaller.
108 6 ← 362 × 3
33 5 ← Difference is less than 362.

⑦

Study Exercise One

A. In each of the following division problems indicate the place value of the first digit of the quotient by placing a box above the corresponding digit of the dividend.

Example: 65)53, 9 46 (box above the 9)

1. 6)825 **2.** 6)372 **3.** 56)983

4. 56)257 **5.** 73)69,582 **6.** 73)84,106

7. 438)972,154 **8.** 438)300,256

B. Find the trial divisor, trial dividend and trial quotient for each of the following:

9. 7)936 **10.** 83)956 **11.** 83)1,026

Study Exercise One (Frame 8, contd.)

12. 28 |36,581 **13.** 28 |87,643 **14.** 562 |243,965

C. For each of the following, the trial quotient needs to be adjusted. Give the correct adjusted trial quotient.

15. 56 |257 **16.** 89 |50,641 **17.** 438 |300,256

(8)

Long Division

Example 1: 842 ÷ 56

Solution:

```
Line (a)        15
Line (b)     56 |842
Line (c)        56↓
Line (d)        282
Line (e)        280
Line (f)          2  ←——— Remainder
```

Therefore, 842 ÷ 56 = 15 R2.

To check the answer, multiply 56 by 15 and add the remainder, 2.

```
  56     840
× 15      +2
 280     842
 56
 840
```

(9)

Example 2: 9,376 ÷ 56

Solution:

```
Line (a)        167
Line (b)     56 |9,376
Line (c)        56 ↓
Line (d)        377
Line (e)        336 ↓
Line (f)         416
Line (g)         392
Line (h)          24  ←——— Remainder
```

Therefore, 9,376 ÷ 56 = 167 R24.

To check the answer, multiply 56 by 167 and add the remainder, 24.

```
  167     9,352
 × 56       +24
1 002     9,376
8 35
9,352
```

(10)

Example 3: 5,692,073 ÷ 734

Solution:

```
Line (a)           7 754
Line (b)     734 |5,692,073
Line (c)          5 138
Line (d)            554 0
Line (e)            513 8
Line (f)             40 27
Line (g)             36 70
Line (h)              3 573
Line (i)              2 936
Line (j)                637  ←—— Remainder
```

Therefore, 5,692,073 ÷ 734 = 7,754 R637.

To check the answer, multiply 734 by 7,754 and add the remainder, 637.

```
      7,754        5,691,436
    × 734        +       637
     31 016        5,692,073
    232 62
  5 427 8
  5,691,436
```

(11)

Study Exercise Two

A. Fill in the boxes with the correct numbers to illustrate the following long division.

```
         □ 3 7 1
1.  39 |9 2, 5 0 3
        7 8
        1 4 □
        1 1 7
          2 8 □
          □ □ □
              7 □
              3 9
              □ □
```

B. Perform the following long divisions and check your answers.

2. 1,892 ÷ 7 **3.** 1,892 ÷ 73 **4.** 142 $\overline{)2,556}$

5. 67 $\overline{)4,796}$ **6.** 188,732 ÷ 887 **7.** 5,719,437 ÷ 37,149

(12)

Difficulties with Zero in Division

Example 1: 1,425 ÷ 7

Long Form

```
Line (a)       203
Line (b)   7 |1,425
Line (c)      1 4↓|
Line (d)       02 |
Line (e)        0 |<———— 7 × 0
Line (f)        25
Line (g)        21
Line (h)         4
```

Therefore, 1,425 ÷ 7 = 203 R4.

Short Form

```
Line (a)       203
Line (b)   7 |1,425
Line (c)      1 4↓↓
Line (d)        25
Line (e)        21
Line (f)         4
```

Therefore, 1,425 ÷ 7 = 203 R4. (13)

Example 2: 84,056 ÷ 42

Long Form

```
Line (a)        2,001
Line (b)   42 |84,056
Line (c)       84 ↓||
Line (d)        0 0||
Line (e)          0↓|        42 × 0
Line (f)          05|
Line (g)           0↓        42 × 0
Line (h)           56
Line (i)           42
Line (j)           14
```

Therefore, 84,056 ÷ 42 = 2,001 R14.

Short Form

```
Line (a)        2,001
Line (b)   42 |84,056
Line (c)       84
Line (d)        0 056
Line (e)           42
Line (f)           14
```

Therefore, 84,056 ÷ 42 = 2,001 R14. (14)

Study Exercise Three

A. Fill in the blanks for the following long form division.

1. 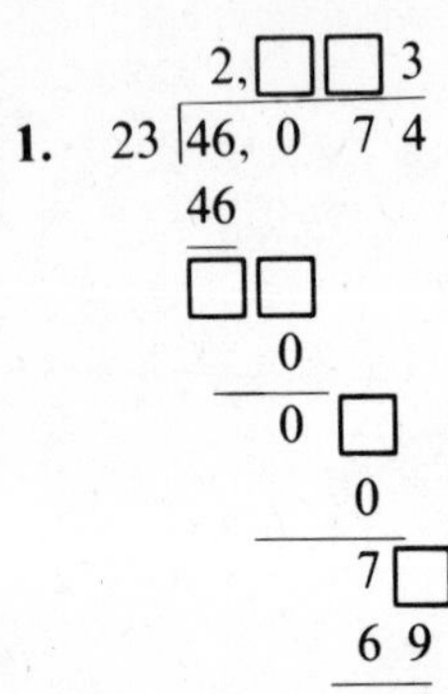

B. Fill in the blanks for the following short form division.

2. 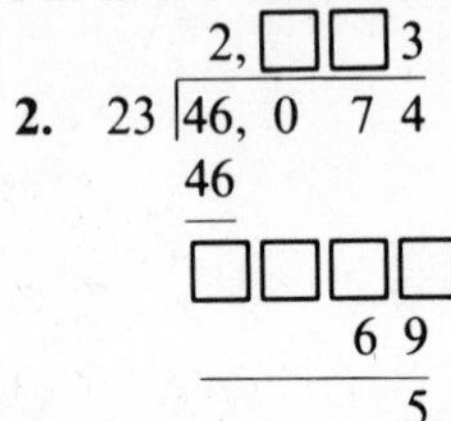

C. Divide the following using short form.

3. 1,535 ÷ 5 **4.** 54,146 ÷ 27 **5.** $\frac{42{,}182}{42}$

6. $\frac{13{,}225}{66}$ **7.** 59 $\overline{)11{,}807}$

(15)

Two Uses of Division

1. Given a certain number of objects and a specified number of groups, division will tell us how many objects need to be placed in each group so that the groups will be the same size.

Example: There are 52 cards in a deck. If all the cards are to be dealt to 4 people in a bridge game, how many cards will each person receive?

Solution: There are 52 objects to be placed in 4 equal groups. Divide 52 by 4.

```
   13
4 |52
   4
   --
   12
   12
   --
    0
```

Each person will receive 13 cards.

(Frame 16, contd.)

2. Given a certain number of objects and a specified number of objects to be placed in a group, division will tell us how many groups are needed.

Example: At a certain college there are 450 students who wish to take an arithmetic course. If each classroom seats 30 students, how many classes should be offered?

Solution: There are 450 objects. Each group can hold 30. Therefore, divide 450 by 30.

$$\begin{array}{r} 15 \\ 30\,\overline{)450} \\ \underline{30} \\ 150 \\ \underline{150} \\ 0 \end{array}$$

15 groups or classes of arithmetic should be offered.

Study Exercise Four

Using division, solve the following word problems.

1. Jim's annual salary is $11,100. How much is his monthly salary?
2. A building contractor estimates he needs 384 cubic yards of dirt fill. If a truck can carry 12 cubic yards of dirt, how many truckloads of fill will he need?
3. On his vacation, Harry traveled 3,290 miles and used 235 gallons of gasoline. On the average, how many miles did he travel for each gallon of gasoline?

Evenly Divisible

A dividend is said to be *evenly divisible* by a divisor providing the remainder is zero.

Examples

1. 34 is evenly divisible by 2.

$$\begin{array}{r} 17 \\ 2\,\overline{)34} \\ \underline{2} \\ 14 \\ \underline{14} \\ 0 \end{array} \leftarrow \text{Remainder is zero.}$$

2. 21 is evenly divisible by 3.

$$\begin{array}{r} 7 \\ 3\,\overline{)21} \\ \underline{21} \\ 0 \end{array} \leftarrow \text{Remainder is zero.}$$

3. 35 is evenly divisible by 5.

$$\begin{array}{r} 7 \\ 5\,\overline{)35} \\ \underline{35} \\ 0 \end{array} \leftarrow \text{Remainder is zero.}$$

4. 90 is evenly divisible by 10.

$$\begin{array}{r} 9 \\ 10\,\overline{)90} \\ \underline{90} \\ 0 \end{array} \leftarrow \text{Remainder is zero.}$$

(Frame 19, contd.)

5. 16 is *not* evenly divisible by 5. $\begin{array}{r} 3 \\ 5\overline{)16} \\ 15 \\ \hline 1 \end{array}$ ← Remainder is *not* zero.

(19)

Study Exercise Five

Answer yes or no.

1. Is 48 evenly divisible by 2?
2. Is 23 evenly divisible by 2?
3. Is 321 evenly divisible by 3?
4. Is 925 evenly divisible by 5?
5. Is 836 evenly divisible by 5?
6. Is 670 evenly divisible by 10?

Even Numbers

Whole numbers which are evenly divisible by 2 are said to be *even numbers*.

0, 2, 4, 6, 8, 10, 12, 14, 16, 18, 20, 22, ...

Note: 0 is an even number because; $\begin{array}{r} 0 \\ 2\overline{)0} \\ 0 \\ \hline 0 \end{array}$ ← Remainder is zero.

(21)

Odd Numbers

Whole numbers which are *not* evenly divisible by 2 are said to be *odd numbers*.

1, 3, 5, 7, 9, 11, 13, 15, 17, 19, 21, ...

(22)

Study Exercise Six

Classify each of the following whole numbers as being *even* or *odd*.

1. 24	**2.** 38	**3.** 49
4. 0	**5.** 283	**6.** 1,904
7. 33,641	**8.** 126,729	**9.** 2,005
10. 531,738		

(23)

Divisibility Tests

Divisibility tests are quick methods of looking at a number and deciding whether it is evenly divisible by a given number.

Divisibility Test for 2

Any whole number is divisible by 2 if the last digit represents an even number.

Example 1: 5,794 is evenly divisible by 2.
↑
Last digit is even.

(Frame 25, contd.)

Example 2: 8,645 is *not* evenly divisible by 2.
↑
Last digit is *not* even.

Divisibility Test for 3

Any whole number is evenly divisible by 3 if the sum of its digits is evenly divisible by 3.

Example 1: 42 is evenly divisible by 3.

$4 + 2 = 6$ [Sum of the digits is 6.]

Example 2: 7,062 is evenly divisible by 3.

$7 + 0 + 6 + 2 = 15$ [Sum of the digits is 15.]

Example 3: 514 is not evenly divisible by 3.

$5 + 1 + 4 = 10$ [Sum of the digits is 10.]

Study Exercise Seven

A. Which of the following whole numbers are evenly divisible by 2? Answer yes or no.

1. 16	**2.** 35	**3.** 146
4. 2,097	**5.** 1,570	**6.** 2,613,536

B. Which of the following whole numbers are evenly divisible by 3? Answer yes or no.

7. 51	**8.** 92	**9.** 132
10. 3,041	**11.** 6,936	**12.** 2,012,541

(27)

Divisibility Test for 5

Any whole number is evenly divisible by 5 if the last digit is either a "5" or a "0".

Example 1: 125 is evenly divisible by 5.
↑
Last digit is a "5".

Example 2: 4,230 is evenly divisible by 5.
↑
Last digit is a "0".

Example 3: 2,057 is *not* evenly divisible by 5.
↑
Last digit is neither "5" nor "0".

Divisibility Test for 10

Any whole number is evenly divisible by 10 if the last digit is a "0"

Example 1: 230 is evenly divisible by 10.
↑
Last digit is a "0".

Example 2: 504 is *not* evenly divisible by 10.
↑
Last digit is not a "0".

Study Exercise Eight

A. Which of the following whole numbers are evenly divisible by 5? Answer yes or no.

1. 235 **2.** 452 **3.** 552 **4.** 1,960

B. Which of the following whole numbers are evenly divisible by 10? Answer yes or no.

5. 675 **6.** 200 **7.** 1,004 **8.** 123,590

C. Examine each number and decide if it is evenly divisible by 2, 3, 5, or 10.

Example: 132 is divisible by 2 and 3.

9. 135 **10.** 720 **11.** 4,233

12. 1,680 **13.** 2,802

(30)

REVIEW EXERCISES

A. Name the dividend, divisor, quotient and remainder.

1.
```
        248
   43 )10,674
        8 6
        2 07
        1 72
          354
          344
           10
```

2. (Fill in the blanks.) To check the previous division problem: 43 × ______ = 10,664 and 10,664 + ______ = 10,674.

B. Perform the following long divisions and check your answers.

3. $6\overline{)95}$ **4.** $7\overline{)832}$ **5.** $42\overline{)8,875}$

6. 5,692 ÷ 21 **7.** 88,241 ÷ 44 **8.** $\dfrac{62,005}{17}$

9. $\dfrac{9,659}{123}$ **10.** 852,561 ÷ 272 **11.** 4,203 ÷ 21

12. $153\overline{)459,062}$

C. Solve the following word problems.

13. A college scholarship fund contains $16,675. If each scholarship is to be worth $725, how many scholarships can be awarded?

14. A 120 foot block wall fence costs $720. What is the cost per running foot?

D. Using the divisibility tests for 2, 3, 5, and 10, determine the numbers for which each of the following is evenly divisible.

15. 141 **16.** 570 **17.** 23,010

18. 5,162 **19.** 485

(31)

SOLUTIONS TO REVIEW EXERCISES

A. **1.** dividend is 10,674; divisor is 43; quotient is 248; remainder is 10.

2. 43 × <u>248</u> = 10,664 and 10,664 + <u>10</u> = 10,674.

B.

	Answer	*Check*
3.	15 R5	15 × 6 = 90 and 90 + 5 = 95
4.	118 R6	118 × 7 = 826 and 826 + 6 = 832
5.	211 R13	211 × 42 = 8,862 and 8,862 + 13 = 8,875
6.	271 R1	271 × 21 = 5,691 and 5,691 + 1 = 5,692
7.	2,005 R21	2,005 × 44 = 88,220 and 88,220 + 21 = 88,241
8.	3,647 R6	3,647 × 17 = 61,999 and 61,999 + 6 = 62,005
9.	78 R65	78 × 123 = 9,594 and 9,594 + 65 = 9,659
10.	3,134 R113	3,134 × 272 = 852,448 and 852,448 + 113 = 852,561
11.	200 R3	200 × 21 = 4,200 and 4,200 + 3 = 4,203
12.	3,000 R62	3,000 × 153 = 459,000 and 459,000 + 62 = 459,062

SOLUTIONS TO REVIEW EXERCISES, CONTD.

(Frame 32, contd.)

C. 13. 23 scholarships
14. \$6 per running foot

D. 15. 3
16. 2, 3, 5, 10
17. 2, 3, 5, 10
18. 2
19. 5

SUPPLEMENTARY PROBLEMS

A. Name the dividend, divisor, quotient and remainder.

1.
```
      124
58 |7,205
    5 8
    1 40
    1 16
      245
      232
       13
```

2. (Fill in the blanks.) To check the preceding problem: 58 × ______ = 7,192 and 7,192 + ______ = ______.

B. Classify each of the following whole numbers as being even or odd.

3. 67 **4.** 132 **5.** 2,001
6. 90 **7.** 5,132 **8.** 140
9. 13 **10.** 0 **11.** 1

C. Using the divisibility tests for 2, 3, 5, and 10, determine for which numbers each of the following is evenly divisible.

12. 213 **13.** 720 **14.** 2,342
15. 93 **16.** 385

D. Solve the following word problems.

17. A stock market investor has \$1,725 which he wishes to invest in a stock costing \$23 a share. Neglecting commissions and taxes, how many shares could he purchase?

18. A contractor builds a 2,350 square foot house for \$32,900. What was the cost per square foot?

E. Perform the following long divisions and check your answers.

19. 3 |423 **20.** 5 |950 **21.** $\frac{720}{8}$
22. 6,642 ÷ 9 **23.** 5,408 ÷ 27 **24.** 329 ÷ 17
25. 2,005 ÷ 13 **26.** $\frac{5{,}212}{21}$ **27.** 98 |1,205
28. 126 |4,658 **29.** 231 |46,251 **30.** 630,000 ÷ 21
31. 2,400 ÷ 80 **32.** 3,600 ÷ 60 **33.** 20,504 ÷ 120

SUPPLEMENTARY PROBLEMS, CONTD.

34. 604 $\overline{)569{,}478}$ **35.** 236 $\overline{)5{,}652{,}058}$ **36.** $\dfrac{12{,}856}{301}$

37. $\dfrac{100{,}261}{92}$ **38.** 120 $\overline{)360{,}051}$ **39.** 765 $\overline{)865{,}042}$

40. 38 $\overline{)152{,}961}$

SOLUTIONS TO STUDY EXERCISES

Study Exercise One (Frame 8)

A.

1. 6 $\overline{)8\ 2\ 5}$ (□ above 8) **2.** 6 $\overline{)3\ 7\ 2}$ (□ above 7) **3.** 5 6 $\overline{)9\ 8\ 3}$ (□ above 3)

4. 5 6 $\overline{)2\ 5\ 7}$ (□ above 7) **5.** 7 3 $\overline{)6\ 9,\ 5\ 8\ 2}$ (□ above 5) **6.** 7 3 $\overline{)8\ 4,\ 1\ 0\ 6}$ (□ above 4)

7. 4 3 8 $\overline{)9\ 7\ 2,\ 1\ 5\ 4}$ (□ above 2) **8.** 4 3 8 $\overline{)3\ 0\ 0,\ 2\ 5\ 6}$ (□ above 2)

B.

	Trial Divisor		*Trial Dividend*		*Trial Quotient*	
9.	7		9		9 ÷ 7	1
10.	8~~3~~	8	9~~5~~	9	9 ÷ 8	1
11.	8~~3~~	8	10~~2~~	10	10 ÷ 8	1
12.	2~~8~~	2	3~~6~~	3	3 ÷ 2	1
13.	2~~8~~	2	8~~7~~	8	8 ÷ 2	4
14.	5~~67~~	5	2,4~~39~~	24	24 ÷ 5	4

C.

	Problem	*Adjusted Trial Quotient*
15.	4 56 $\overline{)257}$ 224 33	4
16.	5 89 $\overline{)50{,}641}$ 44 5 6 1	5
17.	6 438 $\overline{)300{,}256}$ 262 8 37 4	6

8A

SOLUTIONS TO STUDY EXERCISES, CONTD.

Study Exercise Two (Frame 12)

A. **1.**

```
        [2] 3 7 1
   39 | 9 2, 5 0 3
        7 8
        1 4 [5]
        1 1  7
          2 8 [0]
        [2][7][3]
               7 [3]
               3  9
             [3][4]
```

B.

	Problem	*Answer*
2.	7 ⟌1,892 = 270; 14, 49, 49, 02, 0, 2	270 R2
3.	73 ⟌1,892 = 25; 1 46, 432, 365, 67	25 R67
4.	142 ⟌2,556 = 18; 1 42, 1 136, 1 136, 0	18 or 18 R0
5.	67 ⟌4,796 = 71; 4 69, 106, 67, 39	71 R39

SOLUTIONS TO STUDY EXERCISES, CONTD.

Study Exercise Two (Frame 12, contd.)

```
            212
6.  887 |188,732          212 R688
         177 4
          11 33
           8 87
           2 462
           1 774
             688
```

```
                153
7.  37,149 |5,719,437     153 R35,640
            3 714 9
            2 004 53
            1 857 45
              147 087
              111 447
               35 640
```

12A

Study Exercise Three (Frame 15)

```
            2 [0][0] 3
A. 1.  23 |46,  0   7 4
           46
          [0] [0]
                0
               0[7]
                0
                 7 [4]
                 6  9
                    5
```

```
            2 [0][0] 3
B. 2.  23 |46,  0   7 4
           46
          [0][0][7][4]
                 6  9
                    5
```

C. *Problem* *Answer*

```
        307
3.  5 |1,535              307  or  307 R0
       1 5
        035
         35
          0
```

SOLUTIONS TO STUDY EXERCISES, CONTD.

Study Exercise Three (Frame 15, contd.)

```
          2 005
4.  27 |54,146        2,005 R11
        54
         0 146
           135
            11
```

```
          1 004
5.  42 |42,182        1,004 R14
        42
         0 182
           168
            14
```

```
            200
6.  66 |13,225        200 R25
        13 2
           025
             0
            25
```

```
            200
7.  59 |11,807        200 R7
        11 8
           007
             0
             7
```

15A

Study Exercise Four (Frame 17)

```
           925
1.  12 |11,100        Jim's monthly salary is $925.
        10 8
           30
           24
            60
            60
             0
```

```
         32
2.  12 |384           32 truckloads of fill.
        36
         24
         24
          0
```

SOLUTIONS TO STUDY EXERCISES, CONTD.

Study Exercise Four (Frame 17, contd.)

3.
$$\begin{array}{r} 14 \\ 235\overline{)3{,}290} \\ \underline{2\,35} \\ 940 \\ \underline{940} \\ 0 \end{array}$$

14 miles for each gallon of gasoline.

17A

Study Exercise Five (Frame 20)

1. yes **2.** no **3.** yes
4. yes **5.** no **6.** yes

20A

Study Exercise Six (Frame 23)

1. even **2.** even **3.** odd
4. even **5.** odd **6.** even
7. odd **8.** odd **9.** odd
10. even

23A

Study Exercise Seven (Frame 27)

A. **1.** yes **2.** no **3.** yes
4. no **5.** yes **6.** yes

B. **7.** yes **8.** no **9.** yes
10. no **11.** yes **12.** yes

27A

Study Exercise Eight (Frame 30)

A. **1.** yes **2.** no **3.** no
4. yes

B. **5.** no **6.** yes **7.** no
8. yes

C. **9.** 3, 5 **10.** 2, 3, 5, 10 **11.** 3
12. 2, 3, 5, 10 **13.** 2, 3

---

unit

7

Solving Verbal Problems by Reduction and Expansion

Objectives:

By the end of this unit you should be able to:

1. solve reduction problems by division.
2. solve expansion problems by multiplication.
3. solve reduction-expansion problems by using both division and multiplication.

(1)

Reduction

Reduction means to change from *"many to one"* by using the operation of division.

Example 1:

If 12 identical items cost a total of 60 cents, how much does 1 item cost?

Solution:

The total cost must be divided into 12 equal parts in order to find the cost of one item.

$$\begin{array}{r} 5 \\ 12\overline{)60} \\ \underline{60} \\ 0 \end{array}$$

Each item costs 5 cents.

Example 2.

If 5 apples cost 70 cents, what is the price of one apple?

Solution:

The total cost must be divided into 5 equal parts.

$$\begin{array}{r} 14 \\ 5\overline{)70} \\ \underline{5} \\ 20 \end{array}$$

Each apple costs 14 cents.

Example 3:
An automobile can travel 325 miles on 25 gallons of gasoline. How far can it travel on one gallon of gasoline?

Solution:
The total mileage must be divided into 25 equal parts.

$$\begin{array}{r} 13 \\ 25\overline{)325} \\ \underline{25} \\ 75 \\ \underline{75} \\ 0 \end{array}$$

The automobile can travel 13 miles on one gallon of gasoline. 5

Example 4:
How many identical articles will one dollar buy, if 15 dollars will buy 45 articles?

Solution:
Divide 45 into 15 equal parts.

$$\begin{array}{r} 3 \\ 15\overline{)45} \\ \underline{45} \\ 0 \end{array}$$

One dollar will buy 3 articles. 6

Study Exercise One

Solve the following problems by reduction.

1. If 5 candy bars cost 30 cents, how much does one candy bar cost?
2. If 23 identical articles weigh a total of 1,863 pounds, what is the weight of one article?
3. A guided missile travels 5,280 feet in 6 seconds. How far does it travel in one second?
4. How many identical articles will one dollar buy if 21 dollars will buy 63 articles? 7

Expansion

Expansion means to change from *"one to many"* by using the operation of multiplication. 8

Example 1:
If one candy bar costs 7 cents, what is the cost of 12 candy bars?

Solution:
The cost of each is 7 cents, so the cost of 12 may be obtained by multiplying 12 times 7 cents.

$$\begin{array}{r} 12 \\ \underline{\times 7} \\ 84 \end{array}$$

The twelve candy bars cost a total of 84 cents. 9

Example 2:

How far will a train travel in 60 seconds if it is traveling 88 feet each second?

Solution:

Multiply 88 feet by 60.

$$\begin{array}{r} 88 \\ \times 60 \\ \hline 5{,}280 \end{array}$$

The train will travel 5,280 feet in 60 seconds. (10)

Example 3.

If a box weighs 17 pounds, what is the total weight of 9 such boxes?

Solution:

Multiply 17 pounds by 9.

$$\begin{array}{r} 17 \\ \times 9 \\ \hline 153 \end{array}$$

The total weight of 9 boxes is 153 pounds. (11)

Study Exercise Two

Solve the following problems by expansion.

1. If one pound of mixed nuts costs 29 cents, how much is the cost for 3 pounds?
2. If one article weighs 87 pounds, find the total weight of 8 identical articles.
3. If a car can travel 16 miles on one gallon of gasoline, how far can it travel on 20 gallons of gasoline?
4. If the cost of one square yard of carpet is 13 dollars, what would be the cost to carpet a room having an area of 15 square yards? (12)

Reduction and Expansion

1. In doing reduction problems we "reduce to one" by using division.
2. In doing expansion problems we "expand to many" by using multiplication.

We will now combine these steps to solve problems involving both reduction and expansion. 

Example 1:

If 6 similar candy bars cost 24 cents, what is the cost of 9 candy bars?

Solution:

Step (1) Organize the data.
6 bars cost 24 cents.

Step (2) Reduce to the cost of one.
$24 \div 6 = 4$
Each candy bar costs 4 cents.

Step (3) Expand to cost for 9.
$4 \times 9 = 36$
The 9 bars cost 36 cents. (14)

Example 2.

If a car can travel 98 miles on 7 gallons of gasoline, find how far it can travel on 20 gallons of gasoline.

Solution:

Step (1) Organize the data.
98 miles on 7 gallons of gasoline.

Step (2) Reduce to 1 gallon.
$98 \div 7 = 14$
The car will travel 14 miles on 1 gallon of gasoline.

Step (3; Expand to 20 gallons.
$14 \times 20 = 280$
The car will travel 280 miles on 20 gallons of gasoline.

Example 3:

Find the distance a car will travel in 7 seconds, if it travels 150 feet in 3 seconds.

Solution:

Step (1) Organize the data.
150 feet in 3 seconds.

Step (2) Reduce to 1 second.
$150 \div 3 = 50$
The car will travel 50 feet in 1 second.

Step (3) Expand to 7 seconds.
$50 \times 7 = 350$
The car will travel 350 feet in 7 seconds.

Example 4:

If 56 identical articles weigh 28 pounds, how many articles will weigh 7 pounds?

Solution:

Step (1) Organize the data.
56 articles weigh 28 pounds.

Step (2) Reduce to 1 pound.
$56 \div 28 = 2$
2 articles weigh 1 pound.

Step (3) Expand to 7 pounds.
$2 \times 7 = 14$
14 articles weigh a total of 7 pounds.

Study Exercise Three

1. If 7 articles cost 28 cents, find the cost of 21 articles.
2. Find the distance a train will travel in 12 seconds if it travels 180 feet in 2 seconds.
3. If a car travels 105 miles on 7 gallons of gasoline, how far will it travel on 21 gallons?
4. If 60 identical articles weigh 15 pounds, how many articles will weigh 9 pounds?

REVIEW EXERCISES

Solve the following problems using reduction and expansion.

1. If one article weighs 47 pounds, find the weight of 6 such articles.
2. If 8 identical articles weigh 120 pounds, find the weight of one article.
3. If 9 articles weigh 72 pounds, find the weight of 6 articles.
4. If 7 articles weigh 56 pounds, find the weight of 11 articles.
5. If a car travels 42 miles on 2 gallons of gasoline, how far will it travel on 5 gallons?
6. If a car travels 360 miles on 20 gallons of gasoline, how far will it travel on 13 gallons?
7. If 80 feet of blockwall fencing cost 480 dollars, how much will 120 feet cost?
8. If a rocket travels 8,800 feet in 5 seconds, how far will it travel in 60 seconds?
9. If an industrial plant produces 11,200 articles in 28 days, how many would it produce in 5 days.?
10. Find the cost of 7 candy bars if 12 bars sell for 72 cents.
11. 84 identical articles weigh 21 pounds; how many articles will weigh 8 pounds?
12. 96 identical articles cost a total of 12 cents; how many articles may be purchased for 5 cents? 19

SOLUTIONS TO REVIEW EXERCISES

1. $47 \times 6 = 282$
   The 6 articles weigh 282 pounds.
2. $120 \div 8 = 15$
   One article weighs 15 pounds.
3. $72 \div 9 = 8$
   $8 \times 6 = 48$
   The 6 articles weigh 48 pounds.
4. $56 \div 7 = 8$
   $8 \times 11 = 88$
   The 11 articles weigh 88 pounds.
5. $42 \div 2 = 21$
   $21 \times 5 = 105$
   The car will travel 105 miles on 5 gallons of gasoline.
6. $360 \div 20 = 18$
   $18 \times 13 = 234$
   The car will travel 234 miles on 13 gallons of gasoline.
7. $480 \div 80 = 6$
   $6 \times 120 = 720$
   120 feet of fence will cost 720 dollars.
8. $8{,}800 \div 5 = 1{,}760$
   $1{,}760 \times 60 = 105{,}600$
   The rocket will travel 105,600 feet in 60 seconds.
9. $11{,}200 \div 28 = 400$
   $400 \times 5 = 2{,}000$
   2,000 articles would be produced in 5 days.
10. $72 \div 12 = 6$
    $6 \times 7 = 42$
    The 7 candy bars would cost 42 cents.
11. $84 \div 21 = 4$
    $4 \times 8 = 32$
    32 articles would weigh a total of 8 pounds.
12. $96 \div 12 = 8$
    $8 \times 5 = 40$
    40 articles may be purchased for 5 cents.

SUPPLEMENTARY PROBLEMS

1. If one article costs 12 cents, find the cost of 8 identical articles.
2. If 7 articles cost 63 cents, find the cost of one article.
3. If 7 articles cost 35 cents, find the cost of 13 articles.
4. If 5 articles cost 75 cents, find the cost of 2 articles.
5. Find the time for a jet airliner to travel 5 miles if it travels 2 miles in 12 seconds.
6. Find the distance a car will travel in 12 seconds if it travels 300 feet in 6 seconds.
7. If a car travels 60 miles on 4 gallons of gasoline, how far will it travel on 9 gallons?
8. If 96 identical objects weigh 48 pounds, how many objects weigh 16 pounds?
9. If it costs 600 dollars to drill a 200 foot well, how much will it cost to drill a 320 foot well?
10. If 15 identical boxes are required to package a total of 5 cubic feet, how many boxes will be needed to hold 7 cubic feet?
11. If 42 identical articles cost a total of 21 dollars, how many articles will cost 11 dollars?
12. If a car travels 121 miles on 11 gallons of gasoline, how far will it travel on 14 gallons?

SOLUTIONS TO STUDY EXERCISES

Study Exercise One (Frame 7)

1. $30 \div 5 = 6$; each candy bar costs 6 cents.
2. $1{,}863 \div 23 = 81$; one article weighs 81 pounds.
3. $5{,}280 \div 6 = 880$; it travels 880 feet in one second.
4. $63 \div 21 = 3$; one dollar will buy 3 articles.

Study Exercise Two (Frame 12)

1. $29 \times 3 = 87$; 3 pounds cost 87 cents.
2. $87 \times 8 = 696$; 8 articles weigh 696 pounds.
3. $16 \times 20 = 320$; the car can travel 320 miles.
4. $13 \times 15 = 195$; the cost of carpet would be 195 dollars.

Study Exercise Three (Frame 18)

1. $28 \div 7 = 4$
   $4 \times 21 = 84$
   The cost is 84 cents for 21 articles.
2. $180 \div 2 = 90$
   $90 \times 12 = 1{,}080$
   The train will travel 1,080 feet in 12 seconds.
3. $105 \div 7 = 15$
   $15 \times 21 = 315$
   The car will travel 315 miles on 21 gallons.
4. $60 \div 15 = 4$
   $4 \times 9 = 36$
   36 articles will weigh a total of 9 pounds.

---

unit

Exponents, Perfect Squares and Square Roots

Objectives:

By the end of this unit you should be able to:

1. identify the terms *base*, *exponent*, *perfect square*, and *square root*.
2. evaluate exponential expressions.
3. find square roots of perfect squares.

(1)

Numbers related by multiplication are called *factors*.

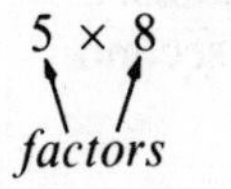

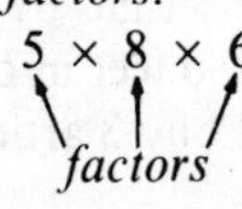

(2)

One number may be used as a factor several times.

Example 1: $2 \times 2 \times 2$

Example 2: $3 \times 3 \times 3 \times 3 \times 3$

(3)

A Simpler Notation

Example 1: $2 \times 2 \times 2 = 2^3$

Example 2: $3 \times 3 \times 3 \times 3 \times 3 = 3^5$

(4)

Exponential Notation

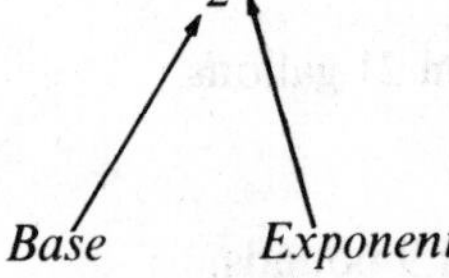

The base, 2, is used as a factor three times. The exponent indicates the number of times the base is used as a factor.

(5)

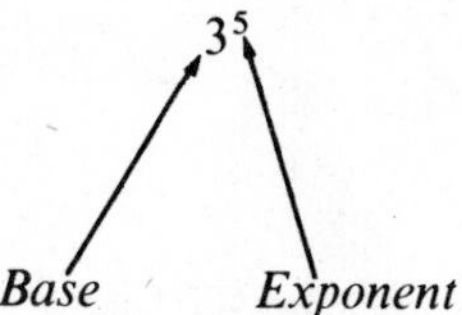

The exponent, 5, indicates that the base, 3, is used as a factor five times. (6)

Reading Exponents

Example 1: 5^2 is read, *"5 to the second"* or *"5 squared."*

Example 2: 2^3 is read, *"2 to the third"* or *"2 cubed."*

Example 3: 7^5 is read, *"7 to the fifth."*

Example 4: $5^2 \times 2^3$ is read, *"5 squared times 2 cubed."*

Example 5: $3^7 \times 5^{12}$ is read, *"3 to the seventh times 5 to the twelfth."* (7)

Study Exercise One

A. Indicate the base and exponent for each of the following.

1. 3^2 **2.** 2^3 **3.** 10^2
4. 5^7 **5.** 10^3 **6.** 3^{10}

B. Write a word statement describing how each of the following is read.

7. 7^2 **8.** 9^3 **9.** $7^2 \times 9^3$
10. $2^3 \times 3^5$ (8)

Changing to Exponential Notation

Example 1: $5 \times 5 = 5^2$

Example 2. $7 \times 7 \times 7 \times 7 = 7^4$

Example 3: $10 \times 10 \times 10 = 10^3$

Example 4: $6 = 6^1$

Example 5: $3 \times 3 \times 5 \times 5 \times 5 \times 5 = 3^2 \times 5^4$

Example 6: $2 \times 2 \times 2 \times 5 \times 5 \times 7 = 2^3 \times 5^2 \times 7^1$ (9)

Study Exercise Two

Change each of the following to exponential notation.

1. 3×3
2. $5 \times 5 \times 5$
3. $2 \times 2 \times 2 \times 7 \times 7$
4. $5 \times 5 \times 5 \times 5 \times 11 \times 11 \times 13 \times 13 \times 13 \times 17$
5. $2 \times 3 \times 3 \times 7 \times 7 \times 7$ (10)

Evaluating Exponential Expressions

Example 1: 3^2

Solution:

Line (a) $3^2 = \underbrace{3 \times 3}$

Line (b) $3^2 = 9$

Example 2: 2^3

Solution:

Line (a) $2^3 = \underbrace{2 \times 2} \times 2$

Line (b) $2^3 = \underbrace{4 \times 2}$

Line (c) $2^3 = 8$

Example 3: 3^4

Solution:

Line (a) $3^4 = \underbrace{3 \times 3} \times \underbrace{3 \times 3}$

Line (b) $3^4 = \underbrace{9 \times 9}$

Line (c) $3^4 = 81$

Example 4: $2^3 \times 3^2$

Solution:

Line (a) $2^3 \times 3^2 = \underbrace{2 \times 2 \times 2} \times \underbrace{3 \times 3}$

Line (b) $2^3 \times 3^2 = \underbrace{8 \times 9}$

Line (c) $2^3 \times 3^2 = 72$

(11)

Study Exercise Three

Evaluate the following.

1. 5^2 **2.** 3^3 **3.** 2^2

4. 7^2 **5.** 1^3 **6.** 0^2

7. 10^2 **8.** 10^3 **9.** $2^3 \times 5^2$

10. $3^2 \times 5^1 \times 7^2$ **11.** $2^4 \times 3^2 \times 5^1$ **12.** $2^1 \times 3^2 \times 5^2$

(12)

Perfect Squares

Any whole number is said to be a perfect square if it can be written using a whole number base with an exponent of two.

Example:

$25 = 5 \times 5$ or 5^2.

Therefore, 25 is a perfect square.

More Examples of Perfect Squares

1. 9 is a perfect square because $9 = 3^2$.
2. 16 is a perfect square because $16 = 4^2$.
3. 36 is a perfect square because $36 = 6^2$.
4. 100 is a perfect square because $100 = 10^2$.

(14)

Study Exercise Four

Decide which of the following are perfect squares, then write using a base with an exponent of 2.

1. 25	**2.** 49	**3.** 12
4. 100	**5.** 10	**6.** 0
7. 1	**8.** 50	**9.** 64
10. 144		

Square Roots of Perfect Squares

A square root of a number is one of its two equal factors.

Example:

A square root of 25 is 5 because $25 = 5 \times 5$.

equal factors

More Examples

1. A square root of 9 is 3 because $9 = 3 \times 3$.
2. A square root of 16 is 4 because $16 = 4 \times 4$.
3. A square root of 1 is 1 because $1 = 1 \times 1$.
4. A square root of 144 is 12 because $144 = 12 \times 12$.

(17)

When a perfect square is written using a base with an exponent of 2, the square root is simply the base.

Example:

$25 = 5^2$

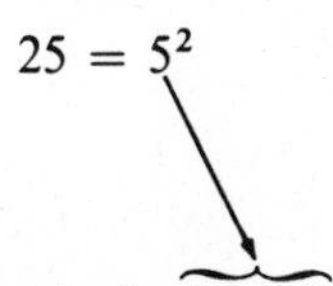

The square root is the base 5.

Relationship of Squaring to Square Rooting

Squaring

Example 1: $5^2 = 25$

Square Rooting

Squaring

Example 2: $7^2 = 49$

Square Rooting

Symbol for Square Root

The symbol, $\sqrt{25}$, says to find the square root of 25.

$$\sqrt{25} = 5$$

(20)

Study Exercise Five

Find the following square roots.

1. $\sqrt{36}$ **2.** $\sqrt{49}$ **3.** $\sqrt{1}$

4. $\sqrt{0}$ **5.** $\sqrt{64}$ **6.** $\sqrt{100}$

7. $\sqrt{121}$

Table of Perfect Squares and Square Roots

Perfect Square	Square Root
0	0
1	1
4	2
9	3
16	4
25	5
36	6
49	7
64	8
81	9
100	10
121	11
144	12

This table should be memorized!

REVIEW EXERCISES

A. Fill in the blanks.

1. Numbers related by multiplication are called __________.
2. For the expression 5^7, 5 is the __________ and 7 is the __________.
3. The expression, 8^2, is read "8 to the __________" or "8 __________."
4. The expression, 7^3, is read "7 to the __________" or "7 __________."
5. The expression, 10^5, is read "10 to the __________."

B. Indicate the base and exponent for each of the following.

6. 1^2 7. 10^5 8. 5^{10} 9. 7^8

C. Change each to exponential notation.

10. 3×3 11. $2 \times 2 \times 2 \times 2$
12. $10 \times 10 \times 10$ 13. 5
14. $2 \times 2 \times 3 \times 3 \times 3$ 15. $5 \times 7 \times 7 \times 11 \times 11 \times 11 \times 11$

D. Evaluate the following.

16. 4^2 17. 5^3 18. 6^1
19. 1^3 20. 7^3 21. 10^3
22. $2^2 \times 5^2$ 23. $3^2 \times 5^1 \times 7^2$

E. Find the following square roots.

24. $\sqrt{4}$ 25. $\sqrt{49}$ 26. $\sqrt{100}$
27. $\sqrt{64}$ 28. $\sqrt{81}$ 29. $\sqrt{1}$
30. $\sqrt{0}$

(23)

SOLUTIONS TO REVIEW EXERCISES

A. 1. factors 2. base, exponent 3. second, squared
4. third, cubed 5. fifth

B.

	Base	*Exponent*
6.	1	2
7.	10	5
8.	5	10
9.	7	8

C. 10. 3^2 11. 2^4 12. 10^3
13. 5^1 14. $2^2 \times 3^3$ 15. $5^1 \times 7^2 \times 11^4$

D. 16. 16 17. 125 18. 6
19. 1 20. 343 21. 1,000
22. 100 23. 2,205

E. 24. 2 25. 7 26. 10
27. 8 28. 9 29. 1
30. 0

(24)

SUPPLEMENTARY PROBLEMS

A. Indicate the base and exponent for each of the following.

1. 2^8 **2.** 8^2 **3.** 1^5 **4.** 5^1

B. Change each to exponential notation.

5. 4×4
6. $3 \times 3 \times 3 \times 3 \times 3 \times 3$
7. 8
8. $2 \times 2 \times 2 \times 3 \times 3$
9. $2 \times 2 \times 3 \times 5 \times 5 \times 7 \times 7 \times 7$

C. Evaluate the following.

10. 2^4 **11.** 3^3 **12.** 5^1
13. 7^2 **14.** $2^2 \times 3^2$ **15.** $2^3 \times 5^2$
16. $2^1 \times 3^3 \times 7^2$ **17.** $2^2 \times 3^1 \times 5^3$ **18.** $10^2 \times 10^3$

D. Find the following square roots.

19. $\sqrt{144}$ **20.** $\sqrt{100}$ **21.** $\sqrt{4}$
22. $\sqrt{36}$ **23.** $\sqrt{0}$ **24.** $\sqrt{121}$
25. $\sqrt{1}$

SOLUTIONS TO STUDY EXERCISES

Study Exercise One (Frame 8)

A.

	Base	*Exponent*
1.	3	2
2.	2	3
3.	10	2
4.	5	7
5.	10	3
6.	3	10

B.
7. "7 to the second" or "7 squared."
8. "9 to the third" or "9 cubed."
9. "7 squared times 9 cubed."
10. "2 cubed times 3 to the fifth." 8A

Study Exercise Two (Frame 10)

1. 3^2 **2.** 5^3
3. $2^3 \times 7^2$ **4.** $5^4 \times 11^2 \times 13^3 \times 17^1$
5. $2^1 \times 3^2 \times 7^3$ 10A

Study Exercise Three (Frame 12)

1. 25 **2.** 27 **3.** 4
4. 49 **5.** 1 **6.** 0
7. 100 **8.** 1,000 **9.** 200
10. 2,205 **11.** 720 **12.** 450 12A

SOLUTIONS TO STUDY EXERCISES, CONTD.

Study Exercise Four (Frame 15)

1. yes; 5^2
2. yes; 7^2
3. no
4. yes; 10^2
5. no
6. yes; 0^2
7. yes; 1^2
8. no
9. yes; 8^2
10. yes; 12^2

15A

Study Exercise Five (Frame 21)

1. 6
2. 7
3. 1
4. 0
5. 8
6. 10
7. 11

21A

unit

9

Primes, Composites and Prime Factoring

Objectives:

By the end of this unit you should understand the meaning of the following terms:

1. natural number
2. prime number
3. composite number

You should also be able to write, in exponential notation, the prime factors of a composite number. (1)

Natural Numbers

1, 2, 3, 4, 5, 6, 7, 8, 9, 10, 11, ...

The natural numbers are all of the whole numbers except zero. (2)

A *prime number* is any natural number which is evenly divisible *only* by itself and one. Remember, "evenly divisible" means that the remainder is zero. One is not considered to be a prime number. (3)

Examples

1. 2 is a prime number.
2. 6 is not a prime number, because 6 is evenly divisible by 2 and 3.
3. 11 is a prime number.
4. 15 is not a prime number, because 15 is evenly divisible by 3 and 5. (4)

A *composite number* is any natural number which is evenly divisible by a number *other* than one or itself.

No number is both prime and composite. (5)

Examples of Composite Numbers

1. 6 is composite because it is evenly divisible by 2 or 3.
2. 15 is composite because it is evenly divisible by 3 or 5.
3. 22 is composite because it is evenly divisible by 2 or 11. (6)

Natural Numbers Separated Into Three Categories

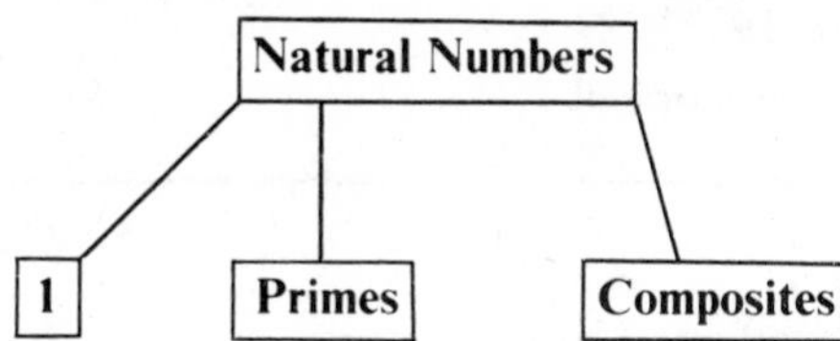

Given any natural number, it is either 1, prime, or composite. 7

Review of the Divisibility Tests

1. Any natural number is evenly divisible by 2 if the last digit is even.

Example: 138 is divisible by 2.

2. Any natural number is evenly divisible by 3 if the sum of its digits is divisible by 3.

Example: 132 is divisible by 3.

3. Any natural number is evenly divisible by 5 if the last digit is either 5 or 0.

Examples: 30 and 145 are both divisible by 5.

4. Any natural number is evenly divisible by 10 if the last digit is zero.

Example: 130 is divisible by 10. 8

The divisibility tests may be used to help determine if a natural number is composite.

Example 1: 25 is composite because it is evenly divisible by 5.

Example 2: 42 is composite because it is evenly divisible by 2 and 3.

Example 3: 5,421 is composite because it is evenly divisible by 3.

Example 4: 77 is composite because it is evenly divisible by 7 and 11. 9

Study Exercise One

Classify each of the following as either prime or composite.

1. 2	**2.** 6	**3.** 7
4. 20	**5.** 16	**6.** 19
7. 141	**8.** 2,675	**9.** 17
10. 121		

10

The Prime Numbers Less Than 30

2, 3, 5, 7, 11, 13, 17, 19, 23, 29

(These should be memorized.) (11)

Prime Factoring

Every composite number can be written as a product of prime factors.

Example:

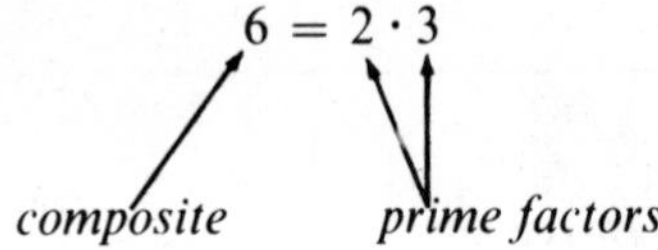

(12)

The Prime Factorization of 30

Line (*a*)	2	30
Line (*b*)	3	15
Line (*c*)		5

$$30 = 2 \times 3 \times 5$$

(13)

The Prime Factorization of 90

Line (*a*)	2	90
Line (*b*)	3	45
Line (*c*)	3	15
Line (*d*)		5

$$90 = 2 \times \underbrace{3 \times 3} \times 5$$

Exponential Form: $90 = 2^1 \times 3^2 \times 5^1$ (14)

The Prime Factorization of 200

Line (*a*)	2	200
Line (*b*)	2	100
Line (*c*)	2	50
Line (*d*)	5	25
Line (*e*)		5

$$200 = \underbrace{2 \times 2 \times 2} \times \underbrace{5 \times 5}$$

Exponential Form: $200 = 2^3 \times 5^2$ (15)

The Prime Factorization of 1,911

Line (a) $3\,\underline{|1{,}911}$
Line (b) $7\,\underline{|637}$
Line (c) $7\,\underline{|91}$
Line (d) 13

$$1{,}911 = 3 \times \underbrace{7 \times 7} \times 13$$

Exponential Form: $1{,}911 = 3^1 \times 7^2 \times 13^1$

16

The Prime Factorization of 297

Line (a) $3\,\underline{|297}$
Line (b) $3\,\underline{|99}$
Line (c) $3\,\underline{|33}$
Line (d) 11

$$297 = \underbrace{3 \times 3 \times 3} \times 11$$

Exponential Form: $297 = 3^3 \times 11^1$

Study Exercise Two

Use the preceding method to give the prime factorization for each of the following. Put your answers in exponential form.

1. 12 **2.** 24 **3.** 108
4. 180 **5.** 92 **6.** 2,205
7. 1,683

REVIEW EXERCISES

A. Fill in the blanks.

1. The natural numbers are all of the whole numbers except __________.

2. "Evenly divisible" means that the remainder is __________.

3. No number is both prime and __________.

4. Given any natural number, it is either __________, __________, or composite.

B. Classify each of the following as either prime or composite.

5. 12   **6.** 23   **7.** 21

8. 49   **9.** 105   **10.** 7

11. 19   **12.** 3,111

C. **13.** List all the prime numbers which are less than 30.

14. List all the composite numbers which are less than 30.

15. Is 1 a prime number?

16. Is 1 a composite number?

D. Give the prime factorizations for each of the following. Put your answers in exponential form.

17. 4   **18.** 9   **19.** 36

20. 35   **21.** 75   **22.** 63

23. 144   **24.** 174   **25.** 4,410 (19)

SOLUTIONS TO REVIEW EXERCISES

A. **1.** zero   **2.** zero

3. composite   **4.** one, prime

B. **5.** composite; divisible by 2 and 3.

6. prime

7. composite; divisible by 3.

8. composite; divisible by 7.

9. composite; divisible by 3 and 5.

10. prime

11. prime

12. composite; divisible by 3.

C. **13.** See Frame 11

14. 4, 6, 8, 9, 10, 12, 14, 15, 16, 18, 20, 21, 22, 24, 25, 26, 27, 28

15. no

16. no

D. **17.** $4 = 2^2$ **18.** $9 = 3^2$

19. $36 = 2^2 \times 3^2$ **20.** $35 = 5^1 \times 7^1$

21. $75 = 3^1 \times 5^2$ **22.** $63 = 3^2 \times 7^1$

23. $144 = 2^4 \times 3^2$ **24.** $174 = 2^1 \times 3^1 \times 29^1$

25. $4{,}410 = 2^1 \times 3^2 \times 5^1 \times 7^2$ (20)

SUPPLEMENTARY PROBLEMS

A. True or False

1. 0 is a natural number.
2. 1 is a prime number.
3. 15 is a composite number.
4. 103 is evenly divisible by 3.
5. 4,263 is evenly divisible by 3.
6. 9,710 is evenly divisible by 2.
7. $2^2 \cdot 21^1$ is the prime factorization of 84 in exponential form.
8. The prime factors of 882 are $2 \times 3 \times 3 \times 7 \times 11$.
9. 72 is both prime and composite.
10. The following group contains all composite numbers: 6, 38, 15, 27, 100.

B. Give the prime factorization for each of the following. Put your answers in exponential form.

11. 14 **12.** 18 **13.** 21
14. 56 **15.** 252 **16.** 378
17. 1,155 **18.** 4,235 **19.** 2,275
20. 3,400

SOLUTIONS TO STUDY EXERCISES

Study Exercise One (Frame 10)

1. 2 is prime.
2. 6 is composite because it is evenly divisible by 2 and 3.
3. 7 is prime.
4. 20 is composite because it is evenly divisible by 2, 5, and 10.
5. 16 is composite because it is evenly divisible by 2.
6. 19 is prime.
7. 141 is composite because it is evenly divisible by 3.
8. 2,675 is composite because it is evenly divisible by 5.
9. 17 is prime.
10. 121 is composite because it is evenly divisible by 11.

10A

Study Exercise Two (Frame 18)

1. 2|12, 2|6, 3

$12 = 2 \times 2 \times 3$
Answer: $12 = 2^2 \times 3^1$

2. 2|24, 2|12, 2|6, 3

$24 = 2 \times 2 \times 2 \times 3$
Answer: $24 = 2^3 \times 3^1$

3. 2|108, 2|54, 3|27, 3|9, 3

$108 = 2 \times 2 \times 3 \times 3 \times 3$
Answer: $108 = 2^2 \times 3^3$

4. 2|180, 2|90, 3|45, 3|15, 5

$180 = 2 \times 2 \times 3 \times 3 \times 5$
Answer: $180 = 2^2 \times 3^2 \times 5^1$

SOLUTIONS TO STUDY EXERCISES, CONTD.

Study Exercise Two (Frame 18, contd.)

5.

$$\begin{array}{r|l} 2 & 92 \\ 2 & 46 \\ & 23 \end{array}$$

$92 = 2 \times 2 \times 23$

Answer: $92 = 2^2 \times 23^1$

6.

$$\begin{array}{r|l} 3 & 2{,}205 \\ 3 & 735 \\ 5 & 245 \\ 7 & 49 \\ & 7 \end{array}$$

$2{,}205 = 3 \times 3 \times 5 \times 7 \times 7$

Answer: $2{,}205 = 3^2 \times 5^1 \times 7^2$

7.

$$\begin{array}{r|l} 3 & 1{,}683 \\ 3 & 561 \\ 11 & 187 \\ & 17 \end{array}$$

$1{,}683 = 3 \times 3 \times 11 \times 17$

Answer: $1{,}683 = 3^2 \times 11^1 \times 17^1$

18A

unit

10

Least Common Multiple

Objectives:

By the end of this unit you should be able to explain the meaning of the following terms:

1. multiple
2. common multiple
3. least common multiple

You should also be able to find the least common multiple of two or more numbers by using prime factors and exponents.

①

Multiple

The product of any natural number and 5 is said to be a *multiple* of 5.

Examples:

1. 15 is a multiple of 5 because $15 = 5 \times 3$.
2. 30 is a multiple of 5 because $30 = 5 \times 6$.
3. 5 is a multiple of 5 because $5 = 5 \times 1$.

6, 9, and 27 are multiples of 3 because:

1. $6 = 3 \times 2$
2. $9 = 3 \times 3$
3. $27 = 3 \times 9$

7, 14, 21, and 28 are multiples of 7 because:

1. $7 = 7 \times 1$
2. $14 = 7 \times 2$
3. $21 = 7 \times 3$
4. $28 = 7 \times 4$

---

1. The *first* multiple of 7 is $7 \times \underline{1}$ or 7.
2. The *second* multiple of 7 is $7 \times \underline{2}$ or 14.
3. The *third* multiple of 7 is $7 \times \underline{3}$ or 21.
4. The *fourth* multiple of 7 is $7 \times \underline{4}$ or 28.
5. The *ninth* multiple of 7 is $7 \times \underline{9}$ or 63.

---

1. The first five multiples of 2 are: 2, 4, 6, 8, and 10.
2. The first five multiples of 3 are: 3, 6, 9, 12, and 15.
3. The first five multiples of 9 are: 9, 18, 27, 36, and 45. ⑥

Study Exercise One

A. Write the first six multiples for each of the following.

1. 4	**2.** 5	**3.** 8
4. 12	**5.** 25	**6.** 100

B.
7. Find the third multiple of 6.
8. Find the sixth multiple of 3.
9. Find the tenth multiple of 7.
10. Find the seventh multiple of 10. ⑦

Common Multiples of 2 and 3

1. The first ten multiples of 2: 2, 4, 6, 8, 10, 12, 14, 16, 18, 20.

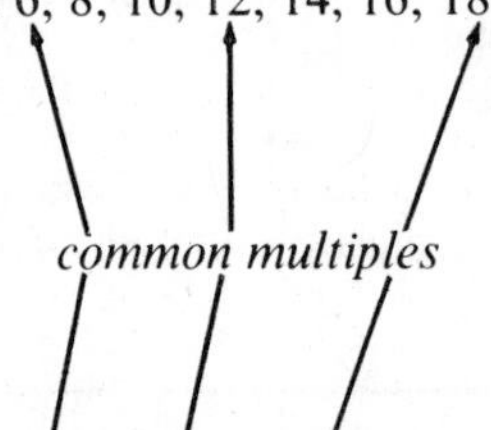

2. The first seven multiples of 3: 3, 6, 9, 12, 15, 18, 21.

The first three *common multiples* of 2 and 3 are 6, 12, and 18. ⑧

The common multiples of 2 and 3 are evenly divisible by both 2 and 3.

Example: 6, 12, and 18 are common multiples of 2 and 3.

Line (*a*) $\begin{array}{r}3\\2\overline{)6}\end{array}$ $\quad$ $\begin{array}{r}2\\3\overline{)6}\end{array}$

Line (*b*) $\begin{array}{r}6\\2\overline{)12}\end{array}$ $\quad$ $\begin{array}{r}4\\3\overline{)12}\end{array}$

Line (*c*) $\begin{array}{r}9\\2\overline{)18}\end{array}$ $\quad$ $\begin{array}{r}6\\3\overline{)18}\end{array}$

6, 12, and 18 are each evenly divisible by 2 and 3. ⑨

Common Multiples of 8 and 12

1. The first six multiples of 8: 8, 16, 24, 32, 40, 48.

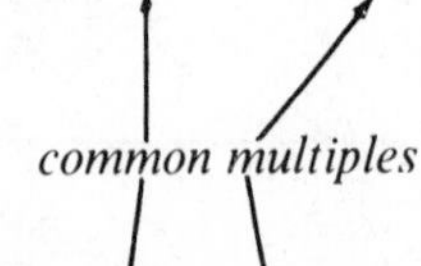

2. The first six multiples of 12: 12, 24, 36, 48, 60, 72.

The first two common multiples of 8 and 12 are 24 and 48. ⑩

The common multiples of 8 and 12 are evenly divisible by both 8 and 12.

Example: 24 and 48 are common multiples of 8 and 12.

Line (a) $8\overset{3}{\overline{)24}}$ $\quad$ $12\overset{2}{\overline{)24}}$

Line (b) $8\overset{6}{\overline{)48}}$ $\quad$ $12\overset{4}{\overline{)48}}$

24 and 48 are evenly divisible by 8 and 12.

(11)

Study Exercise Two

1. List the first six multiples of 9 and the first nine multiples of 6, then find the first three *common* multiples.
2. Find the first three common multiples of 10 and 15.
3. Find the first two common multiples of 2, 4, and 6.

The Least Common Multiple of 6 and 9

Multiples of 6: 6, 12, (18), 24, 30, (36), 42, 48, (54), ...

Multiples of 9: 9, (18), 27, (36), 45, (54), ...

Some common multiples of 6 and 9 are 18, 36, and 54.
The *smallest* common multiple of 6 and 9 is 18.
The smallest common multiple is called the *least common multiple* and is abbreviated *LCM*.
The LCM of 6 and 9 is 18.

The least common multiple (LCM) of 6 and 9 is the smallest number which is evenly divisible by both 6 and 9.

Example: The LCM of 6 and 9 is 18.
18 is the smallest number which is evenly divisible by 6 and 9.

$6\overset{3}{\overline{)18}}$ and $9\overset{2}{\overline{)18}}$

The Least Common Multiple (LCM) of 2, 4, and 6

Multiples of 2: 2, 4, 6, 8, 10, (12), 14, 16, 18, 20, 22, (24), ...

Multiples of 4: 4, 8, (12), 16, 20, (24), 28, ...

Multiples of 6: 6, (12), 18, (24), 30, 36, ...

The smallest common multiple is 12.
The LCM = 12.

The LCM of 2, 4, and 6 is 12.

Therefore, 12 is the smallest number which is evenly divisible by 2, 4, and 6.

$$\begin{array}{r}6\\2\overline{)12}\end{array}\qquad\begin{array}{r}3\\4\overline{)12}\end{array}\qquad\begin{array}{r}2\\6\overline{)12}\end{array}$$

16

Study Exercise Three

1. By listing multiples of 10 and 15, find the least common multiple (LCM).
2. Find the LCM of 3, 6, and 4.
3. Find the smallest number which is evenly divisible by 2, 3, and 4.

Using Exponents to Find the LCM

Example 1: Find the LCM of 60 and 72.

Solution:

Step (1) Write each number in prime factored form.

$60 = 2 \times 2 \times 3 \times 5 \qquad 72 = 2 \times 2 \times 2 \times 3 \times 3$

Step (2) Write in exponential form.

$60 = 2^2 \times 3^1 \times 5^1 \qquad 72 = 2^3 \times 3^2$

Step (3) Write each number that was used as a base, but write it *only once.* (Neglect repetitions.)

$2 \times 3 \times 5$

Step (4) Attach the largest exponent used on each base from Step 2.

$\text{LCM} = 2^3 \times 3^2 \times 5^1$

Step (5) Evaluate the exponential expression.

$\text{LCM} = 8 \times 9 \times 5$ or 360

Example 2: Find the LCM of 12 and 45.

Solution:

Step (1) Write each number in prime factored form.

$12 = 2 \times 2 \times 3 \qquad 45 = 3 \times 3 \times 5$

Step (2) Write in exponential form.

$12 = 2^2 \times 3^1 \qquad 45 = 3^2 \times 5^1$

Step (3) Write each number that was used as a base, but write it *only once.* (Neglect repetitions.)

$2 \times 3 \times 5$

Step (4) Attach the largest exponent used on each base from Step 2.

$\text{LCM} = 2^2 \times 3^2 \times 5^1$

Step (5) Evaluate the exponential expression.

$\text{LCM} = 4 \times 9 \times 5$ or $180.$

Example 3: Find the LCM of 15 and 77.

Solution:

Step (1) Write each number in prime factored form.

$15 = 3 \times 5 \qquad 77 = 7 \times 11$

Step (2) Write in exponential form.

$15 = 3^1 \times 5^1 \qquad 77 = 7^1 \times 11^1$

Step (3) Write each number that was used as a base, but write it only once. (Neglect repetitions.)

$3 \times 5 \times 7 \times 11$

Step (4) Attach the largest exponent used on each base from Step 2.

$\text{LCM} = 3^1 \times 5^1 \times 7^1 \times 11^1$

Step (5) Evaluate the exponential expression.

$\text{LCM} = 1{,}155$

Example 4: Find the LCM of 120, 126 and 900.

Solution:

Step (1) Write each number in prime factored form.

$120 = 2 \times 2 \times 2 \times 3 \times 5 \qquad 126 = 2 \times 3 \times 3 \times 7$

$900 = 2 \times 2 \times 3 \times 3 \times 5 \times 5$

Step (2) Write in exponential form.

$120 = 2^3 \times 3^1 \times 5^1 \qquad 126 = 2^1 \times 3^2 \times 7^1$

$900 = 2^2 \times 3^2 \times 5^2$

Step (3) Write each number that was used as a base, but write it just once. (Neglect repetitions.)

$2 \times 3 \times 5 \times 7$

Step (4) Attach the largest exponent used on each base from Step 2.

$\text{LCM} = 2^3 \times 3^2 \times 5^2 \times 7^1$

Step (5) Evaluate the exponential expression.

$\text{LCM} = \underbrace{8 \times 9}_{} \times \underbrace{25 \times 7}_{}$

$\text{LCM} = 72 \times 175$

$\text{LCM} = 12{,}600$

Study Exercise Four

A. Use exponents to find the LCM.

1. Find the LCM of 12 and 10.
2. Find the LCM of 40, 20, and 28.
3. Find the LCM of 21 and 10.

B.

4. Find the smallest number which is evenly divisible by 15, 45, and 9.
5. The following three numbers are given in prime factored form using exponents:

$3{,}500 = 2^2 \times 5^3 \times 7^1$

$2{,}450 = 2^1 \times 5^2 \times 7^2$

$990 = 2^1 \times 3^2 \times 5^1 \times 11^1$

Find the LCM. You may leave your answer in prime factored form with exponents.

REVIEW EXERCISES

A. Fill in the blanks.

1. The product of any natural number and 7 is said to be a ____________ of 7.
2. The second multiple of 8 is ______.
3. The fifth multiple of 9 is ______.
4. The first multiple of 3 is ______.
5. The first four multiples of 5 are: 5, ______, ______, 20.
6. The first six multiples of 7 are: 7, ______, 21, 28, ______, ______.

B. Find the first three *common* multiples for:

7. 2 and 5
8. 2, 4, 6

C. **9.** Which of the three common multiples in question 7 is the least common multiple?
10. Which of the three common multiples in question 8 is the least common multiple?

D. Use exponents to find the LCM.

11. Find the LCM of 12 and 18.
12. Find the LCM of 40 and 60.
13. Find the LCM of 700, 490, and 196.
14. The following three numbers are given in prime factored form using exponents:

$$4{,}875 = 3^1 \times 5^3 \times 13^1$$
$$450 = 2^1 \times 3^2 \times 5^2$$
$$10{,}584 = 2^3 \times 3^3 \times 7^2$$

Find the LCM. Leave your answer in prime factored form using exponents. 23

SOLUTIONS TO REVIEW EXERCISES

A. **1.** multiple **2.** 16 **3.** 45
4. 3 **5.** 10, 15 **6.** 14, 35, 42

B. **7.** 2, 4, 6, 8, (10), 12, 14, 16, 18, (20), 22, 24, 26, 28, (30), ...

5, (10), 15, (20), 25, (30), 35, 40, ...

8. 2, 4, 6, 8, 10, (12), 14, 16, 18, 20, 22, (24), 26, 28, 30, 32, 34, (36), ...

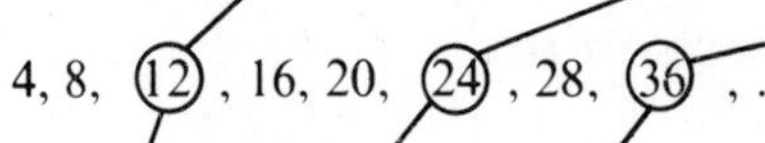

4, 8, (12), 16, 20, (24), 28, (36), ...

6, (12), 18, (24), 30, (36), ...

C. **9.** LCM = 10
10. LCM = 12

D. **11.** *Step (1)* $12 = 2 \times 2 \times 3$ $\quad 18 = 2 \times 3 \times 3$
Step (2) $12 = 2^2 \times 3^1$ $\quad 18 = 2^1 \times 3^2$
Step (3) 2×3
Step (4) $\text{LCM} = 2^2 \times 3^2$
Step (5) $\text{LCM} = 36$

12. *Step (1)* $40 = 2 \times 2 \times 2 \times 5$ $\quad 60 = 2 \times 2 \times 3 \times 5$
Step (2) $40 = 2^3 \times 5^1$ $\quad 60 = 2^2 \times 3^1 \times 5^1$
Step (3) $2 \times 3 \times 5$
Step (4) $\text{LCM} = 2^3 \times 3^1 \times 5^1$
Step (5) $\text{LCM} = 120$

SOLUTIONS TO REVIEW EXERCISES, CONTD.

(Frame 24, contd.)

13. *Step (1)* $700 = 2 \times 2 \times 5 \times 5 \times 7$ $\quad 490 = 2 \times 5 \times 7 \times 7$
$196 = 2 \times 2 \times 7 \times 7$
Step (2) $700 = 2^2 \times 5^2 \times 7^1$ $\quad 490 = 2^1 \times 5^1 \times 7^2$
$196 = 2^2 \times 7^2$
Step (3) $2 \times 5 \times 7$
Step (4) $\text{LCM} = 2^2 \times 5^2 \times 7^2$
Step (5) LCM = 4,900

14. $\text{LCM} = 2^3 \times 3^3 \times 5^3 \times 7^2 \times 13^1$

SUPPLEMENTARY PROBLEMS

A. Fill in the blanks.

1. The smallest number which is evenly divisible by two numbers is said to be the ________ ________ ________ of those numbers.
2. 35 is a multiple of 7 because ______ × 5 = 35.
3. The third multiple of 4 is ______.
4. The seventh multiple of 10 is ______.
5. The first multiple of 7 is ______.
6. The first four multiples of 6 are: 6, ______, 18, ______.
7. The first six multiples of 8 are: ______, 16, ______, ______, 40, ______.

B. Find the first three common multiples for:

8. 2 and 4
9. 2, 4, and 5

C. Use exponents to find the LCM.

10. 50 and 20
11. 24 and 18
12. 126, 108, and 98
13. The following three numbers are given in prime factored form using exponents:

$$3{,}500 = 2^2 \times 5^3 \times 7^1$$
$$4{,}875 = 3^1 \times 5^3 \times 13^1$$
$$10{,}584 = 2^3 \times 3^3 \times 7^2$$

Find the LCM. Leave your answer in prime factored form using exponents.

SOLUTIONS TO STUDY EXERCISES

Study Exercise One (Frame 7)

A.
1. 4, 8, 12, 16, 20, 24
2. 5, 10, 15, 20, 25, 30
3. 8, 16, 24, 32, 40, 48
4. 12, 24, 36, 48, 60, 72
5. 25, 50, 75, 100, 125, 150
6. 100, 200, 300, 400, 500, 600

B.
7. 6 × 3 or 18
8. 3 × 6 or 18
9. 7 × 10 or 70
10. 10 × 7 or 70

SOLUTIONS TO STUDY EXERCISES, CONTD.

Study Exercise Two (Frame 12)

1. 9, (18), 27, (36), 45, (54)

6, 12, (18), 24, 30, (36), 42, 48, (54)

The first three common multiples of 9 and 6 are: 18, 36, and 54.

2. 10, 20, (30), 40, 50, (60), 70, 80, (90)

15, (30), 45, (60), 75, (90)

The first three common multiples of 10 and 15 are: 30, 60, and 90.

3. 2, 4, 6, 8, 10, (12), 14, 16, 18, 20, 22, (24)

4, 8, (12), 16, 20, (24)

6, (12), 18, (24), 30

The first two common multiples of 2, 4, and 6 are: 12 and 24.

(12A)

Study Exercise Three (Frame 17)

1. 10, 20, (30), 40, 50, 60, 70, 80, 90, ...

15, (30), 45, 60, 75, 90, ...

The LCM of 10 and 15 is 30.

2. 3, 6, 9, (12), 15, 18, 21, 24, 27, 30, ...

6, (12), 18, 24, 30, ...

4, 8, (12), 16, 20, 24, 28, 32, ...

The LCM of 3, 6, and 4 is 12.

3. 2, 4, 6, 8, 10, (12), 14, 16, 18, 20, 22, 24, ...

3, 9, (12), 15, 18, 21, 24, 27, 30, ...

4, 8, (12), 16, 20, 24, 28, ...

The LCM is 12. Therefore, 12 is the smallest number which is evenly divisible by 2, 3, and 4.

(17A)

SOLUTIONS TO STUDY EXERCISES, CONTD.

Study Exercise Four (Frame 22)

A. **1.** *Step (1)* $12 = 2 \times 2 \times 3$ $10 = 2 \times 5$
Step (2) $12 = 2^2 \times 3^1$ $10 = 2^1 \times 5^1$
Step (3) $2 \times 3 \times 5$
Step (4) $\text{LCM} = 2^2 \times 3^1 \times 5^1$
Step (5) $\text{LCM} = 60$

2. *Step (1)* $40 = 2 \times 2 \times 2 \times 5$ $20 = 2 \times 2 \times 5$ $28 = 2 \times 2 \times 7$
Step (2) $40 = 2^3 \times 5^1$ $20 = 2^2 \times 5^1$ $28 = 2^2 \times 7^1$
Step (3) $2 \times 5 \times 7$
Step (4) $\text{LCM} = 2^3 \times 5^1 \times 7^1$
Step (5) $\text{LCM} = 280$

3. *Step (1)* $21 = 3 \times 7$ $10 = 2 \times 5$
Step (2) $21 = 3^1 \times 7^1$ $10 = 2^1 \times 5^1$
Step (3) $2 \times 3 \times 5 \times 7$
Step (4) $\text{LCM} = 2^1 \times 3^1 \times 5^1 \times 7^1$
Step (5) $\text{LCM} = 210$

B. **4.** Find the LCM of 15, 45, and 9.
Step (1) $15 = 3 \times 5$ $45 = 3 \times 3 \times 5$ $9 = 3 \times 3$
Step (2) $15 = 3^1 \times 5^1$ $45 = 3^2 \times 5^1$ $9 = 3^2$
Step (3) 3×5
Step (4) $\text{LCM} = 3^2 \times 5^1$
Step (5) $\text{LCM} = 45$
45 is the smallest number which is evenly divisible by 15, 45, and 9.

5. $\text{LCM} = 2^2 \times 3^2 \times 5^3 \times 7^2 \times 11^1$

22A

Practice Test—Whole Numbers—Units 1–10

1. Give the place value for the underlined digit: 26,4<u>5</u>7,053.
2. Write in words the name for 26,041,030.
3. Write the numeral for: eight hundred forty-two thousand, six hundred, nine.
4. Round off the whole number, 1,365,432, to the nearest ten thousand.
5. Add the following by writing each number in place value notation: $2{,}341 + 413 + 35$.
6. Perform the following addition and check your answer.

$$\begin{array}{r} 62{,}948 \\ 408{,}932 \\ \underline{9{,}756} \end{array}$$

7. Perform the following subtraction by writing each number in place value notation: $765 - 341$.
8. Perform the following subtraction and check your answer:

$$\begin{array}{r} 206{,}401 \\ \underline{-58{,}382} \end{array}$$

9. Multiply the following: $3{,}652 \times 1{,}000$.
10. Multiply the following using the short form and check your answer: $6{,}048 \times 708$.
11. List three symbols for showing that 15 is to be divided by 3.
12. Which of the following divisions *cannot* be performed?
    (a) $20 \div 5$ (b) $0 \div 5$ (c) $5 \div 0$
13. Perform the following long division and check your answer: $32{,}607 \div 153$.
14. State the divisibility test for 3.
15. In the exponential expression 5^4, indicate the base and indicate the exponent.
16. Evaluate the following exponential expression: $2^3 \cdot 3^2 \cdot 5^2$.
17. Evaluate: $\sqrt{81}$.
18. List the first five prime numbers.
19. Give the prime factorization of 450.
20. Give the fifth multiple of 12.
21. Find the LCM of 120, 126, and 450.
22. If 6 candy bars cost 42 cents, how much does one candy bar cost?
23. If a box weighs 21 pounds, what is the total weight of 8 identical boxes?
24. Find the cost of 5 candy bars if 12 bars cost 96 cents.
25. If a manufacturing concern produces 6,000 articles in 5 days, how many articles will it produce in 21 days?

Answers—Practice Test—Whole Numbers

1. hundred thousands
2. twenty-six million, forty-one thousand, thirty.
3. 842, 609
4. 1,370,000
5. 2 thousands + 7 hundreds + 8 tens + 9 units
6. 481,636
7. 4 hundreds + 2 tens + 4 units
8. 148,019
9. 3,652,000
10. 4,281,984
11. $15 \div 3$, $3\overline{)15}$, $\frac{15}{3}$
12. (c) $5 \div 0$
13. 213 R18
14. The sum of the digits is evenly divisible by 3.
15. The base is 5 and the exponent is 4.
16. 1,800 **17.** 9 **18.** 2, 3, 5, 7, 11
19. $2^1 \times 3^2 \times 5^2$ **20.** 60 **21.** 12,600
22. 7 cents **23.** 168 pounds **24.** 40 cents
25. 25,200 articles

unit

11

Introduction to Fractions

Objectives:

By the end of this unit you should:

1. know what a fraction is.

2. memorize the Golden Rule of Fractions.

3. know what is meant by numerator, denominator, and terms of a fraction. ①

If a pie is cut into 4 equal portions, then what part of the pie is a portion?

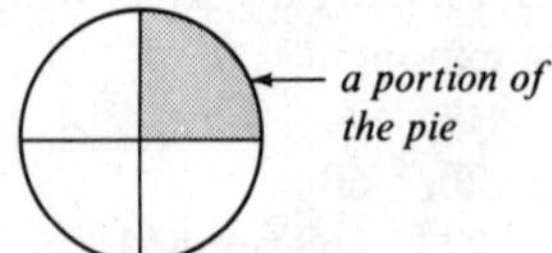

②

Four portions are equal to the whole pie. The portion is a part of the whole pie. Can you compare the part to the whole?

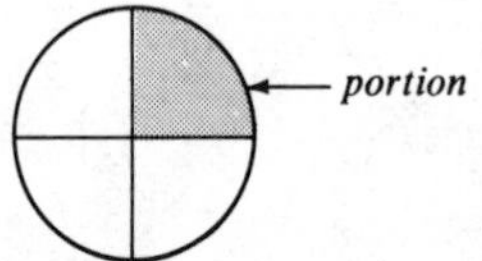

③

We will say that a portion of the pie is to the whole pie as 1 is to 4. The number idea that describes this comparison will be written $\frac{1}{4}$ and will be called a *fraction*.

When a thing is broken into equal parts, each part is called a fraction of the whole thing. ④

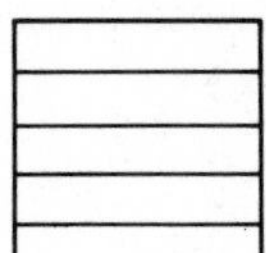

The square has been divided into 5 equal parts. Each part of the square is $\frac{1}{5}$. If two parts are taken, we have two-fifths and write it $\frac{2}{5}$. If three parts are taken, we have $\frac{3}{5}$. (5)

Study Exercise One

In the drawings below, write the fraction that the shaded part represents to the whole.

1. 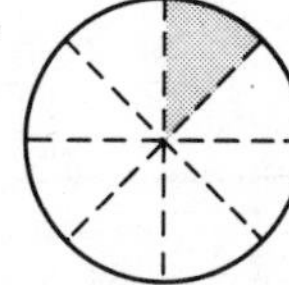
2. 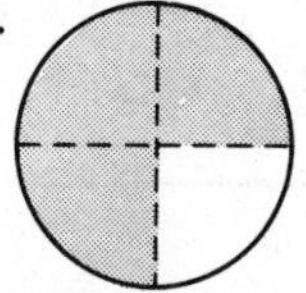
3. 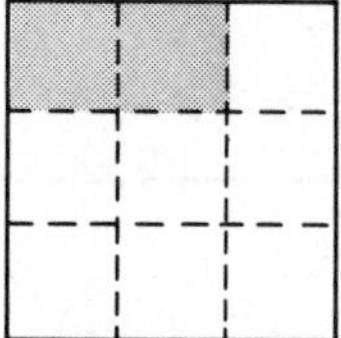
4. 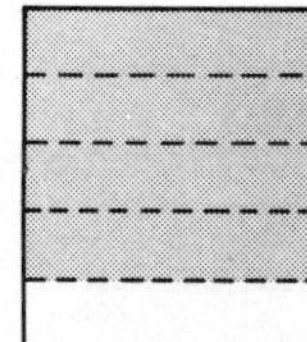
5. 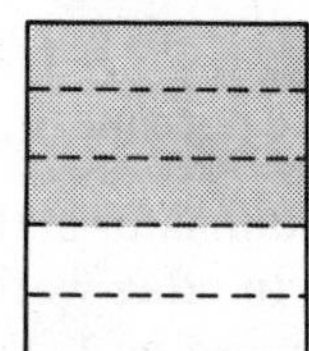
6. 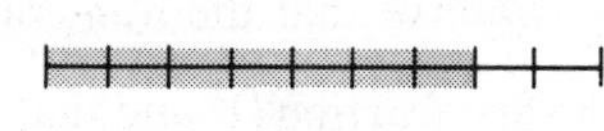

(6)

Terms of a Fraction

We have written fractions by writing one whole number over another.

In the fraction $\frac{3}{5}$, the 3 and 5 are called *terms* of the fraction; the 3 is called the *numerator* and the 5 is called the *denominator*.

$\frac{3}{5}$ 3 ← *numerator*; 5 ← *denominator*

(7)

The same fraction may be written in many different forms.

In the diagrams below, $\frac{2}{4} = \frac{8}{16}, \frac{8}{16} = \frac{1}{2}, \frac{1}{2} = \frac{3}{6}, \frac{2}{4} = \frac{3}{6}$, etc.

1.

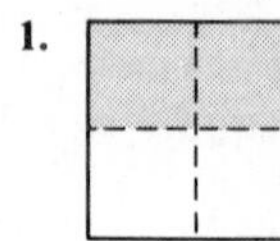

$\frac{2}{4}$

2.

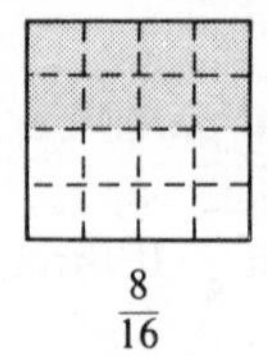

$\frac{8}{16}$

3.

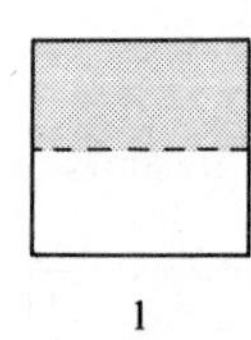

$\frac{1}{2}$

4.

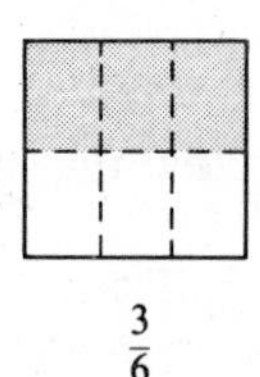

$\frac{3}{6}$

8

In the preceding frame we saw that:

$$\frac{1}{2} = \frac{2}{4} \quad \text{or} \quad \frac{1}{2} = \frac{3}{6} \quad \text{or} \quad \frac{1}{2} = \frac{8}{16}$$

In the statement, $\frac{1}{2} = \frac{3}{6}$, observe that the numerator of the second fraction (3) is three times the numerator of the first fraction (1) and that the denominator of the second fraction (6) is three times the denominator of the first fraction (2). In other words, the numerator and denominator of the first fraction have each been multiplied by 3.

The Golden Rule of Fractions—Part 1

We have discovered in Frame 9 a part of the basic rule of fractions which is sometimes called the Golden Rule of Fractions.

Multiplying both numerator and denominator of a fraction by the same non-zero number does not alter the value of the fraction.

Examples:

1. We take the fraction $\frac{3}{5}$ and multiply numerator and denominator by the number 4.

$$\frac{3}{5} = \frac{3 \times 4}{5 \times 4} = \frac{12}{20}$$

2. We take $\frac{2}{7}$ and multiply numerator and denominator by 6.

$$\frac{2}{7} = \frac{2 \times 6}{7 \times 6} = \frac{12}{42}$$

3. We take $\frac{1}{5}$ and multiply numerator and denominator by 5.

$$\frac{1}{5} = \frac{1 \times 5}{5 \times 5} = \frac{5}{25}$$

Golden Rule of Fractions—Part 2

Since $\frac{1}{2} = \frac{3}{6}$, it is equally true that $\frac{3}{6} = \frac{1}{2}$. Observe that the numerator and denominator of the first fraction have each been divided by 3 to get the second fraction. We thus arrive at the second part of the basic rule of fractions.

Dividing both numerator and denominator of a fraction by the same non-zero number does not alter the value of the fraction. (12)

Golden Rule of Fractions

The two parts are combined to form *The Golden Rule of Fractions:*

Multiplying or dividing both numerator and denominator of a fraction by the same non-zero number does not alter the value of the fraction. (13)

Examples:

1. We take the fraction $\frac{6}{10}$ and divide numerator and denominator by 2.

$$\frac{6}{10} = \frac{6 \div 2}{10 \div 2} = \frac{3}{5}$$

2. We take $\frac{20}{36}$ and divide numerator and denominator by 4.

$$\frac{20}{36} = \frac{20 \div 4}{36 \div 4} = \frac{5}{9}$$

(14)

Study Exercise Two

Take the given fraction and find an equivalent fraction by multiplying or dividing numerator and denominator by the indicated number.

1. $\frac{1}{7}$; multiply by 2

2. $\frac{4}{6}$; divide by 2

3. $\frac{10}{12}$; multiply by 3

4. $\frac{10}{12}$; divide by 2

5. $\frac{8}{9}$; multiply by 4

6. $\frac{12}{21}$; divide by 3

(15)

Changing The Form Of A Fraction

The Golden Rule may be used to change the terms of a fraction.

Problem: Let us change $\frac{1}{4}$ to an equivalent fraction with denominator 12.

$$\frac{1}{4} = \frac{?}{12}$$

Solution: Since $3 \times 4 = 12$, $\frac{1}{4} = \frac{?}{12}$ (denominator $\times 3$)

The denominator was multiplied by 3, so we must also multiply the numerator by 3.

$\frac{1}{4} = \frac{?}{12}$ (numerator $\times 3$, denominator $\times 3$) which gives $\frac{1}{4} = \frac{3}{12}$

16

Example 1: Supply the missing number, $\frac{4}{7} = \frac{?}{21}$

Solution: Since $3 \times 7 = 21$, $\frac{4}{7} = \frac{?}{21}$ (numerator $\times 3$, denominator $\times 3$) which gives $\frac{4}{7} = \frac{12}{21}$

Example 2: Supply the missing number, $\frac{12}{18} = \frac{?}{9}$

Solution: Since $18 \div 2 = 9$, $\frac{12}{18} = \frac{?}{9}$ (numerator $\div 2$, denominator $\div 2$) which gives $\frac{12}{18} = \frac{6}{9}$

17

Study Exercise Three

A. Supply the missing term.

1. $\frac{3}{10} = \frac{?}{30}$

2. $\frac{3}{4} = \frac{?}{12}$

3. $\frac{2}{3} = \frac{?}{27}$

4. $\frac{2}{5} = \frac{?}{25}$

(Frame 18, contd.)

5. $\frac{9}{15} = \frac{?}{5}$

6. $\frac{8}{32} = \frac{?}{8}$

7. $\frac{9}{18} = \frac{?}{6}$

8. $\frac{35}{60} = \frac{?}{12}$

9. $\frac{5}{?} = \frac{25}{40}$

10. $\frac{3}{10} = \frac{15}{?}$

B. True or False

11. $\frac{3}{6} = \frac{1}{2}$

12. $\frac{2}{3} = \frac{36}{48}$

13. $\frac{12}{21} = \frac{4}{7}$

14. $\frac{21}{49} = \frac{3}{7}$

15. $\frac{6}{7} = \frac{48}{42}$

16. $\frac{7}{15} = \frac{21}{60}$

17. $\frac{12}{15} = \frac{4}{5}$

18. $\frac{7}{8} = \frac{28}{36}$

(18)

Whole Numbers As Fractions

A whole number may be written as a fraction with denominator one.

Thus, $3 = \frac{3}{1}$, $5 = \frac{5}{1}$, $8 = \frac{8}{1}$

A whole number may also be written as a fraction with any non-zero number as denominator.

For example, $3 = \frac{3}{1}$ and $\frac{3}{1} = \frac{6}{2}$; thus, $3 = \frac{6}{2}$.

Also, $\frac{3}{1} = \frac{12}{4}$ and so, $3 = \frac{12}{4}$.

Example 1: Write 5 as a fraction with denominator 4.

Solution: $5 = \frac{5}{1}$ and $\frac{5}{1} = \frac{5 \times 4}{1 \times 4} = \frac{20}{4}$. Thus, $5 = \frac{20}{4}$

Example 2: Write 8 as a fraction with denominator 8.

Solution: $8 = \frac{8}{1}$ and $\frac{8}{1} = \frac{8 \times 8}{1 \times 8} = \frac{64}{8}$. Thus, $8 = \frac{64}{8}$

The Number Zero

Zero written as a fraction is $\frac{0}{1}$. By the Golden Rule, we can choose any non-zero number and multiply numerator and denominator to get an equivalent fraction. Let us choose 5.

$$\frac{0}{1} = \frac{0 \times 5}{1 \times 5} = \frac{0}{5}. \qquad \text{Thus } 0 = \frac{0}{5}$$

Study Exercise Four

1. Write 7 as a fraction with denominator of 1.
2. Write 4 as a fraction with denominator of 3.
3. Write 8 as a fraction with denominator of 7.
4. Write 1 as a fraction with denominator of 7.
5. Write 0 as a fraction with denominator of 3.

The Meaning Of A Fraction

A fraction can also be thought of as an indicated division.

Thus $\frac{4}{2}$ means $4 \div 2$ and $\frac{2}{3}$ means $2 \div 3$.

Since division by zero is not allowed, no fraction can be written with a denominator of zero. $\frac{5}{0}$ would mean $5 \div 0$ and is undefined. $\frac{7}{0}, \frac{1}{0}, \frac{6}{0}, \frac{0}{0}$, are all undefined; but $\frac{0}{7}, \frac{0}{1}, \frac{0}{6}$, are defined and each equals zero.

REVIEW EXERCISES

A. Write the fraction that the shaded part represents to the whole.

1. 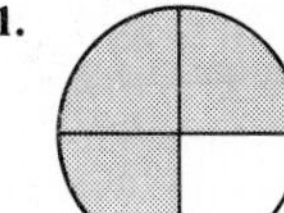

2. 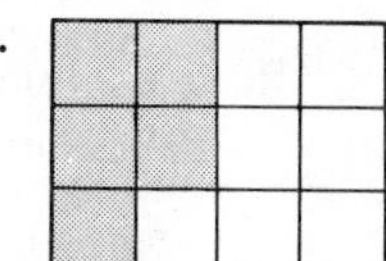

B. **3.** For the fraction $\frac{2}{3}$, we call 2 the __________ and 3 the __________.

4. The fractions $\frac{2}{7}$, and $\frac{2}{5}$ have the same __________.

5. Which of these is not a fraction: $\frac{5}{9}, \frac{1}{3}, \frac{4}{0}$?

C. Find an equivalent fraction by using the Golden Rule as indicated.

6. $\frac{7}{9}$; multiply by 3

7. $\frac{16}{18}$; divide by 2

8. $\frac{12}{20}$; multiply by 4

D. Supply the missing term:

9. $\frac{2}{3} = \frac{6}{?}$

10. $\frac{8}{20} = \frac{?}{5}$

11. $7 = \frac{14}{?}$

12. $0 = \frac{?}{7}$

E. True or False

13. $\frac{3}{10} = \frac{6}{20}$

14. $\frac{24}{36} = \frac{8}{9}$

(24)

SOLUTIONS TO REVIEW EXERCISES

A. **1.** $\frac{3}{4}$ **2.** $\frac{5}{12}$

B. **3.** numerator; denominator
4. numerator
5. $\frac{4}{0}$ is not a fraction since $4 \div 0$ is not allowed.

C. **6.** $\frac{7}{9} = \frac{7 \times 3}{9 \times 3} = \frac{21}{27}$ **7.** $\frac{16}{18} = \frac{16 \div 2}{18 \div 2} = \frac{8}{9}$

8. $\frac{12}{20} = \frac{12 \times 4}{20 \times 4} = \frac{48}{80}$

D. **9.** $\frac{2}{3} = \frac{6}{?} = \frac{6}{9}$ ($\times 3$ above, $\times 3$ below) **10.** $\frac{8}{20} = \frac{?}{5} = \frac{2}{5}$ ($\div 4$ above, $\div 4$ below)

11. $\frac{7}{1} = \frac{14}{?} = \frac{14}{2}$ ($\times 2$ above, $\times 2$ below) **12.** $\frac{0}{1} = \frac{?}{7} = \frac{0}{7}$ ($\times 7$ above, $\times 7$ below)

E. **13.** true
14. false 25

SUPPLEMENTARY PROBLEMS

A. Find the fraction suggested by the following drawings.

1. 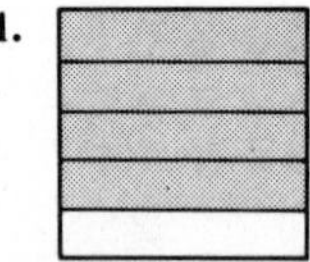 **2.** **3.** 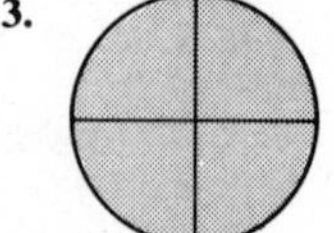

B. **4.** In writing $\frac{3}{5}$, we call 3 the __________ of the fraction and 5 the __________ of the fraction; we call 3 and 5 __________ of the fraction.
5. State the Golden Rule of Fractions

C. Find an equivalent fraction by using the Golden Rule as indicated.

6. $\frac{32}{40}$; divide by 8 **7.** $\frac{25}{120}$; divide by 5

8. $\frac{7}{35}$; divide by 7 **9.** $\frac{3}{8}$; multiply by 8

SUPPLEMENTARY PROBLEMS, CONTD.

10. $\frac{7}{8}$; multiply by 5 **11.** $\frac{0}{4}$; multiply by 5

D. Supply the missing term.

12. $\frac{5}{8} = \frac{?}{56}$ **13.** $\frac{8}{?} = \frac{48}{54}$ **14.** $\frac{1}{6} = \frac{12}{?}$

15. $\frac{?}{7} = \frac{33}{77}$ **16.** $\frac{36}{60} = \frac{?}{15}$ **17.** $\frac{1}{?} = \frac{11}{66}$

18. $\frac{?}{64} = \frac{7}{16}$ **19.** $\frac{1}{8} = \frac{?}{56}$

E. True or False

20. $\frac{4}{5} = \frac{12}{15}$ **21.** $\frac{18}{21} = \frac{6}{7}$ **22.** $4 = \frac{80}{20}$

23. $\frac{7}{84} = \frac{42}{12}$ **24.** $0 = \frac{5}{0}$ **25.** $\frac{55}{80} = \frac{11}{16}$

F. **26.** A baseball team won 11 out of 17 games. Express as a fraction the part of the games won.

27. If John is 16 years old and Bill is 17, express as a fraction the ratio of John's age to Bill's age.

SOLUTIONS TO STUDY EXERCISES

Study Exercise One (Frame 6)

1. $\frac{1}{8}$ **2.** $\frac{3}{4}$ **3.** $\frac{2}{9}$

4. $\frac{4}{5}$ **5.** $\frac{3}{5}$ **6.** $\frac{7}{9}$

6A

Study Exercise Two (Frame 15)

1. $\frac{1 \times 2}{7 \times 2} = \frac{2}{14}$ **2.** $\frac{4 \div 2}{6 \div 2} = \frac{2}{3}$

3. $\frac{10 \times 3}{12 \times 3} = \frac{30}{36}$ **4.** $\frac{10 \div 2}{12 \div 2} = \frac{5}{6}$

5. $\frac{8 \times 4}{9 \times 4} = \frac{32}{36}$ **6.** $\frac{12 \div 3}{21 \div 3} = \frac{4}{7}$

Study Exercise Three (Frame 18)

1. $\frac{3}{10} = \frac{9}{30}$ (×3 numerator, ×3 denominator)

2. $\frac{3}{4} = \frac{9}{12}$ (×3 numerator, ×3 denominator)

3. $\frac{2}{3} = \frac{18}{27}$ (×9 numerator, ×9 denominator)

SOLUTIONS TO STUDY EXERCISES, CONTD.

Study Exercise Three (Frame 18, contd.)

4. $\frac{2}{5} = \frac{10}{25}$ (numerator × 5; denominator × 5)

5. $\frac{9}{15} = \frac{3}{5}$ (numerator ÷ 3; denominator ÷ 3)

6. $\frac{8}{32} = \frac{2}{8}$ (numerator ÷ 4; denominator ÷ 4)

7. $\frac{9}{18} = \frac{3}{6}$ (numerator ÷ 3; denominator ÷ 3)

8. $\frac{35}{60} = \frac{7}{12}$ (numerator ÷ 5; denominator ÷ 5)

9. $\frac{5}{8} = \frac{25}{40}$ (numerator ÷ 5; denominator ÷ 5)

10. $\frac{3}{10} = \frac{15}{50}$ (numerator × 5; denominator × 5)

B. **11.** true **12.** false **13.** true **14.** true
15. false **16.** false **17.** true **18.** false

18A

Study Exercise Four (Frame 22)

1. $7 = \frac{7}{1}$

2. $4 = \frac{4}{1} = \frac{4 \times 3}{1 \times 3} = \frac{12}{3}$

3. $8 = \frac{8}{1} = \frac{8 \times 7}{1 \times 7} = \frac{56}{7}$

4. $1 = \frac{1}{1} = \frac{1 \times 7}{1 \times 7} = \frac{7}{7}$

5. $0 = \frac{0}{1} = \frac{0 \times 3}{1 \times 3} = \frac{0}{3}$

22A

unit

12

Reducing and Comparing Fractions

Objectives:

By the end of this unit you should be able to:

1. reduce a fraction to lowest terms by the Golden Rule or by cancelling.
2. compare fractions.

①

Expanding to Higher Terms

By the Golden Rule, a fraction like $\frac{4}{6}$ may be changed to an equivalent fraction by multiplying numerator and denominator by 3.

Thus, $\frac{4}{6} = \frac{4 \times 3}{6 \times 3} = \frac{12}{18}$

When the application of the Golden Rule results in a larger numerator and denominator, it is called *expanding to higher terms.*

②

Below are examples of fractions expanded to higher terms:

1. $\frac{3}{5}$; $\frac{3 \times 5}{5 \times 5} = \frac{15}{25}$; thus $\frac{3}{5} = \frac{15}{25}$
2. $\frac{1}{7}$; $\frac{1 \times 2}{7 \times 2} = \frac{2}{14}$; thus $\frac{1}{7} = \frac{2}{14}$
3. $\frac{5}{1}$; $\frac{5 \times 3}{1 \times 3} = \frac{15}{3}$; thus $\frac{5}{1} = \frac{15}{3}$

③

Reduction to Lower Terms

When the numerator and denominator of a fraction have a common factor, we can divide by the common factor and reduce the fraction to lower terms.

For example, $\frac{4}{6}$ can be reduced to lower terms since 4 and 6 contain a factor of 2.

$$\frac{4}{6} = \frac{4 \div 2}{6 \div 2} = \frac{2}{3}$$

④

Below are examples of fractions reduced to lower terms:

1. $\frac{12}{27}$; $\frac{12 \div 3}{27 \div 3} = \frac{4}{9}$; thus $\frac{12}{27} = \frac{4}{9}$

2. $\frac{27}{36}$; $\frac{27 \div 9}{36 \div 9} = \frac{3}{4}$; thus $\frac{27}{36} = \frac{3}{4}$

3. $\frac{30}{42}$; $\frac{30 \div 6}{42 \div 6} = \frac{5}{7}$; thus $\frac{30}{42} = \frac{5}{7}$ (5)

Study Exercise One

A. Expand to higher terms by multiplying numerator and denominator by the indicated number.

1. $\frac{3}{7}$; multiply by 4 **2.** $\frac{4}{6}$; multiply by 3 **3.** $\frac{5}{8}$; multiply by 7

B. Reduce to lower terms

4. $\frac{9}{15}$ **5.** $\frac{12}{26}$ **6.** $\frac{4}{18}$ **7.** $\frac{14}{21}$ (6)

Reducing

A fraction may be reduced only when both numerator and denominator have a common factor.

The fraction $\frac{12}{27}$ can be reduced since 12 and 27 have a common factor of 3.

The fraction $\frac{5}{8}$ can not be reduced since 5 and 8 have no common factor. (7)

Many times the terms of a fraction have several common factors. For example, $\frac{16}{24}$ may be reduced to $\frac{8}{12}$ since 2 is a common factor.

$$\frac{16}{24} = \frac{16 \div 2}{24 \div 2} = \frac{8}{12}$$

But, $\frac{8}{12}$ can be reduced to $\frac{4}{6}$

$$\frac{8}{12} = \frac{8 \div 2}{12 \div 2} = \frac{4}{6}$$

But, $\frac{4}{6}$ can be reduced to $\frac{2}{3}$

$$\frac{4}{6} = \frac{4 \div 2}{6 \div 2} = \frac{2}{3}$$

(8)

Reducing to Lowest Terms

If we reduce a fraction until we get to a fraction where there are no common factors in the numerator and denominator, we say the fraction is *reduced to lowest terms*.
In working with mathematics, *all fractions should be reduced to lowest terms.* (9)

Example 1: Reduce $\frac{24}{28}$ to lowest terms

Solution: $\frac{24}{28} = \frac{24 \div 2}{28 \div 2} = \frac{12}{14}$

$\frac{12}{14} = \frac{12 \div 2}{14 \div 2} = \frac{6}{7}$

Thus, $\frac{24}{28}$ reduces to $\frac{6}{7}$

Example 2: Reduce $\frac{26}{39}$ to lowest terms

Solution: $\frac{26}{39} = \frac{26 \div 13}{39 \div 13} = \frac{2}{3}$

Thus, $\frac{26}{39}$ reduces to $\frac{2}{3}$

(10)

Study Exercise Two

Reduce to lowest terms:

1. $\frac{15}{20}$
2. $\frac{18}{24}$
3. $\frac{36}{45}$
4. $\frac{28}{35}$
5. $\frac{8}{24}$
6. $\frac{18}{45}$
7. $\frac{30}{96}$
8. $\frac{54}{60}$

(11)

Another Method For Reducing To Lowest Terms

We can use prime factorization of a number to help us reduce a fraction.

Example: Reduce $\frac{6}{14}$

Solution:

Step (1) Write 6 as a product of primes: 2×3
Step (2) Write 14 as a product of primes: 2×7

$$\frac{6}{14} = \frac{2 \times 3}{2 \times 7}$$

Step (3) Now divide numerator and denominator by 2

$$\frac{2 \times 3 \div 2}{2 \times 7 \div 2} = \frac{1 \times 3}{1 \times 7} = \frac{3}{7}$$

Cancelling

Since $\frac{2 \times 3}{2 \times 7} = \frac{3}{7}$, the 2's have been cancelled.

$$\frac{\overset{1}{\not{2}} \times 3}{\underset{1}{\not{2}} \times 7} = \frac{3}{7}$$

A one is put in their place because we really divided by 2 in each case. (13)

Example 1: Reduce $\frac{12}{16}$

Solution: $12 = 2 \times 2 \times 3; \qquad 16 = 2 \times 2 \times 2 \times 2$

$$\frac{12}{16} = \frac{\overset{1}{\cancel{2}} \times \overset{1}{\cancel{2}} \times 3}{\underset{1}{\cancel{2}} \times \underset{1}{\cancel{2}} \times 2 \times 2}$$

$$= \frac{1 \times 1 \times 3}{1 \times 1 \times 2 \times 2}$$

$$= \frac{3}{4}$$

Example 2: Reduce $\frac{26}{39}$

Solution:
$$\frac{26}{39} = \frac{2 \times 13}{3 \times 13}$$

$$= \frac{2 \times \overset{1}{\cancel{13}}}{3 \times \underset{1}{\cancel{13}}}$$

$$= \frac{2}{3}$$

(14)

Study Exercise Three

Reduce by cancelling:

1. $\frac{4}{22}$ **2.** $\frac{21}{36}$ **3.** $\frac{27}{36}$

4. $\frac{28}{40}$ **5.** $\frac{12}{48}$ **6.** $\frac{30}{36}$

(15)

Like Fractions

If two fractions have the same denominator, they are called *like fractions*.

The fractions $\frac{1}{7}$ and $\frac{3}{7}$ are like fractions since they have the same denominator, seven.

Comparing Fractions

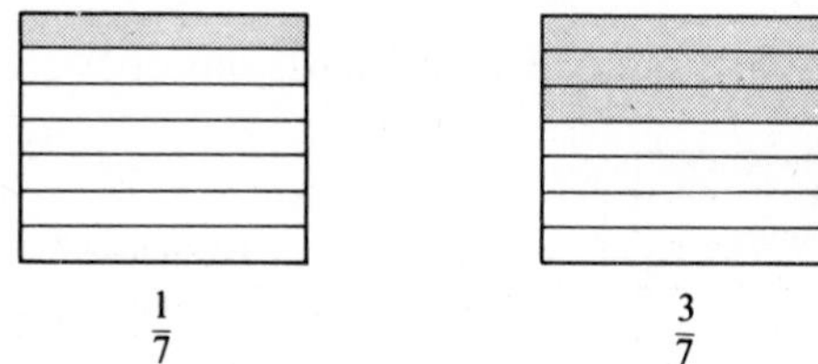

It is easy to compare two fractions with the same denominator. Which do you think is larger, $\frac{1}{7}$ or $\frac{3}{7}$?

Three parts are larger than 1 part out of 7, so $\frac{3}{7}$ is larger than $\frac{1}{7}$.

Example: Which is the largest of $\frac{4}{13}, \frac{7}{13}, \frac{2}{13}$?

Solution: Seven parts are larger than either 4 or 2 parts. Hence, $\frac{7}{13}$ is larger than either $\frac{4}{13}$ or $\frac{2}{13}$.

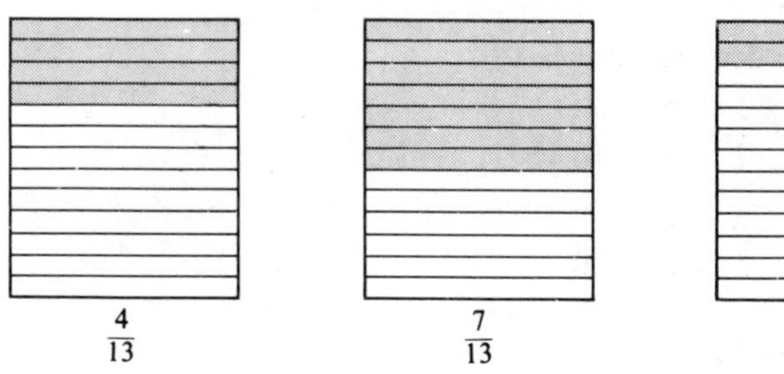

Study Exercise Four

Find the largest of the given group of fractions.

1. $\frac{3}{4}, \frac{2}{4}$

2. $\frac{2}{8}, \frac{5}{8}, \frac{6}{8}$

3. $\frac{7}{13}, \frac{6}{13}, \frac{5}{13}$

Question: Which is larger: $\frac{2}{3}$ or $\frac{3}{5}$?

Answer: It is difficult to compare 2 parts out of 3 with 3 parts out of 5.

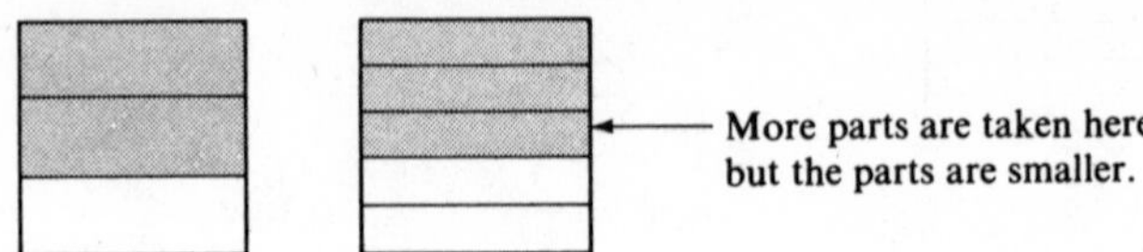
More parts are taken here
but the parts are smaller.

But if the denominators were the same, comparison would be easy. We will expand the denominators to 15.

$$\frac{2}{3} = \frac{2 \times 5}{3 \times 5} = \frac{10}{15}; \qquad \frac{3}{5} = \frac{3 \times 3}{5 \times 3} = \frac{9}{15}$$

$\frac{10}{15}$ is larger than $\frac{9}{15}$; therefore, $\frac{2}{3}$ is larger.

Comparison of Fractions

To compare unlike fractions, expand the fractions so they all have the same denominator. Then compare them as outlined on Frames 17 and 18.

Example: Which is larger, $\frac{1}{2}$ or $\frac{3}{5}$?

Solution: We will expand so each denominator is 10.

$$\frac{1}{2} = \frac{1 \times 5}{2 \times 5} = \frac{5}{10}$$

$$\frac{3}{5} = \frac{3 \times 2}{5 \times 2} = \frac{6}{10}$$

Therefore, $\frac{3}{5}$ is larger than $\frac{1}{2}$.

Study Exercise Five

In each group, find which fraction is larger.

1. $\frac{1}{4}, \frac{2}{3}$ **2.** $\frac{1}{6}, \frac{2}{9}$ **3.** $\frac{3}{5}, \frac{7}{10}$

4. $\frac{3}{4}, \frac{5}{6}$ **5.** $\frac{3}{5}, \frac{1}{10}, \frac{1}{6}$

REVIEW EXERCISES

A. Expand to higher or lower terms by multiplying or dividing numerator and denominator by the indicated number.

1. $\frac{11}{16}$; multiply by 3

2. $\frac{20}{32}$; divide by 4

B. Reduce to lowest terms.

3. $\frac{12}{28}$

4. $\frac{32}{48}$

5. $\frac{15}{27}$

6. $\frac{4}{40}$

C. Determine which fraction is larger.

7. $\frac{3}{5}$ or $\frac{2}{5}$

8. $\frac{2}{5}$ or $\frac{1}{4}$

(23)

SOLUTIONS TO REVIEW EXERCISES

A. 1. $\frac{11}{16} = \frac{11 \times 3}{16 \times 3} = \frac{33}{48}$

2. $\frac{20}{32} = \frac{20 \div 4}{32 \div 4} = \frac{5}{8}$

B. 3. $\frac{12}{28} = \frac{12 \div 4}{28 \div 4} = \frac{3}{7}$

4. $\frac{32}{48} = \frac{\overset{2}{\cancel{32}}}{\underset{3}{\cancel{48}}} = \frac{2}{3}$

5. $\frac{15}{27} = \frac{\overset{1}{\cancel{3}} \times 5}{\underset{1}{\cancel{3}} \times 3 \times 3} = \frac{5}{9}$

6. $\frac{4}{40} = \frac{\overset{1}{\cancel{4}}}{\underset{10}{\cancel{40}}} = \frac{1}{10}$

C. 7. $\frac{3}{5}$ is larger

8. $\frac{2}{5} = \frac{2 \times 4}{5 \times 4} = \frac{8}{20}$

$\frac{1}{4} = \frac{1 \times 5}{4 \times 5} = \frac{5}{20}$

$\frac{8}{20}$ is larger than $\frac{5}{20}$. Therefore, $\frac{2}{5}$ is larger than $\frac{1}{4}$.

SUPPLEMENTARY PROBLEMS

A. Expand to higher terms by multiplying numerator and denominator by the indicated number.

1. $\frac{3}{4}$; multiply by 4 **2.** $\frac{5}{12}$; multiply by 5

3. $\frac{1}{6}$; multiply by 7 **4.** $\frac{5}{6}$; multiply by 2

5. $\frac{15}{16}$; multiply by 4 **6.** $\frac{5}{9}$; multiply by 8

B. Reduce to lowest terms:

7. $\frac{18}{24}$ **8.** $\frac{45}{75}$ **9.** $\frac{16}{36}$ **10.** $\frac{84}{108}$

11. $\frac{8}{20}$ **12.** $\frac{42}{60}$ **13.** $\frac{21}{28}$ **14.** $\frac{57}{76}$

15. $\frac{62}{93}$ **16.** $\frac{40}{200}$

C. True or False:

17. $\frac{43}{86} = \frac{1}{2}$ **18.** $\frac{12}{15} = \frac{3}{5}$ **19.** $\frac{27}{29} = \frac{9}{10}$

20. $\frac{3}{4} = \frac{39}{52}$ **21.** $\frac{24}{28} = \frac{6}{7}$ **22.** $\frac{15}{45} = \frac{1}{3}$

D. Determine which fraction is larger:

23. $\frac{5}{8}$ or $\frac{2}{3}$? **24.** $\frac{4}{5}$ or $\frac{3}{4}$?

25. $\frac{1}{8}$ or $\frac{1}{10}$? 26. $\frac{1}{4}$, $\frac{1}{2}$, or $\frac{1}{6}$?

E. Arrange in order of size (smallest first).

27. $\frac{3}{4}$, $\frac{2}{3}$, $\frac{3}{5}$ **28.** $\frac{1}{2}$, $\frac{1}{5}$, $\frac{1}{3}$

SOLUTIONS TO STUDY EXERCISES

Study Exercise One (Frame 6)

A. 1. $\frac{3}{7} = \frac{3 \times 4}{7 \times 4} = \frac{12}{28}$ **2.** $\frac{4}{6} = \frac{4 \times 3}{6 \times 3} = \frac{12}{18}$ **3.** $\frac{5}{8} = \frac{5 \times 7}{8 \times 7} = \frac{35}{56}$

B. 4. $\frac{9}{15} = \frac{9 \div 3}{15 \div 3} = \frac{3}{5}$ **5.** $\frac{12}{26} = \frac{12 \div 2}{26 \div 2} = \frac{6}{13}$ **6.** $\frac{4}{18} = \frac{4 \div 2}{18 \div 2} = \frac{2}{9}$

7. $\frac{14}{21} = \frac{14 \div 7}{21 \div 7} = \frac{2}{3}$

6A

SOLUTIONS TO STUDY EXERCISES, CONTD.

Study Exercise Two (Frame 11)

1. $\frac{15}{20} = \frac{15 \div 5}{20 \div 5} = \frac{3}{4}$ **2.** $\frac{18}{24} = \frac{18 \div 6}{24 \div 6} = \frac{3}{4}$ **3.** $\frac{36}{45} = \frac{36 \div 9}{45 \div 9} = \frac{4}{5}$

4. $\frac{28}{35} = \frac{28 \div 7}{35 \div 7} = \frac{4}{5}$ **5.** $\frac{8}{24} = \frac{8 \div 8}{24 \div 8} = \frac{1}{3}$ **6.** $\frac{18}{45} = \frac{18 \div 9}{45 \div 9} = \frac{2}{5}$

7. $\frac{30}{96} = \frac{30 \div 6}{96 \div 6} = \frac{5}{16}$ **8.** $\frac{54}{60} = \frac{54 \div 6}{60 \div 6} = \frac{9}{10}$

11A

Study Exercise Three (Frame 15)

1. $\frac{4}{22} = \frac{\overset{1}{\cancel{2}} \times 2}{\underset{1}{\cancel{2}} \times 11} = \frac{2}{11}$ **2.** $\frac{21}{36} = \frac{\overset{1}{\cancel{3}} \times 7}{\underset{1}{\cancel{3}} \times 3 \times 2 \times 2} = \frac{7}{12}$

3. $\frac{27}{36} = \frac{\overset{1}{\cancel{3}} \times \overset{1}{\cancel{3}} \times 3}{\underset{1}{\cancel{3}} \times \underset{1}{\cancel{3}} \times 2 \times 2} = \frac{3}{4}$ **4.** $\frac{28}{40} = \frac{\overset{1}{\cancel{2}} \times \overset{1}{\cancel{2}} \times 7}{\underset{1}{\cancel{2}} \times \underset{1}{\cancel{2}} \times 2 \times 5} = \frac{7}{10}$

5. $\frac{12}{48} = \frac{\overset{1}{\cancel{2}} \times \overset{1}{\cancel{2}} \times \overset{1}{\cancel{3}}}{\underset{1}{\cancel{2}} \times \underset{1}{\cancel{2}} \times 2 \times 2 \times \underset{1}{\cancel{3}}} = \frac{1}{4}$ **6.** $\frac{30}{36} = \frac{\overset{1}{\cancel{2}} \times \overset{1}{\cancel{3}} \times 5}{\underset{1}{\cancel{2}} \times 2 \times \underset{1}{\cancel{3}} \times 3} = \frac{5}{6}$

Study Exercise Four (Frame 19)

1. Since 3 parts out of 4 are larger than 2 parts out of 4, $\frac{3}{4}$ is larger than $\frac{2}{4}$.

2. $\frac{6}{8}$ is the largest

3. $\frac{7}{13}$ is the largest

19A

Study Exercise Five (Frame 22)

1. $\frac{1}{4} = \frac{1 \times 3}{4 \times 3} = \frac{3}{12}$; $\frac{2}{3} = \frac{2 \times 4}{3 \times 4} = \frac{8}{12}$

Since $\frac{8}{12}$ is larger than $\frac{3}{12}$, $\frac{2}{3}$ is larger than $\frac{1}{4}$

2. $\frac{1}{6} = \frac{1 \times 3}{6 \times 3} = \frac{3}{18}$; $\frac{2}{9} = \frac{2 \times 2}{9 \times 2} = \frac{4}{18}$

$\frac{2}{9}$ is larger than $\frac{1}{6}$

3. $\frac{3}{5} = \frac{3 \times 2}{5 \times 2} = \frac{6}{10}$; $\frac{7}{10}$

$\frac{7}{10}$ is larger than $\frac{3}{5}$

SOLUTIONS TO STUDY EXERCISES, CONTD.

Study Exercise Five (Frame 22, contd.)

4. $\frac{3}{4} = \frac{3 \times 3}{4 \times 3} = \frac{9}{12}$; $\frac{5}{6} = \frac{5 \times 2}{6 \times 2} = \frac{10}{12}$

$\frac{5}{6}$ is larger than $\frac{3}{4}$

5. $\frac{3}{5} = \frac{3 \times 6}{5 \times 6} = \frac{18}{30}$; $\frac{1}{10} = \frac{1 \times 3}{10 \times 3} = \frac{3}{30}$; $\frac{1}{6} = \frac{1 \times 5}{6 \times 5} = \frac{5}{30}$

$\frac{3}{5}$ is the largest.

22A

unit

13

Mixed Numerals, Improper Fractions And Lowest Common Denominator

Objectives:

By the end of this unit you should be able to:

1. identify a mixed numeral and an improper fraction.
2. change mixed numerals to improper fractions and improper fractions to mixed numerals.
3. find the LCD.

Mixed Numerals

A *mixed numeral* is a symbol for a number that contains both a whole number and a fraction.

Some mixed numerals include:

$$2\frac{1}{3},\quad 5\frac{2}{7},\quad 1\frac{1}{2}$$

②

Reading Mixed Numerals

$2\frac{1}{3}$ is read *two and one-third.*

$5\frac{2}{7}$ is read *five and two-sevenths.*

$1\frac{1}{2}$ is read *one and one-half.*

③

Meaning of a Mixed Numeral

A mixed numeral such as $3\frac{2}{5}$ actually means $3 + \frac{2}{5}$. It is customary to omit the plus sign and merely write $3\frac{2}{5}$.

Question: What is the meaning of $6\frac{5}{7}$?

Answer: $6\frac{5}{7}$ means $6 + \frac{5}{7}$.

Improper Fractions

A fraction whose numerator is equal to or greater than its denominator is called an *improper fraction.*

Some improper fractions are:

$$\frac{7}{2}, \frac{8}{5}, \frac{3}{3}, \frac{12}{5}$$

Study Exercise One

1. Which of the following are mixed numerals?

$$7, \frac{5}{5}, \frac{2}{3}, 4\frac{1}{3}, 8, \frac{13}{3}, 6\frac{2}{5}$$

2. Which of the numbers in problem one are improper fractions?
3. Read: $5\frac{7}{10}$
4. What is the meaning of $5\frac{7}{10}$?

Changing Improper Fractions To Mixed Numerals

Remember that a fraction indicates division. Now let us change $\frac{34}{5}$ to a mixed numeral.

We find $34 \div 5 = 6$ with remainder 4

$$\begin{array}{r} 6 \\ 5\overline{)34} \\ \underline{30} \\ 4 \end{array}$$

The remainder is put over the divisor.

Thus $34 \div 5 = 6\frac{4}{5}$; or $\frac{34}{5} = 6\frac{4}{5}$.

Example: Write $\frac{38}{8}$ as a mixed numeral.

Solution: $38 \div 8 = 4$ with remainder 6

$38 \div 8 = 4\frac{6}{8}$; or $\frac{38}{8} = 4\frac{6}{8}$

We should reduce the fraction $\frac{6}{8}$

$\frac{6 \div 2}{8 \div 2} = \frac{3}{4}$

Therefore, $\frac{38}{8} = 4\frac{3}{4}$.

(8)

Study Exercise Two

Change to mixed numerals.

1. $\frac{19}{5}$ **2.** $\frac{52}{8}$ **3.** $\frac{16}{6}$

4. $\frac{46}{8}$ **5.** $\frac{37}{3}$

Common Denominators

It will be necessary to change unlike fractions to the same or common denominator.

Let us change $\frac{2}{3}$ and $\frac{4}{7}$ to fractions with the same denominator.

We need a denominator which is a multiple of the two denominators 3 and 7. We take $3 \times 7 = 21$ to be the common denominator.

$$\frac{2}{3} = \frac{2 \times 7}{3 \times 7} = \frac{14}{21}$$

$$\frac{4}{7} = \frac{4 \times 3}{7 \times 3} = \frac{12}{21}$$

Suppose we wanted to change $\frac{1}{6}$ and $\frac{3}{10}$ to fractions with the same denominator.

We could choose $6 \times 10 = 60$ as the common denominator, but there is a smaller number that would also work. That number is 30.

$$\frac{1}{6} = \frac{1 \times 5}{6 \times 5} = \frac{5}{30}$$

$$\frac{3}{10} = \frac{3 \times 3}{10 \times 3} = \frac{9}{30}$$

But how do you determine what is the lowest common denominator?

L C D

In order to keep the numbers we work with as small as possible, we wish to use the lowest common denominator. We will refer to the lowest common denominator as the *LCD*.

Rule: The LCD of two fractions that do not have a common factor in their denominators is the product of the denominators.

Example: The LCD of $\frac{1}{2}$ and $\frac{2}{5}$ is 10 since 2 and 5 have no common factor. (12)

Example: Change $\frac{3}{5}$ and $\frac{1}{6}$ to fractions with the same denominator.

Solution: Since 5 and 6 do not have a common factor, the LCD is $5 \times 6 = 30$.

$$\frac{3}{5} = \frac{3 \times 6}{5 \times 6} = \frac{18}{30}$$

$$\frac{1}{6} = \frac{1 \times 5}{6 \times 5} = \frac{5}{30}$$ (13)

Study Exercise Three

Find the LCD of each of the following groups of fractions:

1. $\frac{1}{3}$ and $\frac{5}{8}$ **2.** $\frac{3}{11}$ and $\frac{4}{7}$

3. $\frac{2}{5}$ and $\frac{4}{9}$ **4.** $\frac{3}{4}$ and $\frac{2}{5}$ (14)

Fractions do not always come in pairs. It may be necessary to find the LCD for three or more fractions. If none of the fractions have a common factor, the LCD is the product of all the denominators.

Example: Find the LCD of $\frac{2}{3}, \frac{3}{4}, \frac{1}{5}$.

Solution: Since 3, 4, and 5 have no common factor, the LCD is $3 \times 4 \times 5 = 60$.

$$\frac{2}{3} \xrightarrow[\times 20]{\times 20} \frac{\quad}{60} = \frac{40}{60}, \qquad \frac{3}{4} \xrightarrow[\times 15]{\times 15} \frac{\quad}{60} = \frac{45}{60}, \qquad \frac{1}{5} \xrightarrow[\times 12]{\times 12} \frac{\quad}{60} = \frac{12}{60}$$ (15)

LCD

Rule: The LCD of two or more fractions is the LCM of their denominators.

Example: Find the LCD of $\frac{5}{8}$ and $\frac{1}{6}$.

Solution:

Step (1) First factor each denominator into prime factors.

$8 = 2^3$

$6 = 2^1 \times 3^1$

Step (2) Then write each factor down: 2, 3.

Step (3) Use the largest exponent that occurs on each. $2^3, 3^1$

Step (4) Multiply together $2^3 \times 3^1 = 8 \times 3 = 24$

Step (5) The LCD is 24.

Another Example: Find the LCD of $\frac{1}{6}$ and $\frac{3}{16}$.

Solution:

Line (a) $6 = 2^1 \times 3^1$

Line (b) $16 = 2^4$

Line (c) 2, 3

Line (d) $2^4, 3^1$

Line (e) $2^4 \times 3^1$

Line (f) $16 \times 3 = 48$

The LCD is 48.

Study Exercise Four

Find the LCD:

1. $\frac{2}{3}, \frac{7}{12}$
2. $\frac{13}{20}, \frac{7}{12}$
3. $\frac{1}{2}, \frac{11}{16}$
4. $\frac{11}{24}, \frac{5}{8}$
5. $\frac{5}{6}, \frac{7}{18}$
6. $\frac{2}{3}, \frac{1}{6}$

The LCD of three or more fractions is found in a similar fashion as with 2 fractions.

Example: Find the LCD of $\frac{1}{6}, \frac{1}{8}, \frac{1}{9}$.

Solution:

$6 = 2^1 \times 3^1$

$8 = 2^3$

$9 = 3^2$

The different factors are 2, 3

$2^3, 3^2$

$2^3 \times 3^2 = 8 \times 9 = 72$

The LCD is 72.

Study Exercise Five

Find the LCD:

1. $\frac{1}{4}, \frac{1}{12}, \frac{5}{6}$ 2. $\frac{7}{15}, \frac{2}{25}, \frac{1}{10}$ 3. $\frac{3}{8}, \frac{3}{10}, \frac{3}{4}, \frac{1}{2}$

Now that we can find the LCD, we should be able to expand fractions to equivalent fractions with the same denominator. That denominator should be the LCD.

Example: Change to fractions with the same denominator: $\frac{1}{6}, \frac{2}{9}$

Solution: First we find the LCD

$$6 = 2^1 \times 3^1, 9 = 3^2$$
$$\text{LCD} = 2^1 \times 3^2 = 18$$
$$\frac{1}{6} = \frac{1 \times 3}{6 \times 3} = \frac{3}{18}$$
$$\frac{2}{9} = \frac{2 \times 2}{9 \times 2} = \frac{4}{18}$$

Study Exercise Six

Change to fractions with the same denominator. Use the LCD.

1. $\frac{1}{16}, \frac{7}{12}$ 2. $\frac{3}{4}, \frac{5}{6}$ 3. $\frac{1}{3}, \frac{2}{15}, \frac{5}{18}$

REVIEW EXERCISES

1. A mixed numeral is a symbol for a number that contains both a ____________ number and a ____________.
2. An improper fraction is a fraction whose numerator is ____________ or ____________ its denominator.
3. The meaning of $2\frac{1}{3}$ is ____________.
4. Change to mixed numerals.
   a) $\frac{26}{3}$ b) $\frac{52}{5}$
5. Find the LCD of the fractions $\frac{8}{9}$ and $\frac{2}{5}$.
6. Find the LCD of $\frac{1}{8}$ and $\frac{5}{12}$.
7. Find the LCD of $\frac{1}{4}, \frac{5}{6}$, and $\frac{3}{10}$.
8. Change $\frac{1}{8}, \frac{3}{20}$ to fractions with the same denominator. Use the LCD.

SOLUTIONS TO REVIEW EXERCISES

1. whole, fraction
2. greater than, equal to
3. two plus one-third $\left(2 + \frac{1}{3}\right)$
4. a) $8\frac{2}{3}$ b) $10\frac{2}{5}$
5. Since 9 and 5 have no common factor, the LCD $= 9 \times 5 = 45$
6. $8 = 2^3, \quad 12 = 2^2 \times 3^1$
   $\text{LCD} = 2^3 \times 3^1 = 8 \times 3 = 24$
7. $4 = 2^2, \quad 6 = 2^1 \times 3^1, \quad 10 = 2^1 \times 5^1$
   $\text{LCD} = 2^2 \times 3^1 \times 5^1 = 4 \times 3 \times 5 = 60$
8. $8 = 2^3, \quad 20 = 2^2 \times 5^1$
   $\text{LCD} = 2^3 \times 5^1 = 8 \times 5 = 40$
   $\frac{1}{8} = \frac{1 \times 5}{8 \times 5} = \frac{5}{40}$
   $\frac{3}{20} = \frac{3 \times 2}{20 \times 2} = \frac{6}{40}$

SUPPLEMENTARY PROBLEMS

1. Select the mixed numerals from the following:
   $\frac{3}{5}, 8, 3\frac{1}{4}, \frac{5}{6}, 2\frac{11}{13}, \frac{7}{8}$
2. Select the improper fractions from the following:
   $\frac{1}{6}, \frac{7}{6}, \frac{3}{3}, \frac{2}{7}, \frac{5}{2}, \frac{8}{8}$

SUPPLEMENTARY PROBLEMS, CONTD.

Change to a whole number or a mixed numeral.

3. $\frac{3}{2}$ **4.** $\frac{12}{5}$ **5.** $\frac{12}{3}$

6. $\frac{22}{7}$ **7.** $\frac{46}{16}$ **8.** $\frac{33}{9}$

9. $\frac{8}{8}$ **10.** $\frac{23}{16}$

Find the LCD of each of the following groups of fractions.

11. $\frac{1}{4}, \frac{3}{14}$ **12.** $\frac{2}{7}, \frac{3}{11}$ **13.** $\frac{7}{15}, \frac{1}{21}$

14. $\frac{4}{18}, \frac{7}{30}$ **15.** $\frac{1}{26}, \frac{3}{14}$ **16.** $\frac{1}{4}, \frac{7}{10}, \frac{5}{12}$

17. $\frac{5}{6}, \frac{1}{15}, \frac{5}{18}$ **18.** $\frac{1}{27}, \frac{1}{63}, \frac{1}{72}$ **19.** $\frac{4}{9}, \frac{5}{6}, \frac{7}{12}, \frac{11}{15}$

20. $\frac{11}{12}, \frac{25}{42}$

Change to fractions with the same denominator. Use the LCD.

21. $\frac{1}{4}, \frac{5}{6}$ **22.** $\frac{1}{6}, \frac{5}{9}$ **23.** $\frac{2}{3}, \frac{3}{5}, \frac{5}{6}$

24. $\frac{8}{21}, \frac{9}{35}$ **25.** $\frac{1}{3}, \frac{3}{4}, \frac{2}{5}$ **26.** $\frac{7}{12}, \frac{5}{8}$

SOLUTIONS TO STUDY EXERCISES

Study Exercise One (Frame 6)

1. $4\frac{1}{3}, 6\frac{2}{5}$ **2.** $\frac{5}{5}, \frac{13}{3}$

3. Five and seven-tenths **4.** $5 + \frac{7}{10}$

6A

Study Exercise Two (Frame 9)

1. $\frac{19}{5} = 19 \div 5 = 3\frac{4}{5}$ **2.** $\frac{52}{8} = 52 \div 8 = 6\frac{4}{8} = 6\frac{1}{2}$

3. $\frac{16}{6} = 16 \div 6 = 2\frac{4}{6} = 2\frac{2}{3}$ **4.** $\frac{46}{8} = 46 \div 8 = 5\frac{6}{8} = 5\frac{3}{4}$

5. $\frac{37}{3} = 37 \div 3 = 12\frac{1}{3}$

SOLUTIONS TO STUDY EXERCISES, CONTD.

Study Exercise Three (Frame 14)

1. Since 3 and 8 have no common factor,
$\text{LCD} = 3 \times 8 = 24$

2. Since 11 and 7 have no common factor,
$\text{LCD} = 11 \times 7 = 77$

3. Since 5 and 9 have no common factor,
$\text{LCD} = 5 \times 9 = 45$

4. Since 4 and 5 have no common factor,
$\text{LCD} = 4 \times 5 = 20.$

14A

Study Exercise Four (Frame 18)

1. $3 = 3^1, \quad 12 = 2^2 \times 3^1$
$\text{LCD} = 2^2 \times 3^1 = 12$

2. $20 = 2^2 \times 5^1, \quad 12 = 2^2 \times 3^1$
$\text{LCD} = 2^2 \times 5^1 \times 3^1 = 60$

3. $2 = 2^1, \quad 16 = 2^4$
$\text{LCD} = 2^4 = 16$

4. $24 = 2^3 \times 3^1, \quad 8 = 2^3$
$\text{LCD} = 2^3 \times 3^1 = 24$

5. $6 = 2^1 \times 3^1, \quad 18 = 2^1 \times 3^2$
$\text{LCD} = 2^1 \times 3^2 = 18$

6. $3 = 3^1, \quad 6 = 2^1 \times 3^1$
$\text{LCD} = 2^1 \times 3^1 = 6$

18A

Study Exercise Five (Frame 20)

1. $4 = 2^2, \quad 12 = 2^2 \times 3^1, \quad 6 = 2^1 \times 3^1$
$\text{LCD} = 2^2 \times 3^1 = 12$

2. $15 = 3^1 \times 5^1, \quad 25 = 5^2, \quad 10 = 2^1 \times 5^1$
$\text{LCD} = 2^1 \times 3^1 \times 5^2 = 150$

3. $8 = 2^3, \quad 10 = 2^1 \times 5^1, \quad 4 = 2^2, \quad 2 = 2^1$
$\text{LCD} = 2^3 \times 5 = 40$

20A

Study Exercise Six (Frame 23)

1. $16 = 2^4, \quad 12 = 2^2 \times 3^1; \quad \text{LCD} = 2^4 \times 3^1 = 48$

$$\frac{1}{16} = \frac{1 \times 3}{16 \times 3} = \frac{3}{48}, \quad \frac{7}{12} = \frac{7 \times 4}{12 \times 4} = \frac{28}{48}$$

2. $4 = 2^2, \quad 6 = 2^1 \times 3^1; \quad \text{LCD} = 2^2 \times 3^1 = 12$

$$\frac{3}{4} = \frac{3 \times 3}{4 \times 3} = \frac{9}{12}, \quad \frac{5}{6} = \frac{5 \times 2}{6 \times 2} = \frac{10}{12}$$

3. $3 = 3^1, \quad 15 = 3^1 \times 5^1, \quad 18 = 2^1 \times 3^2; \quad \text{LCD} = 2^1 \times 3^2 \times 5^1 = 90$

$$\frac{1}{3} = \frac{1 \times 30}{3 \times 30} = \frac{30}{90}, \quad \frac{2}{15} = \frac{2 \times 6}{15 \times 6} = \frac{12}{90}, \quad \frac{5}{18} = \frac{5 \times 5}{18 \times 5} = \frac{25}{90}$$

---

unit

14

Adding and Subtracting Fractions

Objectives:

By the end of this unit you should know how to:

1. add like and unlike fractions.

2. subtract like and unlike fractions. ①

Let us add $\frac{2}{5} + \frac{1}{5}$.

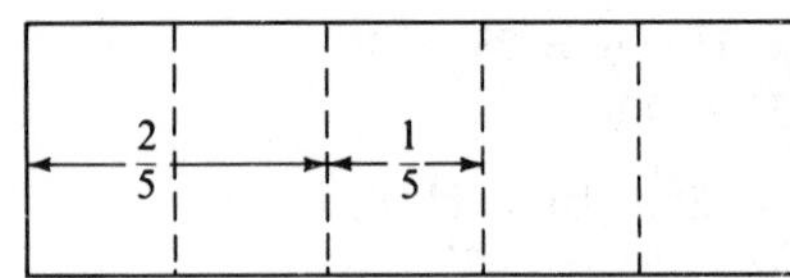

From the diagram we see the result is $\frac{2}{5} + \frac{1}{5} = \frac{3}{5}$.

Notice that when adding $\frac{2}{5} + \frac{1}{5} = \frac{3}{5}$, the numerators are added, but the denominator remains unchanged. ②

Rule: To add two or more fractions with like denominators, add the numerators and put the result over the denominator.

Example 1: $\frac{3}{7} + \frac{1}{7}$

Solution: $\frac{3}{7} + \frac{1}{7} = \frac{3 + 1}{7}$

$= \frac{4}{7}$ ③

$$\frac{3}{7} + \frac{1}{7}$$

Horizontal Format	**Vertical Format**
$\frac{3}{7} + \frac{1}{7} = \frac{4}{7}$	$\begin{array}{r} \frac{3}{7} \\ + \frac{1}{7} \\ \hline \frac{4}{7} \end{array}$

4

Let us add $\frac{2}{5} + \frac{4}{5}$

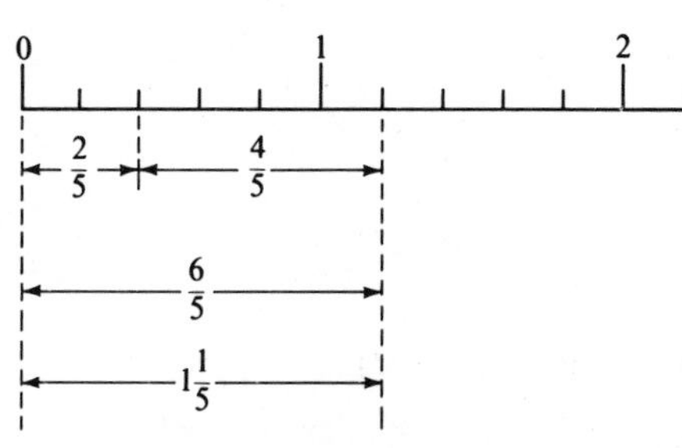

$$\frac{2}{5} + \frac{4}{5} = \frac{2+4}{5}$$
$$= \frac{6}{5}$$
$$= 1\frac{1}{5}$$

The answer must always be expressed in lowest terms. Remember to reduce all answers.

Example: $\frac{7}{8} + \frac{5}{8} = \frac{7+5}{8}$
$$= \frac{12}{8}$$
$$= 1\frac{4}{8}$$
$$= 1\frac{1}{2}$$

Study Exercise One

Add:

1. $\frac{3}{5} + \frac{1}{5}$
2. $\frac{1}{7} + \frac{3}{7}$
3. $\frac{5}{8} + \frac{3}{8}$
4. $\frac{6}{7} + \frac{5}{7}$
5. $\begin{array}{r} \frac{1}{3} \\ +\frac{1}{3} \\ \hline \end{array}$
6. $\begin{array}{r} \frac{3}{10} \\ +\frac{9}{10} \\ \hline \end{array}$

Adding Mixed Numerals

Mixed numerals may be added by combining the whole numbers and then combining the fractions separately.

Example: Add $5\frac{3}{7} + 2\frac{1}{7}$

Solution:

Horizontal Form: $5\frac{3}{7} + 2\frac{1}{7}$

add the whole numbers $(5 + 2)$

add the fractions $\left(\frac{3}{7} + \frac{1}{7}\right)$

$$5\frac{3}{7} + 2\frac{1}{7} = 7 + \frac{4}{7} = 7\frac{4}{7}$$

Vertical Form:

$$\begin{array}{r} 5\frac{3}{7} \\ +2\frac{1}{7} \\ \hline 7\frac{4}{7} \end{array}$$

(8)

Example: $2\frac{7}{8} + 6\frac{5}{8}$

Solution:

$$\begin{aligned} 2\frac{7}{8} + 6\frac{5}{8} &= 8 + \frac{12}{8} \\ &= 8 + 1\frac{4}{8} \\ &= 8 + 1\frac{1}{2} \\ &= 9\frac{1}{2} \end{aligned}$$

Study Exercise Two

Add:

1. $2\frac{3}{11} + 5\frac{6}{11}$
2. $1\frac{15}{16} + 3\frac{13}{16}$
3. $4\frac{8}{12} + 2\frac{6}{12}$
4. $1\frac{13}{16} + 2\frac{4}{16} + \frac{3}{16}$
5. $2\frac{3}{6}$ $+4\frac{5}{6}$
6. $1\frac{3}{5}$ $+\frac{4}{5}$

(10)

Mixed Numerals To Improper Fractions

A mixed numeral such as $3\frac{1}{2}$ can be expressed as an improper fraction.

$3\frac{1}{2}$ means $3 + \frac{1}{2}$. Now $3 = \frac{6}{2}$.

Thus, $3 + \frac{1}{2} = \frac{6}{2} + \frac{1}{2}$

$$= \frac{6 + 1}{2}$$

$$= \frac{7}{2}$$

This process can be shortened as follows:

To change $3\frac{1}{2}$ to an improper fraction, multiply 3×2, then add 1 to the result and divide your answer by 2.

That is, $3\frac{1}{2} = \frac{3 \times 2 + 1}{2} = \frac{7}{2}$

(11)

Example 1: Change $5\frac{1}{3}$ to an improper fraction.

Solution: $5\frac{1}{3} = \frac{5 \times 3 + 1}{3} = \frac{15 + 1}{3} = \frac{16}{3}$

Example 2: Change $4\frac{2}{5}$ to an improper fraction.

Solution: $4\frac{2}{5} = \frac{4 \times 5 + 2}{5} = \frac{20 + 2}{5} = \frac{22}{5}$

(Frame 12, contd.)

Example 3: Add: $2\frac{1}{3} + 1\frac{2}{3}$

Solution: We will change each fraction to an improper fraction.

$$2\frac{1}{3} = \frac{2 \times 3 + 1}{3} = \frac{7}{3}$$

$$1\frac{2}{3} = \frac{1 \times 3 + 2}{3} = \frac{5}{3}$$

$$\frac{7}{3} + \frac{5}{3} = \frac{12}{3} = 4$$

Note: We could also work this problem by adding the whole numbers and the fractions. 

Study Exercise Three

1. Change $8\frac{2}{3}$ to an improper fraction.
2. Change $6\frac{3}{5}$ to an improper fraction.
3. Change $2\frac{1}{2}$ to an improper fraction.
4. Add $8\frac{2}{3} + 6$.
5. Add $5\frac{3}{5} + 6\frac{3}{5}$.

Addition of Unlike Fractions

Fractions can be added when they have the same denominators. For example, $\frac{2}{5} + \frac{1}{5} = \frac{3}{5}$.

To add fractions with unlike denominators, find the LCD and use the Golden Rule.

Example: Add $\frac{3}{4} + \frac{1}{6}$

Solution: $4 = 2^2, \quad 6 = 2^1 \times 3^1$
$\text{LCD} = 2^2 \times 3^1 = 12$

$$\frac{3}{4} = \frac{3 \times 3}{4 \times 3} = \frac{9}{12}$$

$$\frac{1}{6} = \frac{1 \times 2}{6 \times 2} = \frac{2}{12}$$

$$\frac{3}{4} + \frac{1}{6} = \frac{9}{12} + \frac{2}{12} = \frac{11}{12}$$

(14)

Another example: Add: $\frac{1}{16} + \frac{5}{24}$

Solution:

Line (a) $16 = 2^4, \quad 24 = 2^3 \times 3^1$

Line (b) $\text{LCD} = 2^4 \times 3^1 = 48$

Line (c) $\frac{1}{16} = \frac{1 \times 3}{16 \times 3} = \frac{3}{48}, \quad \frac{5}{24} = \frac{5 \times 2}{24 \times 2} = \frac{10}{48}$

Line (d) $\frac{1}{16} + \frac{5}{24} = \frac{3}{48} + \frac{10}{48} = \frac{13}{48}$

Study Exercise Four

Add as indicated:

1. $\frac{3}{5} + \frac{1}{4}$
2. $\frac{5}{8} + \frac{3}{4}$
3. $\frac{3}{8} + \frac{1}{6}$
4. $\frac{5}{14} + \frac{2}{7}$

(16)

Another Method For Adding Mixed Numerals

When adding mixed numerals add the whole numbers first and then add the fractions.

Example: $1\frac{5}{6} + 3\frac{5}{9}$

Solution: $6 = 2^1 \times 3^1, \quad 9 = 3^2$

$\text{LCD} = 2 \times 3^2 = 18$

$$1\frac{5}{6} = 1\frac{15}{18}$$

$$3\frac{5}{9} = 3\frac{10}{18}$$

$$1\frac{5}{6} + 3\frac{5}{9} = 4 + \frac{25}{18}$$

$$= 4 + 1\frac{7}{18}$$

$$= 5\frac{7}{18}$$

(17)

The example in Frame 17 can be done as follows using the horizontal method:

$$1\frac{5}{6} + 3\frac{5}{9} = 1\frac{15}{18} + 3\frac{10}{18}$$
$$= 4\frac{25}{18}$$
$$= 4 + \frac{25}{18}$$
$$= 4 + 1\frac{7}{18}$$
$$= 5\frac{7}{18}$$

Study Exercise Five

Add as indicated:

1. $5\frac{1}{5} + 2\frac{2}{5}$

2. $3\frac{2}{7} + \frac{4}{7}$

3. $1\frac{3}{4} + 3\frac{3}{8}$

4. $6\frac{7}{10} + 4\frac{1}{4}$

5. $4\frac{5}{6} + 2\frac{1}{8}$

6. $1\frac{1}{2} + 1\frac{1}{4} + 1\frac{1}{8}$

Simple Subtraction of Fractions

We wish to subtract $\frac{5}{6} - \frac{3}{6}$

The diagram shows a rectangle divided in 6 parts and 5 parts are shaded to give $\frac{5}{6}$.

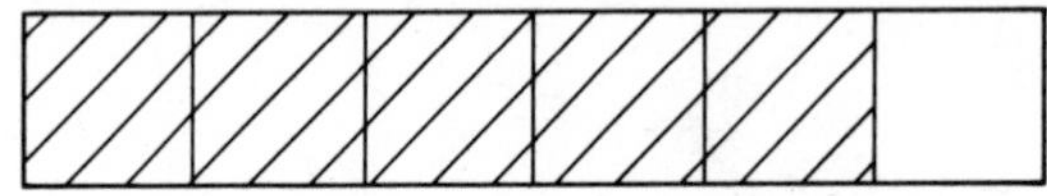

The next diagram shows the remainder when $\frac{3}{6}$ are taken from $\frac{5}{6}$.

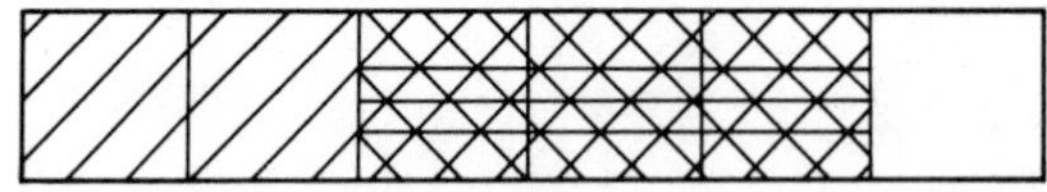

$\frac{2}{6}$ remain and $\frac{2}{6} = \frac{1}{3}$

Thus $\frac{5}{6} - \frac{3}{6} = \frac{2}{6} = \frac{1}{3}$

To subtract fractions, we subtract the numerators and keep the denominators the same.

Example 1: $\frac{5}{6} - \frac{4}{6} = \frac{5-4}{6} = \frac{1}{6}$

Example 2: $\frac{8}{11} - \frac{2}{11} = \frac{8-2}{11} = \frac{6}{11}$

Example 3: $\frac{5}{8} - \frac{3}{8} = \frac{5-3}{8} = \frac{2}{8} = \frac{1}{4}$

In subtracting fractions, if the fractions do not have the same denominator, we will use the LCD to write them with the same denominator.

Example: Subtract $\frac{2}{3} - \frac{1}{12}$.

Solution: $3 = 3^1$, $12 = 2^2 \times 3^1$

The LCD is $2^2 \times 3 = 12$

$\frac{2}{3} = \frac{2 \times 4}{3 \times 4} = \frac{8}{12}$

Thus, $\frac{8}{12} - \frac{1}{12} = \frac{7}{12}$

Study Exercise Six

Subtract as indicated:

1. $\frac{7}{10} - \frac{3}{10}$
2. $\frac{5}{8} - \frac{3}{8}$
3. $\frac{5}{6} - \frac{1}{4}$
4. $\frac{7}{8} - \frac{7}{10}$
5. $\frac{3}{4} - \frac{1}{2}$
6. $\frac{1}{4} - \frac{1}{5}$

Now let us subtract $5 - \frac{3}{8}$. There are two methods that can be used.

Method 1

Change 5 into $4\frac{8}{8}$ and then subtract $4\frac{8}{8} - \frac{3}{8} = 4\frac{5}{8}$. This method is called borrowing.

We borrow 1 from the 5 and change the 1 into $\frac{8}{8}$.

Method 2

Change 5 into an improper fraction with denominator 8, $5 = \frac{40}{8}$ and then subtract.

$$\frac{40}{8} - \frac{3}{8} = \frac{37}{8}$$

$$= 4\frac{5}{8}$$

(24)

Example 1: Subtract $4\frac{4}{9} - 1\frac{8}{9}$

Solution: We will change both fractions into improper fractions.

$$4\frac{4}{9} - 1\frac{8}{9} = \frac{40}{9} - \frac{17}{9} = \frac{23}{9} = 2\frac{5}{9}$$

Example 2: Subtract $3\frac{3}{7} - 2\frac{4}{7}$

Solution: We will use borrowing. Borrow 1 from the 3 and change the 1 to $\frac{7}{7}$.

$$3\frac{3}{7} = 2 + 1\frac{3}{7} = 2 + \frac{10}{7} = 2\frac{10}{7}$$

So, $3\frac{3}{7} - 2\frac{4}{7} = 2\frac{10}{7} - 2\frac{4}{7} = \frac{6}{7}$

(25)

Study Exercise Seven

Subtract as indicated:

1. $3 - \frac{1}{4}$
2. $7 - \frac{3}{5}$
3. $5 - 1\frac{2}{3}$
4. $4 - 3\frac{1}{8}$
5. $3\frac{1}{3} - 1\frac{2}{3}$
6. $6\frac{3}{10} - 3\frac{7}{10}$

(26)

Subtraction of Unlike Fractions

Let us subtract $5\frac{1}{4} - 3\frac{5}{8}$. The LCD is 8.

$$5\frac{1}{4} = 5\frac{1 \times 2}{4 \times 2} = 5\frac{2}{8}$$

Thus, $5\frac{1}{4} - 3\frac{5}{8} = 5\frac{2}{8} - 3\frac{5}{8}$.

Now borrow 1 from the 5.

$$5\frac{2}{8} - 3\frac{5}{8} = 4\frac{10}{8} - 3\frac{5}{8}$$

$$= 1\frac{5}{8}$$

Example: Subtract $6\frac{1}{6} - 2\frac{3}{8}$

Solution:

Line (a) $6 = 2^1 \times 3^1, \quad 8 = 2^3$

Line (b) The LCD is $2^3 \times 3^1 = 24$

Line (c) $6\frac{1}{6} = 6\frac{1 \times 4}{6 \times 4} = 6\frac{4}{24}$

Line (d) $2\frac{3}{8} = 2\frac{3 \times 3}{8 \times 3} = 2\frac{9}{24}$

Line (e) $6\frac{1}{6} - 2\frac{3}{8} = 6\frac{4}{24} - 2\frac{9}{24}$

Line (f) Now borrow 1 from the 6

$$6\frac{4}{24} - 2\frac{9}{24} = 5\frac{28}{24} - 2\frac{9}{24} = 3\frac{19}{24}$$

Study Exercise Eight

Subtract as indicated:

1. $4\frac{1}{6} - 2\frac{3}{4}$ **2.** $10\frac{1}{2} - 2\frac{5}{6}$ **3.** $11\frac{1}{10} - 3\frac{1}{12}$

REVIEW EXERCISES

Add or subtract as indicated:

1. $\frac{3}{8} + \frac{1}{8}$ 2. $\frac{3}{4} + \frac{2}{4}$ 3. $4\frac{4}{5} + \frac{2}{5}$

4. $1\frac{2}{7} + 3\frac{6}{7}$ 5. $6\frac{3}{4} + 2$ 6. $\frac{5}{9} + \frac{7}{12}$

7. $\frac{1}{8} + \frac{3}{16}$ 8. $1\frac{5}{9} + 2\frac{1}{3}$ 9. $\frac{5}{8} - \frac{3}{8}$

10. $\frac{5}{6} - \frac{7}{9}$ 11. $4\frac{1}{4} - 3\frac{3}{8}$ (30)

SOLUTIONS TO REVIEW EXERCISES

1. $\frac{3}{8} + \frac{1}{8} = \frac{3+1}{8} = \frac{4}{8} = \frac{1}{2}$

2. $\frac{3}{4} + \frac{2}{4} = \frac{3+2}{4} = \frac{5}{4} = 1\frac{1}{4}$

3. $4\frac{4}{5} + \frac{2}{5} = 4 + \frac{4+2}{5} = 4 + \frac{6}{5} = 4 + 1\frac{1}{5} = 5\frac{1}{5}$

4. $1\frac{2}{7} + 3\frac{6}{7} = 4 + \frac{2+6}{7} = 4 + \frac{8}{7} = 4 + 1\frac{1}{7} = 5\frac{1}{7}$

5. $6\frac{3}{4} + 2 = 8 + \frac{3}{4} = 8\frac{3}{4}$

6. $9 = 3^2,\quad 12 = 2^2 \times 3^1,\quad \text{LCD} = 2^2 \times 3^2 = 36$

   $\frac{5}{9} = \frac{5 \times 4}{9 \times 4} = \frac{20}{36},\quad \frac{7}{12} = \frac{7 \times 3}{12 \times 3} = \frac{21}{36}$

   $\frac{5}{9} + \frac{7}{12} = \frac{20}{36} + \frac{21}{36} = \frac{41}{36} = 1\frac{5}{36}$

7. $8 = 2^3,\quad 16 = 2^4,\quad \text{LCD} = 2^4 = 16$

   $\frac{1}{8} + \frac{3}{16} = \frac{2}{16} + \frac{3}{16} = \frac{5}{16}$

8. $9 = 3^2,\quad 3 = 3^1,\quad \text{LCD} = 3^2 = 9$

   $1\frac{5}{9} + 2\frac{1}{3} = 1\frac{5}{9} + 2\frac{3}{9} = 3 + \frac{5+3}{9} = 3 + \frac{8}{9} = 3\frac{8}{9}$

9. $\frac{5}{8} - \frac{3}{8} = \frac{5-3}{8} = \frac{2}{8} = \frac{1}{4}$

10. $6 = 2^1 \times 3^1,\quad 9 = 3^2,\quad \text{LCD} = 2^1 \times 3^2 = 18$

    $\frac{5}{6} = \frac{5 \times 3}{6 \times 3} = \frac{15}{18},\quad \frac{7}{9} = \frac{7 \times 2}{9 \times 2} = \frac{14}{18}$

    $\frac{5}{6} - \frac{7}{9} = \frac{15}{18} - \frac{14}{18} = \frac{1}{18}$

11. $4 = 2^2,\quad 8 = 2^3,\quad \text{LCD} = 2^3 = 8$

    $4\frac{1}{4} - 3\frac{3}{8} = 4\frac{2}{8} - 3\frac{3}{8} = 3\frac{10}{8} - 3\frac{3}{8} = \frac{7}{8}$ (31)

SUPPLEMENTARY PROBLEMS

Add or subtract as indicated. Leave answers in reduced form.

1. $\frac{1}{8} + \frac{5}{8}$ **2.** $\frac{3}{12} + \frac{5}{12}$ **3.** $2\frac{1}{6} + \frac{5}{6}$

4. $3\frac{1}{12} + 1\frac{7}{12}$ **5.** $1\frac{7}{12} + \frac{11}{12}$ **6.** $3\frac{7}{9} + 1\frac{5}{9}$

7. $\frac{3}{10} + \frac{4}{5}$ **8.** $\frac{5}{16} + \frac{3}{4}$ **9.** $1\frac{1}{8} + \frac{3}{4}$

10. $\frac{3}{10} + 2\frac{1}{6}$ **11.** $1\frac{3}{10} + 3\frac{4}{5}$ **12.** $1\frac{5}{16} + \frac{3}{4}$

13. $3\frac{7}{10} + 2\frac{1}{2}$ **14.** $2\frac{7}{8} + 1\frac{3}{4}$ **15.** $\frac{7}{10} + 1\frac{3}{4} + \frac{1}{2}$

16. $\frac{2}{5} + \frac{1}{2} + \frac{9}{10}$ **17.** $3\frac{1}{3} + 1\frac{1}{4} + 4\frac{7}{8}$ **18.** $3\frac{2}{5} + \frac{3}{4} + 1\frac{7}{10}$

19. $\frac{5}{8} - \frac{3}{8}$ **20.** $4\frac{7}{12} - 1\frac{5}{12}$ **21.** $4\frac{1}{5} - 2\frac{3}{5}$

22. $5\frac{3}{8} - \frac{7}{8}$ **23.** $\frac{7}{12} - \frac{1}{3}$ **24.** $\frac{5}{6} - \frac{3}{4}$

25. $6\frac{7}{10} - \frac{1}{2}$ **26.** $2\frac{5}{6} - 1\frac{4}{9}$ **27.** $8\frac{1}{3} - 3\frac{5}{6}$

28. $2\frac{1}{4} - \frac{1}{2}$ **29.** $5\frac{1}{4} - 2\frac{5}{6}$ **30.** $1\frac{1}{2} - \frac{3}{4}$

31. $3\frac{1}{6} - 1\frac{5}{8}$ **32.** $2 - 1\frac{3}{8}$ **33.** Take $\frac{9}{10}$ from $1\frac{4}{5}$.

34. Find the difference between $4\frac{5}{8}$ and $2\frac{9}{10}$.

35. Find the sum of $2 + \frac{1}{2} + \frac{1}{4} + \frac{7}{8}$.

SOLUTIONS TO STUDY EXERCISES

Study Exercise One (Frame 7)

1. $\frac{3}{5} + \frac{1}{5} = \frac{3+1}{5} = \frac{4}{5}$

2. $\frac{1}{7} + \frac{3}{7} = \frac{1+3}{7} = \frac{4}{7}$

3. $\frac{5}{8} + \frac{3}{8} = \frac{5+3}{8} = \frac{8}{8} = 1$

4. $\frac{6}{7} + \frac{5}{7} = \frac{6+5}{7} = \frac{11}{7} = 1\frac{4}{7}$

5.
$$\begin{array}{r} \frac{1}{3} \\ +\frac{1}{3} \\ \hline \frac{1+1}{3} \end{array} = \frac{2}{3}$$

6.
$$\begin{array}{r} \frac{3}{10} \\ +\frac{9}{10} \\ \hline \frac{3+9}{10} \end{array} = \frac{12}{10} = 1\frac{2}{10} = 1\frac{1}{5}$$

7A

SOLUTIONS TO STUDY EXERCISES, CONTD.

Study Exercise Two (Frame 10)

1. $2\frac{3}{11} + 5\frac{6}{11} = 7 + \frac{9}{11} = 7\frac{9}{11}$

2. $1\frac{15}{16} + 3\frac{13}{16} = 4 + \frac{28}{16} = 4 + 1\frac{12}{16} = 5\frac{12}{16} = 5\frac{3}{4}$

3. $4\frac{8}{12} + 2\frac{6}{12} = 6 + \frac{14}{12} = 6 + 1\frac{2}{12} = 7\frac{2}{12} = 7\frac{1}{6}$

4. $1\frac{13}{16} + 2\frac{4}{16} + \frac{3}{16} = 3 + \frac{20}{16} = 3 + 1\frac{4}{16} = 4\frac{4}{16} = 4\frac{1}{4}$

5. $$\begin{array}{r} 2\frac{3}{6} \\ +4\frac{5}{6} \\ \hline 6\frac{8}{6} \end{array} = 6 + 1\frac{2}{6} = 7\frac{2}{6} = 7\frac{1}{3}$$

6. $$\begin{array}{r} 1\frac{3}{5} \\ +\ \frac{4}{5} \\ \hline 1\frac{7}{5} \end{array} = 1 + 1\frac{2}{5} = 2\frac{2}{5}$$

10A

Study Exercise Three (Frame 13)

1. $8\frac{2}{3} = \frac{8 \times 3 + 2}{3} = \frac{24 + 2}{3} = \frac{26}{3}$

2. $6\frac{3}{5} = \frac{6 \times 5 + 3}{5} = \frac{30 + 3}{5} = \frac{33}{5}$

3. $2\frac{1}{2} = \frac{2 \times 2 + 1}{2} = \frac{4 + 1}{2} = \frac{5}{2}$

4. $8\frac{2}{3} + 6 = \frac{8 \times 3 + 2}{3} + \frac{6 \times 3}{3}$
$= \frac{26}{3} + \frac{18}{3}$
$= \frac{44}{3}$
$= 14\frac{2}{3}$

5. $5\frac{3}{5} + 6\frac{3}{5} = \frac{5 \times 5 + 3}{5} + \frac{6 \times 5 + 3}{5}$
$= \frac{28}{5} + \frac{33}{5}$
$= \frac{61}{5}$
$= 12\frac{1}{5}$

13A

SOLUTIONS TO STUDY EXERCISES, CONTD.

Study Exercise Four (Frame 16)

1. $5 = 5^1,\ \ 4 = 2^2,\ \ \text{LCD} = 5^1 \times 2^2 = 20$

$\frac{3}{5} = \frac{3 \times 4}{5 \times 4} = \frac{12}{20},\ \ \frac{1}{4} = \frac{1 \times 5}{4 \times 5} = \frac{5}{20}$

$\frac{3}{5} + \frac{1}{4} = \frac{12}{20} + \frac{5}{20} = \frac{17}{20}$

2. $8 = 2^3,\ \ 4 = 2^2,\ \ \text{LCD} = 2^3 = 8$

$\frac{5}{8} + \frac{3}{4} = \frac{5}{8} + \frac{6}{8} = \frac{11}{8} = 1\frac{3}{8}$

3. $8 = 2^3,\ \ 6 = 2^1 \times 3^1,\ \ \text{LCD} = 2^3 \times 3^1 = 24$

$\frac{3}{8} = \frac{3 \times 3}{8 \times 3} = \frac{9}{24},\ \ \frac{1}{6} = \frac{1 \times 4}{6 \times 4} = \frac{4}{24}$

$\frac{3}{8} + \frac{1}{6} = \frac{9}{24} + \frac{4}{24} = \frac{13}{24}$

4. $14 = 2^1 \times 7^1,\ \ 7 = 7^1,\ \ \text{LCD} = 2^1 \times 7^1 = 14$

$\frac{5}{14} + \frac{2}{7} = \frac{5}{14} + \frac{4}{14} = \frac{9}{14}$

16A

Study Exercise Five (Frame 19)

1. $5\frac{1}{5} + 2\frac{2}{5} = 7 + \frac{1 + 2}{5} = 7 + \frac{3}{5} = 7\frac{3}{5}$

2. $3\frac{2}{7} + \frac{4}{7} = 3 + \frac{2 + 4}{7} = 3 + \frac{6}{7} = 3\frac{6}{7}$

3. $4 = 2^2,\ \ 8 = 2^3,\ \ \text{LCD} = 8$

$1\frac{3}{4} + 3\frac{3}{8} = 1\frac{6}{8} + 3\frac{3}{8} = 4 + \frac{9}{8} = 4 + 1\frac{1}{8} = 5\frac{1}{8}$

4. $10 = 2^1 \times 5^1,\ \ 4 = 2^2,\ \ \text{LCD} = 2^2 \times 5^1 = 20$

$6\frac{7}{10} = 6\frac{7 \times 2}{10 \times 2} = 6\frac{14}{20},\ \ 4\frac{1}{4} = 4\frac{1 \times 5}{4 \times 5} = 4\frac{5}{20}$

$6\frac{7}{10} + 4\frac{1}{4} = 6\frac{14}{20} + 4\frac{5}{20} = 10 + \frac{14 + 5}{20} = 10 + \frac{19}{20} = 10\frac{19}{20}$

5. $6 = 2^1 \times 3^1,\ \ 8 = 2^3,\ \ \text{LCD} = 2^3 \times 3^1 = 24$

$4\frac{5}{6} = 4\frac{5 \times 4}{6 \times 4} = 4\frac{20}{24},\ \ 2\frac{1}{8} = 2\frac{1 \times 3}{8 \times 3} = 2\frac{3}{24}$

$4\frac{5}{6} + 2\frac{1}{8} = 4\frac{20}{24} + 2\frac{3}{24} = 6 + \frac{20 + 3}{24} = 6 + \frac{23}{24} = 6\frac{23}{24}$

6. $2 = 2^1,\ \ 4 = 2^2,\ \ 8 = 2^3,\ \ \text{LCD} = 2^3 = 8$

$1\frac{1}{2} + 1\frac{1}{4} + 1\frac{1}{8} = 1\frac{4}{8} + 1\frac{2}{8} + 1\frac{1}{8} = 3 + \frac{4 + 2 + 1}{8}$

$= 3 + \frac{7}{8} = 3\frac{7}{8}$

19A

SOLUTIONS TO STUDY EXERCISES, CONTD.

Study Exercise Six (Frame 23)

1. $\frac{7}{10} - \frac{3}{10} = \frac{7-3}{10} = \frac{4}{10} = \frac{2}{5}$

2. $\frac{5}{8} - \frac{3}{8} = \frac{5-3}{8} = \frac{2}{8} = \frac{1}{4}$

3. LCD is 12; $\frac{5}{6} = \frac{5 \times 2}{6 \times 2} = \frac{10}{12}$, $\frac{1}{4} = \frac{1 \times 3}{4 \times 3} = \frac{3}{12}$

$\frac{5}{6} - \frac{1}{4} = \frac{10}{12} - \frac{3}{12} = \frac{7}{12}$

4. LCD = 40; $\frac{7}{8} = \frac{7 \times 5}{8 \times 5} = \frac{35}{40}$, $\frac{7}{10} = \frac{7 \times 4}{10 \times 4} = \frac{28}{40}$

$\frac{7}{8} - \frac{7}{10} = \frac{35}{40} - \frac{28}{40} = \frac{7}{40}$

5. LCD = 4; $\frac{1}{2} = \frac{2}{4}$

$\frac{3}{4} - \frac{1}{2} = \frac{3}{4} - \frac{2}{4} = \frac{1}{4}$

6. LCD = 20; $\frac{1}{4} = \frac{1 \times 5}{4 \times 5} = \frac{5}{20}$, $\frac{1}{5} = \frac{1 \times 4}{5 \times 4} = \frac{4}{20}$

$\frac{1}{4} - \frac{1}{5} = \frac{5}{20} - \frac{4}{20} = \frac{1}{20}$ 23A

Study Exercise Seven (Frame 26)

1. $3 - \frac{1}{4} = 2\frac{4}{4} - \frac{1}{4} = 2\frac{3}{4}$

2. $7 - \frac{3}{5} = 6\frac{5}{5} - \frac{3}{5} = 6\frac{2}{5}$

3. $5 - 1\frac{2}{3} = 4\frac{3}{3} - 1\frac{2}{3} = 3\frac{1}{3}$

4. $4 - 3\frac{1}{8} = 3\frac{8}{8} - 3\frac{1}{8} = \frac{7}{8}$

5. $3\frac{1}{3} - 1\frac{2}{3} = 2\frac{4}{3} - 1\frac{2}{3} = 1\frac{2}{3}$

6. $6\frac{3}{10} - 3\frac{7}{10} = 5\frac{13}{10} - 3\frac{7}{10} = 2\frac{6}{10} = 2\frac{3}{5}$ 26A

SOLUTIONS TO STUDY EXERCISES, CONTD.

Study Exercise Eight (Frame 29)

1. $6 = 2^1 \times 3^1$, $4 = 2^2$, $\text{LCD} = 2^2 \times 3^1 = 12$

$4\frac{1}{6} = 4\frac{1 \times 2}{6 \times 2} = 4\frac{2}{12}$, $2\frac{3}{4} = 2\frac{3 \times 3}{4 \times 3} = 2\frac{9}{12}$

$$4\frac{1}{6} - 2\frac{3}{4} = 4\frac{2}{12} - 2\frac{9}{12}$$
$$= 3\frac{14}{12} - 2\frac{9}{12}$$
$$= 1\frac{14 - 9}{12}$$
$$= 1\frac{5}{12}$$

2. $2 = 2^1$, $6 = 2^1 \times 3^1$, $\text{LCD} = 2^1 \times 3^1 = 6$

$$10\frac{1}{2} - 2\frac{5}{6} = 10\frac{3}{6} - 2\frac{5}{6}$$
$$= 9\frac{9}{6} - 2\frac{5}{6}$$
$$= 7\frac{9 - 5}{6}$$
$$= 7\frac{4}{6}$$
$$= 7\frac{2}{3}$$

3. $10 = 2^1 \times 5^1$, $12 = 2^2 \times 3^1$, $\text{LCD} = 2^2 \times 3^1 \times 5^1 = 60$

$11\frac{1}{10} = 11\frac{1 \times 6}{10 \times 6} = 11\frac{6}{60}$, $3\frac{1}{12} = 3\frac{1 \times 5}{12 \times 5} = 3\frac{5}{60}$

$$11\frac{1}{10} - 3\frac{1}{12} = 11\frac{6}{60} - 3\frac{5}{60}$$
$$= 8\frac{6 - 5}{60}$$
$$= 8\frac{1}{60}$$

29A

unit

15

Multiplying Fractions

Objectives:

By the end of this unit you should be able to:

1. multiply fractions:

(a) a whole number by a fraction.

(b) two fractions.

(c) a mixed number by a whole number, fraction, or another mixed number.

(d) three fractions.

2. raise fractions to powers. ①

In the unit on whole numbers we learned that multiplication is *repetitive addition*. For example, 3×5 means 5 is added three times. That is, $5 + 5 + 5 = 15$. Also, 4×6 means $6 + 6 + 6 + 6$ ②

Now let us multiply $4 \times \frac{1}{6}$ by the method of addition.

$$4 \times \frac{1}{6} \quad \text{means} \quad \frac{1}{6} + \frac{1}{6} + \frac{1}{6} + \frac{1}{6}$$

We know $\frac{1}{6} + \frac{1}{6} + \frac{1}{6} + \frac{1}{6} = \frac{4}{6} = \frac{2}{3}$

Also, if we interchange the numbers, the product remains the same.

$$\frac{1}{6} \times 4 = \frac{1}{6} + \frac{1}{6} + \frac{1}{6} + \frac{1}{6} = \frac{4}{6} = \frac{2}{3}$$

③

Multiply $3 \times \frac{5}{7}$

$$3 \times \frac{5}{7} = \frac{5}{7} + \frac{5}{7} + \frac{5}{7}$$
$$= \frac{15}{7}$$
$$= 2\frac{1}{7}$$

Since we add 3 fives in the numerator, we can merely multiply 3×5.

$$3 \times \frac{5}{7} = \frac{3 \times 5}{7} = \frac{15}{7} = 2\frac{1}{7}$$

Rule: *To multiply a fraction by a whole number, multiply the numerator by the whole number and keep the denominator unchanged.*

Example 1: $7 \times \frac{2}{3}$

Solution: $7 \times \frac{2}{3} = \frac{7 \times 2}{3} = \frac{14}{3} = 4\frac{2}{3}$

Example 2: $\frac{2}{5} \times 6$

Solution: $\frac{2}{5} \times 6 = \frac{2 \times 6}{5} = \frac{12}{5} = 2\frac{2}{5}$

Example 3: $\frac{2}{3}$ of 8

Solution: Remember that "of" means multiply

$\frac{2}{3}$ of $8 = \frac{2}{3} \times 8 = \frac{2 \times 8}{3} = \frac{16}{3} = 5\frac{1}{3}$

Study Exercise One

1. $\frac{2}{7} \times 3$ **2.** $6 \times \frac{3}{8}$ **3.** $2 \times \frac{1}{2}$ **4.** $\frac{2}{3}$ of 12

By our rule, to multiply $8 \times \frac{5}{6}$, we write:

$$\frac{8 \times 5}{6} = \frac{40}{6} = \frac{20}{3} = 6\frac{2}{3}$$

Notice we must reduce our answer by dividing numerator and denominator by 2.
The work can be made easier by reducing before multiplying (that is, cancelling before multiplying).

$$\overset{4}{\cancel{8}} \times \frac{5}{\underset{3}{\cancel{6}}} = \frac{20}{3}$$

$$= 6\frac{2}{3}$$

(7)

Example 1: Find $25 \times \frac{7}{15}$

Solution: $\overset{5}{\cancel{25}} \times \frac{7}{\underset{3}{\cancel{15}}} = \frac{35}{3}$

$= 11\frac{2}{3}$

Example 2: Find $\frac{5}{18} \times 20$

Solution: $\frac{5}{\underset{9}{\cancel{18}}} \times \overset{10}{\cancel{20}} = \frac{50}{9}$

$= 5\frac{5}{9}$

(8)

Study Exercise Two

Find the products by first cancelling if possible and then multiplying.

1. $\frac{3}{4} \times 22$ **2.** $18 \times \frac{4}{9}$ **3.** $\frac{5}{8} \times 3$ **4.** $\frac{3}{4}$ of 6

(9)

Multiplying Two Fractions

Let us find $\frac{1}{2} \times \frac{4}{5}$.

We know $\frac{1}{2} \times 4 = 2$; so $\frac{1}{2} \times \frac{4}{5}$ must equal $\frac{2}{5}$.

That is $\frac{1}{2} \times \frac{4}{5} = \frac{2}{5}$.

Rule: *To multiply two fractions, multiply the numerators and then multiply the denominators.*

Example 1: $\frac{3}{4} \times \frac{6}{11}$

Solution: $\frac{3}{4} \times \frac{6}{11} = \frac{3 \times 6}{4 \times 11} = \frac{18}{44} = \frac{9}{22}$

Example 2: $\frac{5}{8} \times \frac{4}{15}$

Solution: $\frac{5}{8} \times \frac{4}{15} = \frac{5 \times 4}{8 \times 15} = \frac{20}{120} = \frac{1}{6}$

To make the work easier, we can cancel first, if we wish. We must cancel the same factor from both numerator and denominator.

$$\frac{\overset{1}{\cancel{5}}}{\underset{2}{\cancel{8}}} \times \frac{\overset{1}{\cancel{4}}}{\underset{3}{\cancel{15}}} = \frac{1}{6}$$

(11)

Study Exercise Three

Multiply as indicated and reduce all answers:

1. $\frac{5}{6} \times \frac{2}{3}$ **2.** $\frac{4}{15} \times \frac{3}{14}$ **3.** $\frac{2}{3} \times \frac{5}{7}$

4. $\frac{2}{3} \times \frac{5}{12}$ **5.** $\frac{3}{5} \times \frac{25}{30}$ **6.** $\frac{1}{4} \times \frac{12}{16}$

(12)

Multiplying Mixed Numbers

Procedure:

1. *First change mixed numbers to improper fractions.*
2. *Then proceed in the same way as multiplication of fractions.*

Example: $3\frac{1}{2} \times 1\frac{4}{5}$

Solution: $3\frac{1}{2} \times 1\frac{4}{5} = \frac{7}{2} \times \frac{9}{5} = \frac{63}{10} = 6\frac{3}{10}$

(13)

Example 1: $2\frac{1}{2} \times 5\frac{2}{3}$

Solution: $2\frac{1}{2} \times 5\frac{2}{3} = \frac{5}{2} \times \frac{17}{3} = \frac{85}{6} = 14\frac{1}{6}$

Example 2: $3\frac{1}{3} \times 6\frac{1}{2}$

Solution: $3\frac{1}{3} \times 6\frac{1}{2} = \frac{\overset{5}{\cancel{10}}}{3} \times \frac{13}{\underset{1}{\cancel{2}}} = \frac{65}{3} = 21\frac{2}{3}$

Cancelling should be used when possible to make the work easier.

(14)

Study Exercise Four

1. $3\frac{1}{4} \times \frac{2}{5}$
2. $4\frac{1}{4} \times \frac{8}{9}$
3. $3\frac{2}{3} \times 2\frac{1}{4}$
4. $1\frac{4}{5} \times 3\frac{1}{3}$
5. $2\frac{6}{7} \times 1\frac{1}{4}$
6. $4\frac{3}{5} \times 8$
7. $\frac{5}{8} \times 2\frac{2}{5}$

(15)

Raising Fractions to Powers

Recall that $2^3 = 2 \times 2 \times 2 = 8$

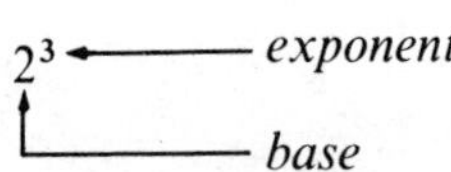

Similarly, $\left(\frac{3}{4}\right)^2 = \frac{3}{4} \times \frac{3}{4} = \frac{9}{16}$

$\left(\frac{3}{4}\right)^2$ ← exponent

↑ base

(16)

Example 1: Find $\left(\frac{2}{5}\right)^2$

Solution: $\left(\frac{2}{5}\right)^2 = \frac{2}{5} \times \frac{2}{5} = \frac{4}{25}$

Example 2: Find $\left(1\frac{1}{3}\right)^2$

Solution: $\left(1\frac{1}{3}\right)^2 = \left(\frac{4}{3}\right)^2 = \frac{4}{3} \times \frac{4}{3} = \frac{16}{9} = 1\frac{7}{9}$

(17)

Study Exercise Five

Find:

1. $\left(\frac{1}{2}\right)^2$
2. $\left(\frac{7}{8}\right)^2$
3. $\left(2\frac{2}{3}\right)^2$
4. $\left(1\frac{1}{5}\right)^2$

(18)

Multiplying Several Fractions

The same method used to multiply two fractions can be used to multiply more than two fractions.

Let us find the product of $\frac{3}{10} \times \frac{1}{6} \times \frac{9}{24}$.

$$\frac{\overset{1}{\cancel{3}}}{10} \times \frac{1}{\underset{2}{\cancel{6}}} \times \frac{\overset{3}{\cancel{9}}}{\underset{8}{\cancel{24}}} = \frac{1 \times 1 \times 3}{10 \times 2 \times 8} = \frac{3}{160}$$

(19)

Example 1: $\frac{2}{7} \times \frac{5}{6} \times \frac{8}{25}$

Solution: $\frac{\overset{1}{\cancel{2}}}{7} \times \frac{\overset{1}{\cancel{5}}}{\underset{3}{\cancel{6}}} \times \frac{8}{\underset{5}{\cancel{25}}} = \frac{1 \times 1 \times 8}{7 \times 3 \times 5} = \frac{8}{105}$

Example 2: $2\frac{1}{3} \times 1\frac{1}{5} \times 3$

Solution: Change to improper fractions:

$$2\frac{1}{3} \times 1\frac{1}{5} \times 3 = \frac{7}{\underset{1}{\cancel{3}}} \times \frac{6}{5} \times \frac{\overset{1}{\cancel{3}}}{1} = \frac{7 \times 6 \times 1}{1 \times 5 \times 1} = \frac{42}{5} = 8\frac{2}{5}$$

(20)

Study Exercise Six

Multiply as indicated and reduce all answers:

1. $\frac{1}{2} \times \frac{8}{15} \times \frac{5}{6}$ **2.** $\frac{2}{5} \times \frac{3}{4} \times \frac{15}{16}$ **3.** $1\frac{1}{2} \times \frac{4}{5} \times 3\frac{1}{6}$

4. $2\frac{3}{4} \times 2\frac{2}{3} \times 3\frac{1}{7}$ **5.** $2\frac{1}{2} \times \frac{4}{5} \times 2\frac{1}{6}$

Square Roots

A *square root* of a number is one of two equal factors of a number.

$3 \times 3 = 9$

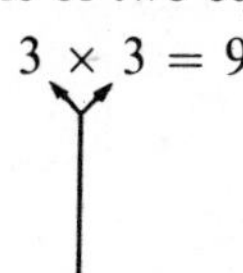

The two equal factors of 9

Thus 3 is a square root of 9.

Recall that the square root of 9 is written $\sqrt{9}$.

$\sqrt{9} = 3$ since $3 \times 3 = 9$

Perfect Square Roots

Study the following square roots until you know them.

$\sqrt{1} = 1$	$\sqrt{25} = 5$	$\sqrt{81} = 9$
$\sqrt{4} = 2$	$\sqrt{36} = 6$	$\sqrt{100} = 10$
$\sqrt{9} = 3$	$\sqrt{49} = 7$	$\sqrt{121} = 11$
$\sqrt{16} = 4$	$\sqrt{64} = 8$	$\sqrt{144} = 12$

(23)

Square Roots of Fractions

$\sqrt{\frac{4}{9}} = \frac{2}{3}$ $\left(\textit{Check:} \ \frac{2}{3} \times \frac{2}{3} = \frac{4}{9}\right)$

$\sqrt{\frac{25}{36}} = \frac{5}{6}$ $\left(\textit{Check:} \ \frac{5}{6} \times \frac{5}{6} = \frac{25}{36}\right)$

Notice that to find the square root of a fraction, find the square root of the numerator and denominator separately.

Square Roots of Mixed Numbers

First change to an improper fraction. Then find the square root of the fraction.

Example: $\sqrt{7\frac{1}{9}}$

Solution: $\sqrt{7\frac{1}{9}} = \sqrt{\frac{64}{9}}$

$= \frac{8}{3} = 2\frac{2}{3}$

Check: $\frac{8}{3} \times \frac{8}{3} = \frac{64}{9} = 7\frac{1}{9}$.

Study Exercise Seven

Find the indicated square roots:

1. $\sqrt{\frac{4}{25}}$ 2. $\sqrt{\frac{49}{100}}$ 3. $\sqrt{\frac{1}{9}}$

4. $\sqrt{1\frac{7}{9}}$ 5. $\sqrt{3\frac{1}{16}}$ 6. $\sqrt{2\frac{7}{9}}$

Summary

To multiply two fractions, multiply the numerators and then multiply the denominators.

Example: $\frac{2}{3} \times \frac{5}{7} = \frac{10}{21}$

Before multiplying be sure any whole number or mixed number is changed to an improper fraction.

Example: $3 \times \frac{7}{11} = \frac{3}{1} \times \frac{7}{11} = \frac{21}{11} = 1\frac{10}{11}$

Example: $2\frac{1}{2} \times \frac{1}{3} = \frac{5}{2} \times \frac{1}{3} = \frac{5}{6}$

Final answers should be reduced. Cancelling can be used to make the work easier.

Example: $\frac{2}{7} \times \frac{5}{8} = \frac{\overset{1}{\cancel{2}}}{7} \times \frac{5}{\underset{4}{\cancel{8}}} = \frac{1 \times 5}{7 \times 4} = \frac{5}{28}$

REVIEW EXERCISES

1. $2 \times \frac{3}{5}$ **2.** $\frac{4}{5}$ of 20 **3.** $\frac{4}{21} \times \frac{7}{8}$

4. $\frac{10}{9} \times \frac{15}{8}$ **5.** $2\frac{1}{3} \times \frac{3}{16}$ **6.** $1\frac{7}{16} \times 1\frac{5}{9}$

7. $\left(\frac{3}{5}\right)^2$ **8.** $\left(3\frac{2}{3}\right)^2$ **9.** $\frac{3}{4} \times \frac{4}{7} \times \frac{1}{2}$

10. $4\frac{1}{6} \times 3\frac{1}{5} \times 1\frac{1}{10}$ **11.** $\sqrt{\frac{81}{121}}$ **12.** $\sqrt{12\frac{1}{4}}$

(28)

SOLUTIONS TO REVIEW EXERCISES

1. $2 \times \frac{3}{5} = \frac{2 \times 3}{5} = \frac{6}{5} = 1\frac{1}{5}$

2. $\frac{4}{5}$ of $20 = \frac{4}{5} \times 20 = \frac{4 \times \overset{4}{\cancel{20}}}{\underset{1}{\cancel{5}}} = 16$

3. $\frac{\overset{1}{\cancel{4}}}{\underset{3}{\cancel{21}}} \times \frac{\overset{1}{\cancel{7}}}{\underset{2}{\cancel{8}}} = \frac{1 \times 1}{3 \times 2} = \frac{1}{6}$

4. $\frac{\overset{5}{\cancel{10}}}{\underset{3}{\cancel{9}}} \times \frac{\overset{5}{\cancel{15}}}{\underset{4}{\cancel{8}}} = \frac{5 \times 5}{3 \times 4} = \frac{25}{12} = 2\frac{1}{12}$

5. $2\frac{1}{3} \times \frac{3}{16} = \frac{7}{\underset{1}{\cancel{3}}} \times \frac{\overset{1}{\cancel{3}}}{16} = \frac{7 \times 1}{1 \times 16} = \frac{7}{16}$

6. $1\frac{7}{16} \times 1\frac{5}{9} = \frac{23}{\underset{8}{\cancel{16}}} \times \frac{\overset{7}{\cancel{14}}}{9} = \frac{23 \times 7}{8 \times 9} = \frac{161}{72} = 2\frac{17}{72}$

7. $\left(\frac{3}{5}\right)^2 = \frac{3}{5} \times \frac{3}{5} = \frac{9}{25}$

8. $\left(3\frac{2}{3}\right)^2 = \left(\frac{11}{3}\right)^2 = \frac{11}{3} \times \frac{11}{3} = \frac{11 \times 11}{3 \times 3} = \frac{121}{9} = 13\frac{4}{9}$

9. $\frac{3}{\underset{1}{\cancel{4}}} \times \frac{\overset{1}{\cancel{4}}}{7} \times \frac{1}{2} = \frac{3 \times 1 \times 1}{1 \times 7 \times 2} = \frac{3}{14}$

SOLUTIONS TO REVIEW EXERCISES, CONTD.

(Frame 29, contd.)

10. $$4\frac{1}{6}\times 3\frac{1}{5}\times 1\frac{1}{10} = \frac{25}{6}\times\frac{\overset{8}{\cancel{16}}}{5}\times\frac{11}{\underset{5}{\cancel{10}}}$$

$$=\frac{\overset{5}{\cancel{25}}}{\underset{3}{\cancel{6}}}\times\frac{\overset{4}{\cancel{\overset{8}{\cancel{16}}}}}{5}\times\frac{11}{\underset{1}{\cancel{\underset{5}{\cancel{10}}}}}$$

$$=\frac{\overset{1}{\cancel{\overset{5}{\cancel{25}}}}}{\underset{3}{\cancel{6}}}\times\frac{\overset{4}{\cancel{\overset{8}{\cancel{16}}}}}{\underset{1}{\cancel{5}}}\times\frac{11}{\underset{1}{\cancel{\underset{5}{\cancel{10}}}}}$$

$$=\frac{1\times 4\times 11}{3\times 1\times 1}$$

$$=\frac{44}{3}$$

$$=14\frac{2}{3}$$

11. $\sqrt{\frac{81}{121}}=\frac{9}{11}$

12. $\sqrt{12\frac{1}{4}}=\sqrt{\frac{49}{4}}=\frac{7}{2}=3\frac{1}{2}$ (29)

SUPPLEMENTARY PROBLEMS

Perform the indicated operations and reduce all fractions.

1. $4\times\frac{5}{6}$ **2.** $12\times\frac{3}{8}$ **3.** $6\times 2\frac{1}{2}$

4. $\frac{5}{6}\times 10$ **5.** $\frac{3}{5}\times 5$ **6.** $\frac{1}{4}\times\frac{2}{5}$

7. $\frac{3}{8}\times\frac{4}{15}$ **8.** $\frac{1}{3}\times 6\frac{1}{5}$ **9.** $12\times 2\frac{2}{3}$

10. $\frac{5}{8}\times\frac{5}{11}$ **11.** $\frac{10}{9}\times\frac{15}{8}$ **12.** $2\frac{1}{2}\times 1\frac{1}{5}$

13. $\frac{9}{16}\times 1\frac{1}{3}$ **14.** $2\frac{9}{16}\times 1\frac{1}{4}$ **15.** $\frac{4}{21}\times\frac{7}{8}$

16. $3\frac{3}{4}\times 3\frac{1}{5}$ **17.** $\frac{1}{2}\times\frac{2}{3}\times 8$ **18.** $\frac{2}{3}\times\frac{7}{8}\times\frac{3}{10}$

19. $\frac{3}{4}\times\frac{3}{5}\times\frac{3}{6}$ **20.** $1\frac{1}{8}\times 1\frac{1}{4}\times 1\frac{1}{3}$ **21.** $3\frac{3}{5}\times 1\frac{3}{8}\times\frac{10}{11}$

22. $\left(\frac{2}{5}\right)^2$ **23.** $\left(\frac{7}{8}\right)^2$ **24.** $\left(\frac{1}{6}\right)^2$

25. $\left(3\frac{1}{2}\right)^2$ **26.** $\left(2\frac{3}{4}\right)^2$ **27.** $\left(5\frac{1}{5}\right)^2$

28. $\left(1\frac{1}{2}\right)^2\times\frac{6}{7}$ **29.** $\left(\frac{1}{3}\right)^2\times 9$ **30.** $\frac{2^2}{5}\times\frac{1}{2}$

SUPPLEMENTARY PROBLEMS, CONTD.

31. $\frac{3}{4^2} \times \frac{2^2}{6}$ **32.** $\left(\frac{2}{3}\right)^3$ **33.** $\left(1\frac{1}{4}\right)^3$

34. $\sqrt{\frac{64}{121}}$ **35.** $\sqrt{\frac{25}{36}}$ **36.** $\sqrt{\frac{9}{144}}$

37. $\sqrt{1\frac{15}{49}}$ **38.** $\sqrt{7\frac{1}{9}}$ **39.** $\sqrt{1\frac{11}{25}}$

40. $\sqrt{\frac{1}{100}}$

SOLUTIONS TO STUDY EXERCISES

Study Exercise One (Frame 6)

1. $\frac{2}{7} \times 3 = \frac{2 \times 3}{7} = \frac{6}{7}$

2. $6 \times \frac{3}{8} = \frac{6 \times 3}{8} = \frac{18}{8} = 2\frac{2}{8} = 2\frac{1}{4}$

3. $2 \times \frac{1}{2} = \frac{2 \times 1}{2} = \frac{2}{2} = 1$

4. $\frac{2}{3}$ of $12 = \frac{2 \times 12}{3} = \frac{24}{3} = 8$

Study Exercise Two (Frame 9)

1. $\frac{3}{4} \times 22 = \frac{3}{\underset{2}{\cancel{4}}} \times \overset{11}{\cancel{22}} = \frac{33}{2} = 16\frac{1}{2}$

2. $18 \times \frac{4}{9} = \overset{2}{\cancel{18}} \times \frac{4}{\underset{1}{\cancel{9}}} = \frac{8}{1} = 8$

3. $\frac{5}{8} \times 3 = \frac{5 \times 3}{8} = \frac{15}{8} = 1\frac{7}{8}$

4. $\frac{3}{4} \times 6 = \frac{3}{\underset{2}{\cancel{4}}} \times \overset{3}{\cancel{6}} = \frac{9}{2} = 4\frac{1}{2}$

9A

Study Exercise Three (Frame 12)

1. $\frac{5}{\underset{3}{\cancel{6}}} \times \frac{\overset{1}{\cancel{2}}}{3} = \frac{5}{9}$

2. $\frac{\overset{2}{\cancel{4}}}{\underset{5}{\cancel{15}}} \times \frac{\overset{1}{\cancel{3}}}{\underset{7}{\cancel{14}}} = \frac{2}{35}$

3. $\frac{2}{3} \times \frac{5}{7} = \frac{10}{21}$

4. $\frac{\overset{1}{\cancel{2}}}{3} \times \frac{5}{\underset{6}{\cancel{12}}} = \frac{5}{18}$

SOLUTIONS TO STUDY EXERCISES, CONTD.

Study Exercise Three (Frame 12, contd.)

5. $\frac{\overset{1}{\cancel{3}}}{\underset{1}{\cancel{5}}} \times \frac{\overset{\overset{1}{\cancel{5}}}{\cancel{25}}}{\underset{\underset{2}{\cancel{10}}}{\cancel{30}}} = \frac{1}{2}$

6. $\frac{1}{\underset{1}{\cancel{4}}} \times \frac{\overset{3}{\cancel{12}}}{16} = \frac{3}{16}$

12A

Study Exercise Four (Frame 15)

1. $3\frac{1}{4} \times \frac{2}{5} = \frac{13}{\underset{2}{\cancel{4}}} \times \frac{\overset{1}{\cancel{2}}}{5} = \frac{13}{10} = 1\frac{3}{10}$

2. $4\frac{1}{4} \times \frac{8}{9} = \frac{17}{\underset{1}{\cancel{4}}} \times \frac{\overset{2}{\cancel{8}}}{9} = \frac{34}{9} = 3\frac{7}{9}$

3. $3\frac{2}{3} \times 2\frac{1}{4} = \frac{11}{\underset{1}{\cancel{3}}} \times \frac{\overset{3}{\cancel{9}}}{4} = \frac{33}{4} = 8\frac{1}{4}$

4. $1\frac{4}{5} \times 3\frac{1}{3} = \frac{\overset{3}{\cancel{9}}}{\underset{1}{\cancel{5}}} \times \frac{\overset{2}{\cancel{10}}}{\underset{1}{\cancel{3}}} = \frac{6}{1} = 6$

5. $2\frac{6}{7} \times 1\frac{1}{4} = \frac{\overset{5}{\cancel{20}}}{7} \times \frac{5}{\underset{1}{\cancel{4}}} = \frac{25}{7} = 3\frac{4}{7}$

6. $4\frac{3}{5} \times 8 = \frac{23}{5} \times \frac{8}{1} = \frac{184}{5} = 36\frac{4}{5}$

7. $\frac{5}{8} \times 2\frac{2}{5} = \frac{\overset{1}{\cancel{5}}}{\underset{2}{\cancel{8}}} \times \frac{\overset{3}{\cancel{12}}}{\underset{1}{\cancel{5}}} = \frac{3}{2} = 1\frac{1}{2}$

15A

Study Exercise Five (Frame 18)

1. $\left(\frac{1}{2}\right)^2 = \frac{1}{2} \times \frac{1}{2} = \frac{1 \times 1}{2 \times 2} = \frac{1}{4}$

2. $\left(\frac{7}{8}\right)^2 = \frac{7}{8} \times \frac{7}{8} = \frac{7 \times 7}{8 \times 8} = \frac{49}{64}$

3. $\left(2\frac{2}{3}\right)^2 = \left(\frac{8}{3}\right)^2 = \frac{8}{3} \times \frac{8}{3} = \frac{8 \times 8}{3 \times 3} = \frac{64}{9} = 7\frac{1}{9}$

4. $\left(1\frac{1}{5}\right)^2 = \left(\frac{6}{5}\right)^2 = \frac{6}{5} \times \frac{6}{5} = \frac{6 \times 6}{5 \times 5} = \frac{36}{25} = 1\frac{11}{25}$

18A

Study Exercise Six (Frame 21)

1. $\frac{1}{\underset{1}{\cancel{2}}} \times \frac{\overset{4}{\cancel{8}}}{\underset{3}{\cancel{15}}} \times \frac{\overset{1}{\cancel{5}}}{6} = \frac{1}{\underset{1}{\cancel{2}}} \times \frac{\overset{\overset{2}{\cancel{4}}}{\cancel{8}}}{\underset{3}{\cancel{15}}} \times \frac{\overset{1}{\cancel{5}}}{\underset{3}{\cancel{6}}} = \frac{1 \times 2 \times 1}{1 \times 3 \times 3} = \frac{2}{9}$

2. $\frac{\overset{1}{\cancel{2}}}{\underset{1}{\cancel{5}}} \times \frac{3}{\underset{2}{\cancel{4}}} \times \frac{\overset{3}{\cancel{15}}}{16} = \frac{1 \times 3 \times 3}{1 \times 2 \times 16} = \frac{9}{32}$

3. $1\frac{1}{2} \times \frac{4}{5} \times 3\frac{1}{6} = \frac{\overset{1}{\cancel{3}}}{\underset{1}{\cancel{2}}} \times \frac{\overset{\overset{1}{\cancel{2}}}{\cancel{4}}}{5} \times \frac{19}{\underset{\underset{1}{\cancel{2}}}{\cancel{6}}} = \frac{1 \times 1 \times 19}{1 \times 5 \times 1} = \frac{19}{5} = 3\frac{4}{5}$

SOLUTIONS TO STUDY EXERCISES, CONTD.

Study Exercise Six (Frame 21, contd.)

4. $2\frac{3}{4} \times 2\frac{2}{3} \times 3\frac{1}{7} = \frac{11}{\cancel{4}_1} \times \frac{\cancel{8}^2}{3} \times \frac{22}{7} = \frac{11 \times 2 \times 22}{1 \times 3 \times 7} = \frac{484}{21} = 23\frac{1}{21}$

5. $2\frac{1}{2} \times \frac{4}{5} \times 2\frac{1}{6} = \frac{\cancel{5}^1}{\cancel{2}_1} \times \frac{\cancel{4}^{\cancel{2}^1}}{\cancel{5}_1} \times \frac{13}{\cancel{6}_3} = \frac{1 \times 1 \times 13}{1 \times 1 \times 3} = \frac{13}{3} = 4\frac{1}{3}$

21A

Study Exercise Seven (Frame 26)

1. $\sqrt{\frac{4}{25}} = \frac{2}{5}$

2. $\sqrt{\frac{49}{100}} = \frac{7}{10}$

3. $\sqrt{\frac{1}{9}} = \frac{1}{3}$

4. $\sqrt{1\frac{7}{9}} = \sqrt{\frac{16}{9}} = \frac{4}{3} = 1\frac{1}{3}$

5. $\sqrt{3\frac{1}{16}} = \sqrt{\frac{49}{16}} = \frac{7}{4} = 1\frac{3}{4}$

6. $\sqrt{2\frac{7}{9}} = \sqrt{\frac{25}{9}} = \frac{5}{3} = 1\frac{2}{3}$

26A

unit

16

Division of Fractions

Objectives:

By the end of this unit you should:

1. be able to find the reciprocal of a number.
2. know how to simplify complex fractions.
3. know how to divide fractions.

(1)

Reciprocal

The *reciprocal* of a fraction is found by inverting the fraction.

Fraction	Reciprocal
$\frac{2}{5}$	$\frac{5}{2}$ or $2\frac{1}{2}$
$\frac{1}{3}$	$\frac{3}{1}$ or 3

(2)

The reciprocal of a whole number or a mixed number is found by first changing to an improper fraction and then inverting.

Number	Improper Fraction	Reciprocal
3	$\frac{3}{1}$	$\frac{1}{3}$
$2\frac{2}{5}$	$\frac{12}{5}$	$\frac{5}{12}$

(3)

Product of a Number By Its Reciprocal

Number	Reciprocal	Number × Reciprocal
↓	↓	↓
3	$\frac{1}{3}$	$3 \times \frac{1}{3} = 1$
$\frac{2}{5}$	$\frac{5}{2}$	$\frac{2}{5} \times \frac{5}{2} = 1$
$2\frac{2}{5}$ or $\frac{12}{5}$	$\frac{5}{12}$	$\frac{12}{5} \times \frac{5}{12} = 1$

Rule: *The product of a number by its reciprocal is 1.*

We can also state that *whenever the product of two numbers is 1, each number is the reciprocal of the other.*

For example, in $\frac{2}{5} \times ? = 1$, the question mark must stand for the reciprocal of $\frac{2}{5}$ or $\frac{5}{2}$, since $\frac{2}{5} \times \frac{5}{2} = 1$.

Instead of a question mark, the letter n can be used.

If $\frac{2}{5} \times n = 1$, then $n = \frac{5}{2}$.

Study Exercise One

Find the reciprocal of each of these numbers:

1. $\frac{3}{11}$ **2.** $\frac{4}{5}$ **3.** 7 **4.** $4\frac{2}{3}$

Find what number the question mark or n represents:

5. $\frac{2}{3} \times ? = 1$ **6.** $\frac{5}{3} \times ? = 1$ **7.** $\frac{3}{7} \times n = 1$ **8.** $\frac{12}{11} \times n = 1$

(6)

Complex Fractions

Examples of complex fractions include:

$$\frac{\frac{2}{3}}{\frac{1}{6}}$$ ← the numerator is $\frac{2}{3}$; ← the denominator is $\frac{1}{6}$

$$\frac{8}{\frac{2}{5}}$$ ← the numerator is 8; ← the denominator is $\frac{2}{5}$

$$\frac{\frac{3}{8}}{4}$$ ← the numerator is $\frac{3}{8}$; ← the denominator is 4

A *complex fraction* is a fraction whose numerator or denominator or both are also fractions.

$$\frac{\frac{2}{3}}{\frac{1}{6}} \qquad \frac{8}{\frac{2}{5}} \qquad \frac{\frac{3}{8}}{4}$$

It is important to draw a longer line between numerator and denominator. (8)

Example 1: What is the denominator of $\frac{\frac{3}{5}}{6}$?

Solution: 6

Example 2: What is the numerator of $\frac{\frac{1}{3}}{\frac{3}{4}}$?

Solution: $\frac{1}{3}$

Example 3: Which of these fractions is a complex fraction?

$$\frac{12}{5}, \quad \frac{\frac{1}{3}}{6}, \quad 2\frac{1}{3}$$

Solution: $\frac{\frac{1}{3}}{6}$

Simplifying a Complex Fraction

Recall that the product of a number by its reciprocal is 1.

For example, $\frac{3}{8} \times \frac{8}{3} = 1$.

We can simplify complex fractions by using the property of reciprocals and the Golden Rule of Fractions.
Do you remember the Golden Rule of Fractions?

Division By One

In Unit 11 we learned three ways to indicate division by one.
For example, six divided by one can be written:

$$6 \div 1, \qquad 1\overline{)6}, \qquad \frac{6}{1}$$

In each case, six divided by one is 6.

Similarly $\frac{\frac{2}{3}}{1} = \frac{2}{3}$ since $1 \times \frac{2}{3} = \frac{2}{3}$

Rule: *Division by 1 does not change a number.* (11)

Simplifying Complex Fractions

Example: Simplify $\dfrac{\frac{2}{3}}{6}$.

Solution: We will make the denominator one by multiplying numerator and denominator by the reciprocal of 6.

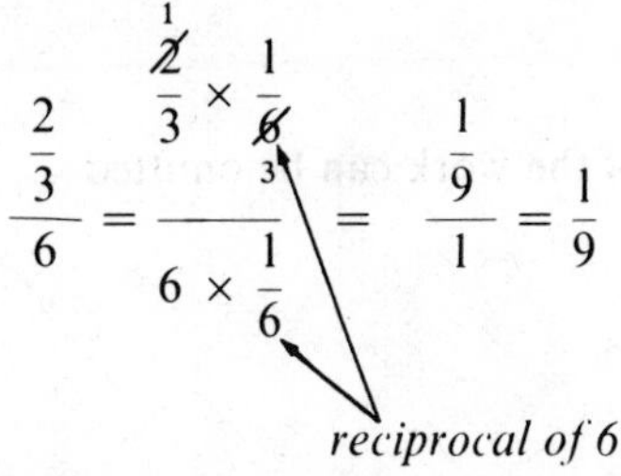

$$\frac{\frac{2}{3}}{6} = \frac{\frac{2}{3} \times \frac{1}{6}}{6 \times \frac{1}{6}} = \frac{\frac{1}{9}}{1} = \frac{1}{9}$$

(12)

Rule: *To simplify a complex fraction, multiply numerator and denominator by the reciprocal of the denominator.*

Example 1: Simplify $\dfrac{3}{\frac{3}{4}}$

Solution: $\dfrac{3}{\frac{3}{4}} = \dfrac{3 \times \frac{4}{3}}{\frac{3}{4} \times \frac{4}{3}} = \dfrac{4}{1} = 4$

Example 2: Simplify $\dfrac{\frac{5}{8}}{\frac{2}{3}}$

Solution: $\dfrac{\frac{5}{8}}{\frac{2}{3}} = \dfrac{\frac{5}{8} \times \frac{3}{2}}{\frac{2}{3} \times \frac{3}{2}} = \dfrac{\frac{15}{16}}{1} = \dfrac{15}{16}$

Study Exercise Two

1. What is the numerator of $\dfrac{8}{\frac{2}{3}}$?

2. What is the denominator of $\dfrac{\frac{1}{6}}{\frac{4}{7}}$?

3. Simplify $\dfrac{\frac{3}{5}}{1}$

4. Simplify $\dfrac{\frac{2}{3}}{4}$

5. Simplify $\dfrac{2}{\frac{3}{4}}$

6. Simplify $\dfrac{8}{1\frac{1}{3}}$

(Frame 14, contd.)

7. Simplify $\dfrac{\frac{5}{9}}{\frac{4}{5}}$ **8.** Simplify $\dfrac{\frac{3}{8}}{\frac{1}{4}}$

(14)

In simplifying a complex fraction like $\dfrac{\frac{3}{8}}{\frac{1}{4}}$ much of the work can be omitted.

$$\frac{\frac{3}{8}}{\frac{1}{4}} = \frac{\frac{3}{\underset{2}{\cancel{8}}} \times \frac{\overset{1}{\cancel{4}}}{1}}{\frac{1}{4} \times \frac{4}{1}} = \frac{\frac{3}{2}}{1} = \frac{3}{2} = 1\frac{1}{2}$$

We can omit the work in between and write:

$$\frac{\frac{3}{8}}{\frac{1}{4}} = \frac{3}{\underset{2}{\cancel{8}}} \times \frac{\overset{1}{\cancel{4}}}{1} = \frac{3}{2} = 1\frac{1}{2}$$

Thus, to divide $\frac{3}{8}$ by $\frac{1}{4}$, we multiply $\frac{3}{8}$ by the reciprocal of $\frac{1}{4}$.

(15)

Rule: *To divide any number by a fraction, multiply by the reciprocal of the fraction.*

$$\frac{6}{18} \div \frac{2}{3} = \frac{\overset{1}{\cancel{6}}}{\underset{\underset{1}{\cancel{6}}}{\cancel{18}}} \times \frac{\overset{1}{\cancel{3}}}{2} = \frac{1}{2}$$

reciprocal of the divisor $\frac{2}{3}$

Remember, to divide you must multiply by the reciprocal of the divisor.

(16)

Example 1: $\frac{4}{10} \div 5$

Solution: $\frac{4}{10} \div 5 = \frac{\overset{2}{\cancel{4}}}{\underset{5}{\cancel{10}}} \times \frac{1}{5} = \frac{2}{25}$

Example 2: $\dfrac{\frac{5}{6}}{\frac{2}{3}}$

Solution: $\dfrac{\frac{5}{6}}{\frac{2}{3}} = \frac{5}{\underset{2}{\cancel{6}}} \times \frac{\overset{1}{\cancel{3}}}{2} = \frac{5}{4} = 1\frac{1}{4}$

Example 3: $6 \div \frac{2}{3}$

Solution: $6 \div \frac{2}{3} = \frac{\overset{3}{\cancel{6}}}{1} \times \frac{3}{\underset{1}{\cancel{2}}} = \frac{9}{1} = 9$

(17)

Study Exercise Three

Divide as indicated and reduce all answers.

1. $\frac{4}{5} \div \frac{3}{10}$
2. $8 \div \frac{2}{3}$
3. $\dfrac{\frac{5}{9}}{\frac{3}{4}}$
4. $\dfrac{\frac{6}{7}}{3}$
5. $\frac{3}{8} \div \frac{6}{7}$
6. $\frac{1}{6} \div \frac{1}{3}$
7. $\dfrac{4}{\frac{2}{3}}$
8. $\dfrac{\frac{5}{9}}{\frac{2}{5}}$

Mixed Numbers in Division Problems

Mixed numbers should first be changed to improper fractions. Then divide by multiplying by the reciprocal of the divisor.

$$2\frac{2}{3} \div \frac{2}{5} = \frac{8}{3} \div \frac{2}{5}$$
$$= \frac{\overset{4}{\cancel{8}}}{3} \times \frac{5}{\underset{1}{\cancel{2}}}$$
$$= \frac{20}{3}$$
$$= 6\frac{2}{3}$$

Example: Find $5\frac{1}{3} \div 2\frac{3}{4}$

Solution: $5\frac{1}{3} \div 2\frac{3}{4} = \frac{16}{3} \div \frac{11}{4}$

$$= \frac{16}{3} \times \frac{4}{11}$$
$$= \frac{64}{33}$$
$$= 1\frac{31}{33}$$

Study Exercise Four

Divide as indicated and reduce all answers.

1. $3\frac{1}{2} \div 2\frac{2}{3}$
2. $4\frac{2}{3} \div \frac{1}{3}$
3. $\dfrac{1\frac{2}{3}}{2\frac{1}{2}}$
4. $\dfrac{6\frac{1}{2}}{2}$
5. $1\frac{3}{4} \div 5\frac{1}{4}$

(21)

REVIEW EXERCISES

1. The reciprocal of $\frac{5}{7}$ is ______.

2. The reciprocal of $6\frac{3}{5}$ is ______.

3. The product of a number and its reciprocal is ___________.

4. Find the number n if $\frac{5}{7} \times n = 1$.

5. Which of the following is a complex fraction?

$$\frac{2}{7}, \quad 1\frac{3}{5}, \quad \frac{5}{\frac{2}{3}}, \quad \frac{\frac{2}{3}}{\frac{1}{5}}$$

6. Simplify $\dfrac{\frac{7}{8}}{1}$

7. Simplify $\dfrac{5}{\frac{3}{4}}$

8. Simplify $\dfrac{\frac{2}{3}}{\frac{1}{4}}$

9. Find $\frac{3}{8} \div \frac{1}{6}$

10. Find $\dfrac{1\frac{2}{3}}{3\frac{5}{9}}$ 22

SOLUTIONS TO REVIEW EXERCISES

1. The reciprocal of $\frac{5}{7}$ is $\underline{\frac{7}{5} \text{ or } 1\frac{2}{5}}$.

2. The reciprocal of $6\frac{3}{5}$ is $\underline{\frac{5}{33}}$.

3. The product of a number and its reciprocal is 1.

4. $\frac{5}{7} \times n = 1$

n is the reciprocal of $\frac{5}{7}$; thus $n = \frac{7}{5}$ or $1\frac{2}{5}$.

5. $\dfrac{5}{\frac{2}{3}}$ and $\dfrac{\frac{2}{3}}{\frac{1}{5}}$ are complex fractions.

6. $\dfrac{\frac{7}{8}}{1} = \frac{7}{8}$

SOLUTIONS TO REVIEW EXERCISES, CONTD.

(Frame 23, contd.)

7. $\dfrac{5}{\frac{3}{4}} = 5 \div \dfrac{3}{4} = \dfrac{5}{1} \times \dfrac{4}{3} = \dfrac{20}{3} = 6\dfrac{2}{3}$

8. $\dfrac{\frac{2}{3}}{\frac{1}{4}} = \dfrac{2}{3} \div \dfrac{1}{4} = \dfrac{2}{3} \times \dfrac{4}{1} = \dfrac{8}{3} = 2\dfrac{2}{3}$

9. $\dfrac{3}{8} \div \dfrac{1}{6} = \dfrac{3}{\underset{4}{\cancel{8}}} \times \dfrac{\overset{3}{\cancel{6}}}{1} = \dfrac{9}{4} = 2\dfrac{1}{4}$

10. $\dfrac{1\frac{2}{3}}{3\frac{5}{9}} = 1\dfrac{2}{3} \div 3\dfrac{5}{9}$

$= \dfrac{5}{3} \div \dfrac{32}{9}$

$= \dfrac{5}{\underset{1}{\cancel{3}}} \times \dfrac{\overset{3}{\cancel{9}}}{32} = \dfrac{15}{32}$

(23)

SUPPLEMENTARY PROBLEMS

A. Find the reciprocals of:

1. $\dfrac{8}{13}$ **2.** $\dfrac{1}{8}$ **3.** $\dfrac{18}{5}$

4. $2\dfrac{3}{4}$ **5.** 7 **6.** $1\dfrac{1}{8}$

B. Find n such that each of the following will be true.

7. $\dfrac{1}{3} \times n = 1$ **8.** $\dfrac{9}{2} \times n = 1$ **9.** $\dfrac{1}{25} \times n = 1$

10. $\dfrac{3}{5} \times n = 1$

C. **11.** If the product of two numbers is 1, then each is a __________ of the other.

12. What is the numerator of $\dfrac{\frac{2}{3}}{5}$?

13. $\dfrac{5}{8} \div 15$ **14.** $\dfrac{2}{9} \div 6$ **15.** $5 \div \dfrac{3}{4}$

16. $8 \div \dfrac{2}{3}$ **17.** $\dfrac{1}{\frac{5}{12}}$ **18.** $\dfrac{\frac{3}{4}}{\frac{3}{8}}$

SUPPLEMENTARY PROBLEMS, CONTD.

19. $\dfrac{2\frac{1}{4}}{\frac{5}{8}}$ **20.** $\dfrac{1\frac{1}{2}}{1\frac{1}{4}}$ **21.** $8\frac{3}{4} \div 2\frac{1}{6}$

22. $2\frac{5}{8} \div \frac{5}{8}$ **23.** $\frac{3}{20} \div \frac{4}{5}$ **24.** $1\frac{1}{6} \div 2\frac{2}{3}$

25. $6\frac{1}{4} \div 1\frac{3}{5}$ **26.** $1\frac{6}{7} \div 1\frac{6}{7}$ **27.** $4\frac{3}{8} \div 14$

28. $9 \div 1\frac{2}{3}$

SOLUTIONS TO STUDY EXERCISES

Study Exercise One (Frame 6)

1. $\frac{11}{3}$ **2.** $\frac{5}{4}$ **3.** $\frac{1}{7}$ **4.** $\frac{3}{14}$

5. The question mark stands for the reciprocal of $\frac{2}{3} : \frac{3}{2}$.

6. The question mark stands for the reciprocal of $\frac{5}{3} : \frac{3}{5}$.

7. n represents the reciprocal of $\frac{3}{7} : \frac{7}{3} = 2\frac{1}{3}$.

8. n represents the reciprocal of $\frac{12}{11} : \frac{11}{12}$.

6A

Study Exercise Two (Frame 14)

1. The numerator of $\dfrac{8}{\frac{2}{3}}$ is 8.

2. The denominator of $\dfrac{\frac{1}{6}}{\frac{4}{7}}$ is $\frac{4}{7}$.

3. $\dfrac{\frac{3}{5}}{1} = \frac{3}{5}$

4. $\dfrac{\frac{2}{3}}{4} = \dfrac{\frac{\not{2}^{1}}{3} \times \frac{1}{\not{4}_{2}}}{4 \times \frac{1}{4}} = \dfrac{\frac{1}{6}}{1} = \frac{1}{6}$

5. $\dfrac{2}{\frac{3}{4}} = \dfrac{2 \times \frac{4}{3}}{\frac{3}{4} \times \frac{4}{3}} = \dfrac{\frac{8}{3}}{1} = \frac{8}{3} = 2\frac{2}{3}$

SOLUTIONS TO STUDY EXERCISES, CONTD.

Study Exercise Two (Frame 14, contd.)

6. $\dfrac{8}{1\frac{1}{3}} = \dfrac{8}{\frac{4}{3}} = \dfrac{\overset{2}{\cancel{8}} \times \frac{3}{\cancel{4}_1}}{\frac{4}{3} \times \frac{3}{4}} = \dfrac{\frac{6}{1}}{1} = 6$

7. $\dfrac{\frac{5}{9}}{\frac{4}{5}} = \dfrac{\frac{5}{9} \times \frac{5}{4}}{\frac{4}{5} \times \frac{5}{4}} = \dfrac{\frac{25}{36}}{1} = \dfrac{25}{36}$

8. $\dfrac{\frac{3}{8}}{\frac{1}{4}} = \dfrac{\frac{3}{\cancel{8}_2} \times \frac{\overset{1}{\cancel{4}}}{1}}{\frac{1}{4} \times \frac{4}{1}} = \dfrac{\frac{3}{2}}{1} = \dfrac{3}{2} = 1\frac{1}{2}$

(14A)

Study Exercise Three (Frame 18)

1. $\dfrac{4}{5} \div \dfrac{3}{10} = \dfrac{4}{\underset{1}{\cancel{5}}} \times \dfrac{\overset{2}{\cancel{10}}}{3} = \dfrac{8}{3} = 2\frac{2}{3}$

2. $8 \div \dfrac{2}{3} = \dfrac{\overset{4}{\cancel{8}}}{1} \times \dfrac{3}{\underset{1}{\cancel{2}}} = \dfrac{12}{1} = 12$

3. $\dfrac{\frac{5}{9}}{\frac{3}{4}} = \dfrac{5}{9} \times \dfrac{4}{3} = \dfrac{20}{27}$

4. $\dfrac{\frac{6}{7}}{3} = \dfrac{\overset{2}{\cancel{6}}}{7} \times \dfrac{1}{\underset{1}{\cancel{3}}} = \dfrac{2}{7}$

5. $\dfrac{3}{8} \div \dfrac{6}{7} = \dfrac{\overset{1}{\cancel{3}}}{8} \times \dfrac{7}{\underset{2}{\cancel{6}}} = \dfrac{7}{16}$

6. $\dfrac{1}{6} \div \dfrac{1}{3} = \dfrac{1}{\underset{2}{\cancel{6}}} \times \dfrac{\overset{1}{\cancel{3}}}{1} = \dfrac{1}{2}$

7. $\dfrac{4}{\frac{2}{3}} = \dfrac{\overset{2}{\cancel{4}}}{1} \times \dfrac{3}{\underset{1}{\cancel{2}}} = \dfrac{6}{1} = 6$

8. $\dfrac{\frac{5}{9}}{\frac{2}{5}} = \dfrac{5}{9} \times \dfrac{5}{2} = \dfrac{25}{18} = 1\frac{7}{18}$

(18A)

SOLUTIONS TO STUDY EXERCISES, CONTD.

Study Exercise Four (Frame 21)

1. $3\frac{1}{2} \div 2\frac{2}{3} = \frac{7}{2} \div \frac{8}{3} = \frac{7}{2} \times \frac{3}{8} = \frac{21}{16} = 1\frac{5}{16}$

2. $4\frac{2}{3} \div \frac{1}{3} = \frac{14}{3} \div \frac{1}{3} = \frac{14}{\overset{}{\underset{1}{\cancel{3}}}} \times \frac{\overset{1}{\cancel{3}}}{1} = \frac{14}{1} = 14$

3. $\dfrac{1\frac{2}{3}}{2\frac{1}{2}} = 1\frac{2}{3} \div 2\frac{1}{2} = \frac{5}{3} \div \frac{5}{2} = \frac{\overset{1}{\cancel{5}}}{3} \times \frac{2}{\underset{1}{\cancel{5}}} = \frac{2}{3}$

4. $\dfrac{6\frac{1}{2}}{2} = 6\frac{1}{2} \div 2 = \frac{13}{2} \div \frac{2}{1} = \frac{13}{2} \times \frac{1}{2} = \frac{13}{4} = 3\frac{1}{4}$

5. $1\frac{3}{4} \div 5\frac{1}{4} = \frac{7}{4} \div \frac{21}{4} = \frac{\overset{1}{\cancel{7}}}{\underset{1}{\cancel{4}}} \times \frac{\overset{1}{\cancel{4}}}{\underset{3}{\cancel{21}}} = \frac{1}{3}$ 21A

unit

17

Working With Fractions

Objectives:

By the end of this unit you should be able to:

1. find what part one number is of another.
2. work problems in reduction and expansion.
3. read, interpret, and work verbal problems.

Fractional Parts of 100

Certain parts of 100 are used so often they should be memorized. The halves and fourths of 100:

$$\frac{1}{2} \text{ of } 100 = 50 \text{ since } \frac{1}{\cancel{2}_{1}} \times \frac{\cancel{100}^{50}}{1} = 50$$

$$\frac{1}{4} \text{ of } 100 = 25 \text{ since } \frac{1}{\cancel{4}_{1}} \times \frac{\cancel{100}^{25}}{1} = 25$$

$$\frac{3}{4} \text{ of } 100 = 75 \text{ since } \frac{3}{\cancel{4}_{1}} \times \frac{\cancel{100}^{25}}{1} = 75$$

The Eighths of 100

$$\frac{1}{8} \text{ of } 100 = 12\frac{1}{2} \text{ since } \frac{1}{\cancel{8}_{2}} \times \frac{\cancel{100}^{25}}{1} = \frac{25}{2} = 12\frac{1}{2}$$

$$\frac{2}{8} \text{ of } 100 = \frac{1}{4} \text{ of } 100 = 25 \text{ since } \frac{2}{8} = \frac{1}{4}$$

Other eighths of 100 are figured in the same way.

The Thirds of 100

$$\frac{1}{3} \text{ of } 100 = \frac{1}{3} \times \frac{100}{1} = \frac{100}{3} = 33\frac{1}{3}$$

$$\frac{2}{3} \text{ of } 100 = \frac{2}{3} \times \frac{100}{1} = \frac{200}{3} = 66\frac{2}{3}$$

Find:

Study Exercise One

1. $\frac{1}{5}$ of 100 **2.** $\frac{3}{4}$ of 100 **3.** $\frac{1}{6}$ of 100

4. $\frac{1}{3}$ of 100 **5.** $\frac{3}{10}$ of 100 **6.** $\frac{2}{5}$ of 100

④

To Find What Part One Number Is Of Another

With whole numbers, to find what fraction of 6 is 5, we write $\frac{5}{6}$.

The fraction is $\frac{\text{"is" number}}{\text{"of" number}}$

Example: What part of 30 is 23?

Solution: $\frac{\text{"is" number}}{\text{"of" number}} = \frac{23}{30}$

⑤

Example 1: What fraction of 3 is $\frac{5}{6}$?

Solution:

$$\frac{\text{"is" number}}{\text{"of" number}} = \frac{\frac{5}{6}}{3}$$

$$= \frac{\frac{5}{6} \times 6}{3 \times 6}$$

$$= \frac{5}{18}$$

Example 2: What part of $3\frac{3}{4}$ is $5\frac{1}{2}$?

Solution:

$$\frac{\text{"is" number}}{\text{"of" number}} = \frac{5\frac{1}{2}}{3\frac{3}{4}}$$

$$= \frac{\frac{11}{2}}{\frac{15}{4}}$$

$$= \frac{\frac{11}{2} \times 4}{\frac{15}{4} \times 4} = \frac{22}{15} = 1\frac{7}{15}$$

⑥

Study Exercise Two

1. What fraction of 8 is $\frac{2}{3}$?

2. What fraction of $\frac{3}{4}$ is $\frac{2}{3}$?

3. What fraction of $3\frac{1}{4}$ is 5?

4. What fraction of $1\frac{5}{7}$ is $\frac{2}{3}$?

5. What fraction of $3\frac{1}{4}$ is $2\frac{1}{2}$?

(7)

Finding a Number When a Fractional Part Is Known

Suppose we know that $\frac{1}{5}$ of a number is 7.

$\frac{5}{5}$ of the number equals the number.

Thus, we find the number by multiplying by 5.

$\frac{1}{5}$ of the number is 7

$\frac{5}{5}$ of the number is $7 \times 5 = 35$

Notice that we are expanding from 1-fifth to 5-fifths.

(8)

Example 1: Find the number if $\frac{1}{8}$ of the number is 4.

Solution: $\frac{1}{8}$ of the number is 4

$\frac{8}{8}$ of the number is $4 \times 8 = 32$

The number is 32.

Example 2: Find the number the question mark represents if $\frac{1}{5} \times ? = 15$.

Solution: $\frac{1}{5}$ of the number is 15

$\frac{5}{5}$ of the number is $15 \times 5 = 75$

The question mark represents the number 75.

(9)

Study Exercise Three

1. Find the number if $\frac{1}{4}$ of the number is 15.

2. Find the number if $\frac{1}{9}$ of the number is 9.

3. Find the number the question mark represents if $\frac{1}{6} \times ? = 16$

4. Find the number the question mark represents if $\frac{1}{5} \times ? = 2\frac{1}{2}$

(10)

Example: Now let's find the number if $\frac{5}{6}$ of the number is $\frac{2}{3}$.

Solution: $\frac{5}{6}$ of the number is $\frac{2}{3}$

$\frac{1}{6}$ of the number is $\frac{2}{3} \div 5 = \frac{2}{3} \times \frac{1}{5} = \frac{2}{15}$

$\frac{6}{6}$ of the number is $\overset{2}{\cancel{6}} \times \frac{2}{\underset{5}{\cancel{15}}} = \frac{4}{5}$

The number is $\frac{4}{5}$

(11)

Example: What is the number if $\frac{5}{8}$ of the number is $2\frac{1}{4}$?

Solution: $\frac{5}{8}$ of the number is $2\frac{1}{4}$

$\frac{5}{8}$ of the number is $\frac{9}{4}$

$\frac{1}{8}$ of the number is $\frac{9}{4} \div 5 = \frac{9}{4} \times \frac{1}{5} = \frac{9}{20}$

$\frac{8}{8}$ of the number is $\frac{9}{20} \times 8 = \frac{9}{\underset{5}{\cancel{20}}} \times \frac{\overset{2}{\cancel{8}}}{1} = \frac{18}{5} = 3\frac{3}{5}$

The number is $3\frac{3}{5}$

(12)

Study Exercise Four

1. Find the number if $\frac{5}{6}$ of the number is $\frac{3}{4}$.

2. Find the number if $3\frac{1}{2}$ times the number is $6\frac{1}{2}$.

3. What is the number if $4\frac{1}{4}$ times the number is 6?

(13)

The Problem of Expansion

Expansion means changing from one to many.

Example: If 1 lb. of candy costs 38¢, find the cost of $2\frac{1}{4}$ lb. of candy.

Solution: Write down the known fact:
1 lb. costs 38¢

Expand from 1 to $2\frac{1}{4}$ by multiplying.

$$2\frac{1}{4} \text{ lb. costs } 38¢ \times 2\frac{1}{4} = \frac{\overset{19}{\cancel{38}}}{1} \times \frac{9}{\underset{2}{\cancel{4}}} = \frac{171}{2} = 85\frac{1}{2}¢$$

(14)

The Problem of Reduction

Reduction means changing from several back to one.

Example: If $2\frac{3}{4}$ lb. cost 48¢, find the cost of 1 lb.

Solution: Write the known fact.
$2\frac{3}{4}$ lb. cost 48¢
Reduce to 1 by dividing

$$1 \text{ lb. costs } 48¢ \div 2\frac{3}{4} = \frac{48}{1} \times \frac{4}{11} = \frac{192}{11} = 17\frac{5}{11}¢$$

Study Exercise Five

1. If 1 article costs $6\frac{3}{4}$¢, find the cost of 6 articles.

2. If eggs cost 58¢ per dozen, what is the cost of $1\frac{1}{2}$ dozen?

3. A house worth $27,000 is assessed at $\frac{2}{3}$ of its value. Find the assessed value of the house.

4. If a missile travels 1 mile in $1\frac{1}{4}$ seconds, how long will it take to travel $16\frac{1}{2}$ miles?

5. If $\frac{3}{4}$ lb. costs $14\frac{1}{2}$¢, find the cost of 1 lb.

Verbal Problems

Addition

Example: Sue made a 2-piece dress requiring $1\frac{1}{4}$ yards for one part and $2\frac{5}{8}$ yards for the other. How much material did she use?

Solution:

$$1\frac{1}{4} + 2\frac{5}{8} = \frac{5}{4} + \frac{21}{8}$$
$$= \frac{5 \times 2}{4 \times 2} + \frac{21}{8}$$
$$= \frac{10}{8} + \frac{21}{8}$$
$$= \frac{31}{8}$$
$$= 3\frac{7}{8}$$

Sue used $3\frac{7}{8}$ yards.

(17)

Verbal Problems

Subtraction

Example: Two months ago, Joe weighed $143\frac{1}{4}$ lb. Now he weighs $137\frac{3}{8}$ lb. How much did he lose?

Solution:

$$143\frac{1}{4} - 137\frac{3}{8} = 143\frac{2}{8} - 137\frac{3}{8}$$
$$= 142\frac{10}{8} - 137\frac{3}{8}$$
$$= 5\frac{7}{8}$$

Joe lost $5\frac{7}{8}$ lb.

Verbal Problems

Multiplication

Example: If 1 cu. ft. of water weighs $62\frac{1}{2}$ lb., find the weight of a column of water containing $2\frac{1}{3}$ cubic feet.

Solution:

$$62\frac{1}{2} \times 2\frac{1}{3} = \frac{125}{2} \times \frac{7}{3}$$

$$= \frac{875}{6}$$

$$= 145\frac{5}{6}$$

The column of water weighs $145\frac{5}{6}$ lb.

(19)

Verbal Problems

Division

Example: A bus is scheduled to go a distance of $66\frac{1}{2}$ miles in $1\frac{1}{2}$ hours.

What average speed must be maintained to arrive on schedule?

Solution:

$$66\frac{1}{2} \div 1\frac{1}{2} = \frac{133}{2} \div \frac{3}{2}$$

$$= \frac{133}{\overset{}{\underset{1}{\cancel{2}}}} \times \frac{\overset{1}{\cancel{2}}}{3}$$

$$= \frac{133}{3}$$

$$= 44\frac{1}{3}$$

The average speed must be $44\frac{1}{3}$ miles per hour.

Study Exercise Six

1. How many feet of wood are needed to make 8 shelves each $6\frac{1}{3}$ feet long?
2. Find the net change in a stock if its high was $20\frac{3}{8}$ and its low $5\frac{1}{4}$.
3. Bob worked $3\frac{3}{4}$ hours on Monday, $2\frac{1}{8}$ hours on Tuesday, and $3\frac{1}{2}$ hours on Wednesday. How many hours did he work altogether?
4. How much does a pound of apples cost if $3\frac{5}{8}$ lb. of apples cost 58¢?

REVIEW EXERCISES

1. Find $\frac{1}{8}$ of 100.

2. Find $\frac{2}{3}$ of 100.

3. What fraction of 6 is $\frac{1}{3}$?

4. What fraction of $1\frac{1}{2}$ is $\frac{2}{3}$?

5. Find the number if $\frac{2}{5}$ of the number is $\frac{1}{4}$.

6. If $\frac{3}{8}$ lb. cost 21¢, find the cost of 1 lb.

7. If a costume requires $3\frac{1}{3}$ yards of material, how many costumes can be made from a 30-yard bolt of material?

SOLUTIONS TO REVIEW EXERCISES

1. $\frac{1}{8} \times 100 = \frac{1}{\cancel{8}_{2}} \times \frac{\cancel{100}^{25}}{1} = \frac{25}{2} = 12\frac{1}{2}$

2. $\frac{2}{3} \times 100 = \frac{200}{3} = 66\frac{2}{3}$

3. $\frac{\text{"is" number}}{\text{"of" number}} = \frac{\frac{1}{3}}{6} = \frac{\frac{1}{3} \times 3}{6 \times 3} = \frac{1}{18}$

4. $\frac{\text{"is" number}}{\text{"of" number}} = \frac{\frac{2}{3}}{1\frac{1}{2}} = \frac{\frac{2}{\cancel{3}_1} \times \cancel{6}^{2}}{\frac{3}{\cancel{2}_1} \times \cancel{6}^{3}} = \frac{4}{9}$

5. $\frac{2}{5}$ of the number is $\frac{1}{4}$

$\frac{1}{5}$ of the number is $\frac{1}{4} \div 2 = \frac{1}{4} \times \frac{1}{2} = \frac{1}{8}$

$\frac{5}{5}$ of the number is $\frac{1}{8} \times 5 = \frac{5}{8}$

the number is $\frac{5}{8}$

6. $\frac{3}{8}$ lb. cost 21¢

$\frac{1}{8}$ lb. cost $21 \div 3 = 21 \times \frac{1}{3} = 7$

$\frac{8}{8}$ lb. cost $7 \times 8 = 56$¢

the cost of 1 lb. is 56¢

7. $3\frac{1}{3}$ yards give 1 costume

$\frac{10}{3}$ yards give 1 costume

$\frac{1}{3}$ yard gives $1 \div 10 = \frac{1}{10}$

$\frac{3}{3}$ yard gives $\frac{1}{10} \times 3 = \frac{3}{10}$

30 yards give $\frac{3}{10} \times 30 = 9$

9 costumes can be made

Alternate Solution

$30 \div 3\frac{1}{3} = \frac{30}{1} \div \frac{10}{3}$

$= \frac{\cancel{30}^{3}}{1} \times \frac{3}{\cancel{10}_{1}}$

$= 9$

SUPPLEMENTARY PROBLEMS

A. Find:

1. $\frac{2}{5}$ of 100

2. $\frac{1}{8}$ of 100

3. $\frac{3}{4}$ of 100

4. $\frac{4}{5}$ of 100

5. $\frac{3}{8}$ of 100

6. $\frac{1}{10}$ of 100

7. $\frac{2}{3}$ of 100

8. $\frac{5}{8}$ of 100

9. $\frac{3}{7}$ of 100

10. $\frac{3}{50}$ of 100

B. **11.** What part of $\frac{1}{4}$ is $\frac{5}{16}$?

12. What part of $\frac{3}{4}$ is $\frac{2}{3}$?

13. What part of $3\frac{1}{2}$ is $2\frac{1}{3}$?

14. What part of 4 is $\frac{5}{6}$?

15. What part of $1\frac{3}{4}$ is $2\frac{1}{3}$?

16. What part of 100 is $3\frac{1}{8}$?

C. **17.** Find the number if $\frac{7}{10}$ of the number is 63.

18. Find the number if $\frac{4}{5}$ of the number is 12.

19. Find the number if $\frac{2}{3}$ of the number is $1\frac{1}{5}$.

20. Find the number if $1\frac{1}{5}$ of the number is 9.

21. Find the number if $3\frac{1}{2}$ times the number is $6\frac{1}{4}$.

22. Find the number the question mark represents if $\frac{5}{16} \times ? = 4\frac{1}{4}$.

23. Find the number the question mark represents if $1\frac{3}{14} \times ? = 4\frac{1}{6}$.

24. Find the number the question mark represents if $3\frac{1}{9} \times ? = \frac{5}{6}$.

D. **25.** If 1 article costs $4\frac{2}{3}$¢, find the cost of 8 articles.

26. If $\frac{3}{4}$ lb. costs $14\frac{1}{2}$¢, find the cost of 1 lb.

27. If a car travels $5\frac{1}{4}$ miles in $9\frac{3}{5}$ minutes, how far will it travel in 12 minutes?

28. If 1 lb. costs $6\frac{1}{2}$¢, how many lbs. can you buy for 16¢?

SUPPLEMENTARY PROBLEMS, CONTD.

29. If a factory produces $8\frac{1}{3}$ articles in $5\frac{1}{4}$ days, how many will it produce in $10\frac{1}{2}$ days.

30. A brass rod was cut into pieces $7\frac{5}{32}$ inches, $4\frac{3}{4}$ inches, and $2\frac{1}{8}$ inches long. How long was the rod if $\frac{1}{16}$ inch was wasted in each cut?

31. How much heavier is a $5\frac{1}{3}$ lb. package than a $2\frac{5}{6}$ lb. package?

32. What is the cost of $2\frac{3}{4}$ pounds of meat at \$1.29 a pound?

33. If $2\frac{1}{4}$ yards of material is needed for a suit, how many suits can be cut from a roll of material 60 yards long?

SOLUTIONS TO STUDY EXERCISES

Study Exercise One (Frame 4)

1. $\frac{1}{\cancel{5}_{1}} \times \frac{\cancel{100}^{20}}{1} = 20$

2. $\frac{3}{\cancel{4}_{1}} \times \frac{\cancel{100}^{25}}{1} = 75$

3. $\frac{1}{\cancel{6}_{3}} \times \frac{\cancel{100}^{50}}{1} = \frac{50}{3} = 16\frac{2}{3}$

4. $\frac{1}{3} \times \frac{100}{1} = \frac{100}{3} = 33\frac{1}{3}$

5. $\frac{3}{\cancel{10}_{1}} \times \frac{\cancel{100}^{10}}{1} = 30$

6. $\frac{2}{\cancel{5}_{1}} \times \frac{\cancel{100}^{20}}{1} = 40$

4A

Study Exercise Two (Frame 7)

1. $\dfrac{\frac{2}{3}}{8} = \dfrac{\frac{2}{3} \times 3}{8 \times 3} = \dfrac{2}{24} = \dfrac{1}{12}$

2. $\dfrac{\frac{2}{3}}{\frac{3}{4}} = \dfrac{\frac{2}{\cancel{3}_{1}} \times \cancel{12}^{4}}{\frac{3}{\cancel{4}_{1}} \times \cancel{12}^{3}} = \dfrac{2 \times 4}{3 \times 3} = \dfrac{8}{9}$

3. $\dfrac{5}{3\frac{1}{4}} = \dfrac{5}{\frac{13}{4}} = \dfrac{5 \times 4}{\frac{13}{4} \times 4} = \dfrac{20}{13} = 1\dfrac{7}{13}$

4. $\dfrac{\frac{2}{3}}{1\frac{5}{7}} = \dfrac{\frac{2}{3}}{\frac{12}{7}} = \dfrac{\frac{2}{\cancel{3}_{1}} \times \cancel{21}^{7}}{\frac{12}{\cancel{7}_{1}} \times \cancel{21}^{3}} = \dfrac{2 \times 7}{12 \times 7} = \dfrac{14}{36} = \dfrac{7}{18}$

5. $\dfrac{2\frac{1}{2}}{3\frac{1}{4}} = \dfrac{\frac{5}{2}}{\frac{13}{4}} = \dfrac{\frac{5}{\cancel{2}_{1}} \times \cancel{4}^{2}}{\frac{13}{\cancel{4}_{1}} \times \cancel{4}^{1}} = \dfrac{5 \times 2}{13 \times 1} = \dfrac{10}{13}$

7A

SOLUTIONS TO STUDY EXERCISES, CONTD.

Study Exercise Three (Frame 10)

1. $\frac{1}{4}$ of the number is 15

$\frac{4}{4}$ of the number is $15 \times 4 = 60$

the number is 60

2. $\frac{1}{9}$ of the number is 9

$\frac{9}{9}$ of the number is $9 \times 9 = 81$

the number is 81

3. $\frac{1}{6}$ of the number is 16

$\frac{6}{6}$ of the number is $16 \times 6 = 96$

the number is 96

4. $\frac{1}{5}$ of the number is $2\frac{1}{2}$

$\frac{5}{5}$ of the number is $2\frac{1}{2} \times 5 = \frac{5}{2} \times 5 = \frac{25}{2} = 12\frac{1}{2}$

the number is $12\frac{1}{2}$

(10A)

Study Exercise Four (Frame 13)

1. $\frac{5}{6}$ of the number is $\frac{3}{4}$

$\frac{1}{6}$ of the number is $\frac{3}{4} \div 5 = \frac{3}{4} \times \frac{1}{5} = \frac{3}{20}$

$\frac{6}{6}$ of the number is $\frac{3}{\cancel{20}_{10}} \times \frac{\cancel{6}^{3}}{1} = \frac{9}{10}$

the number is $\frac{9}{10}$

2. $3\frac{1}{2}$ times the number is $6\frac{1}{2}$

$\frac{7}{2}$ times the number is $\frac{13}{2}$

$\frac{1}{2}$ times the number is $\frac{13}{2} \div 7 = \frac{13}{2} \times \frac{1}{7} = \frac{13}{14}$

$\frac{2}{2}$ times the number is $\frac{13}{\cancel{14}_{7}} \times \frac{\cancel{2}^{1}}{1} = \frac{13}{7} = 1\frac{6}{7}$

the number is $1\frac{6}{7}$

SOLUTIONS TO STUDY EXERCISES, CONTD.

Study Exercise Four (Frame 13, contd.)

3. $4\frac{1}{4}$ times the number is 6

$\frac{17}{4}$ times the number is $\frac{6}{1}$

$\frac{1}{4}$ times the number is $\frac{6}{1} \div 17 = \frac{6}{1} \times \frac{1}{17} = \frac{6}{17}$

$\frac{4}{4}$ times the number is $\frac{6}{17} \times 4 = \frac{24}{17} = 1\frac{7}{17}$

the number is $1\frac{7}{17}$

13A

Study Exercise Five (Frame 16)

1. 1 article costs $6\frac{3}{4}$¢

6 articles cost $6\frac{3}{4}¢ \times 6 = \frac{27}{\cancel{4}_2} \times \cancel{6}^3 = \frac{81}{2} = 40\frac{1}{2}$

The cost of 6 articles is $40\frac{1}{2}$¢

2. 1 dozen eggs costs 58¢

$1\frac{1}{2}$ dozen cost $58 \times 1\frac{1}{2} = \cancel{58}^{29} \times \frac{3}{\cancel{2}_1} = 87$¢

3. $\frac{2}{\cancel{3}_1} \times \cancel{27{,}000}^{9{,}000} = 18{,}000$

The assessed value is \$18,000

4. 1 mile in $1\frac{1}{4}$ seconds

$16\frac{1}{2}$ miles in $1\frac{1}{4} \times 16\frac{1}{2} = \frac{5}{4} \times \frac{33}{2} = \frac{165}{8} = 20\frac{5}{8}$

It will take $20\frac{5}{8}$ seconds

5. $\frac{3}{4}$ lb. costs $14\frac{1}{2}$¢

$\frac{1}{4}$ lb. costs $14\frac{1}{2} \div 3 = \frac{29}{2} \times \frac{1}{3} = \frac{29}{6} = 4\frac{5}{6}$

$\frac{4}{4}$ lb. costs $4\frac{5}{6} \times 4 = \frac{29}{\cancel{6}_3} \times \frac{\cancel{4}^2}{1} = \frac{58}{3} = 19\frac{1}{3}$

The cost of 1 lb. is $19\frac{1}{3}$¢

16A

SOLUTIONS TO STUDY EXERCISES, CONTD.

Study Exercise Six (Frame 21)

1. $8 \times 6\frac{1}{3} = \frac{8}{1} \times \frac{19}{3} = \frac{152}{3} = 50\frac{2}{3}$

$50\frac{2}{3}$ feet of wood are needed

2. $20\frac{3}{8} - 5\frac{1}{4} = 20\frac{3}{8} - 5\frac{2}{8} = 15\frac{1}{8}$

The net change is $15\frac{1}{8}$

3. $3\frac{3}{4} + 2\frac{1}{8} + 3\frac{1}{2} = 3\frac{6}{8} + 2\frac{1}{8} + 3\frac{4}{8}$

$= 8\frac{11}{8} = 8 + 1\frac{3}{8} = 9\frac{3}{8}$

Bob worked $9\frac{3}{8}$ hours

4. $58 \div 3\frac{5}{8} = \frac{58}{1} \div \frac{29}{8} = \frac{\overset{2}{\cancel{58}}}{1} \times \frac{8}{\underset{1}{\cancel{29}}} = 16$

1 lb. costs 16¢

21A

Practice Test—Fractions—Units 11–17

1. State the Golden Rule of Fractions.
2. Change $\frac{4}{5}$ to an equivalent fraction by multiplying by 5.
3. Reduce to lowest terms: (a) $\frac{54}{60}$ (b) $\frac{35}{16}$
4. Which is largest: $\frac{3}{4}$, $\frac{5}{12}$, or $\frac{2}{3}$?
5. What is the LCD of $\frac{2}{3}$, $\frac{1}{4}$, $\frac{3}{8}$?
6. Add $\frac{1}{8} + \frac{5}{12} + \frac{1}{6}$
7. Add $2\frac{5}{9} + 3\frac{2}{3}$
8. Subtract $\frac{5}{6} - \frac{3}{8}$
9. Subtract $3\frac{3}{4} - 1\frac{7}{6}$
10. Multiply $\frac{3}{5} \times \frac{20}{36}$
11. Multiply $2\frac{2}{5} \times 1\frac{1}{10}$
12. Divide $\frac{3}{8} \div \frac{5}{6}$
13. Divide $6\frac{2}{3} \div 1\frac{3}{4}$
14. Simplify $\dfrac{\frac{2}{9}}{\frac{4}{5}}$
15. What is the reciprocal of $1\frac{3}{7}$?
16. Find $\left(2\frac{1}{3}\right)^2$
17. Find $\sqrt{12\frac{1}{4}}$

Practice Test—Fractions, contd.

18. What fraction of $2\frac{3}{4}$ is $1\frac{1}{3}$?

19. Find the number if $1\frac{5}{16}$ of the number is 7.

20. From the sum of $4\frac{3}{4}$ and $1\frac{1}{6}$ subtract the sum of $2\frac{1}{3}$ and $1\frac{11}{12}$.

21. Simplify: $\dfrac{1\frac{2}{3}}{6}$

22. From $\frac{3}{4}$, $\frac{1}{2}$, $3\frac{2}{3}$, $\frac{15}{6}$, $\frac{4}{4}$, $\dfrac{\frac{1}{2}}{3}$, select all:

(a) mixed numbers (b) improper fractions

(c) complex fractions

23. Cathy typed 255 words in $7\frac{1}{2}$ minutes. How many words did she average per minute?

24. How many pieces $\frac{1}{4}$ inch long can be cut from a piece of metal $12\frac{3}{4}$ inches long?

25. To change Fahrenheit temperature to centigrade temperature, first subtract 32 degrees from the Fahrenheit reading, then take $\frac{5}{9}$ of the answer.

(a) change 68°F to centigrade (b) change 212°F to centigrade

Answers—Practice Test—Fractions

1. The numerator and denominator of a fraction may be multiplied or divided by the same non-zero number without altering the value of the fraction.

2. $\frac{4}{5} = \frac{4 \times 5}{5 \times 5} = \frac{20}{25}$

3. (a) $\frac{9}{10}$ (b) $2\frac{3}{16}$

4. $\frac{3}{4}$

5. LCD is 24

6. $\frac{17}{24}$

7. $6\frac{2}{9}$

8. $\frac{11}{24}$

9. $1\frac{7}{12}$

10. $\frac{1}{3}$

11. $2\frac{16}{25}$

12. $\frac{9}{20}$

13. $3\frac{17}{21}$

14. $\frac{5}{18}$

15. $\frac{7}{10}$

16. $5\frac{4}{9}$

17. $3\frac{1}{2}$

18. $\frac{16}{33}$

19. $5\frac{1}{3}$

20. $1\frac{2}{3}$

21. $\frac{5}{18}$

22. (a) $3\frac{2}{3}$ (b) $\frac{15}{6}, \frac{4}{4}$ (c) $\frac{\frac{1}{2}}{3}$

23. 34 words

24. 51 pieces

25. (a) 20° centigrade
    (b) 100° centigrade

unit

18

Decimal Numerals

Objectives:

By the end of this unit you should:

1. understand decimal place value

2. be able to read decimals

①

Place Value

In Unit One we discussed the place value of whole numbers. Study the place values below for the number 4,367,285

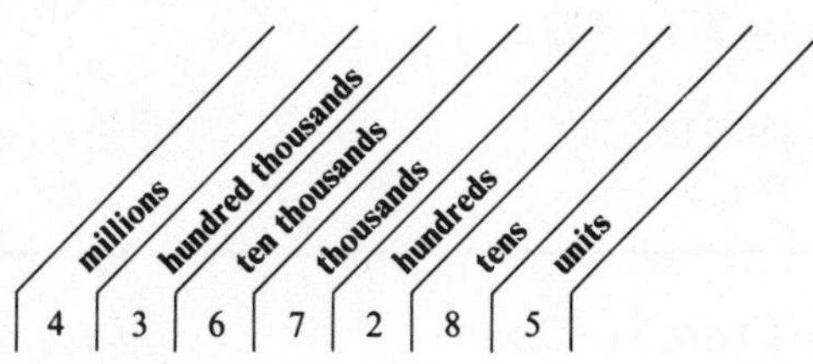

The place value of the digit 3 is *hundred thousands*.
The place value of the digit 2 is *hundreds*.
What is the place value of the digit 7?

②

---

thousands	hundreds	tens	units

Each place value is ten times the place value on its right.

1 thousand = 10 hundreds
1 hundred = 10 tens
1 ten = 10 units
1 unit = 10?

③

---

thousands	hundreds	tens	units	tenths

1 unit = 10 tenths

We will place a period between the units' digit and the tenths' digit. This period is called a decimal point and is read "and."

For example, 142.5 is read *one hundred forty-two and 5 tenths.* (4)

If we keep moving to the right, each place value is $\frac{1}{10}$ of the place value before it.

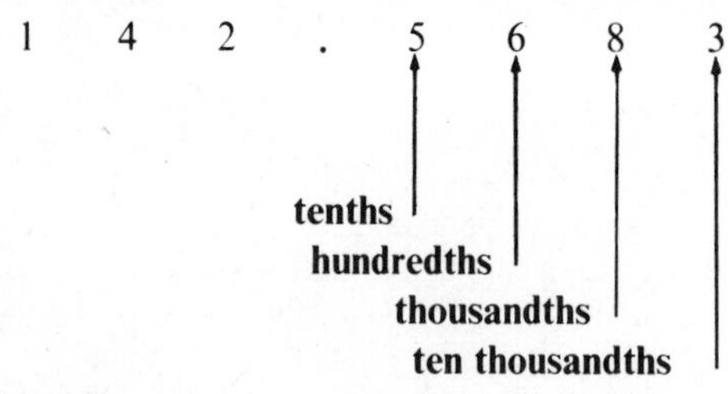

(5)

---

tenths	hundredths	thousandths	ten thousandths	hundred thousandths

Examples:

1. What is the place value of the digit 3 in 215.638?
   Solution: hundredths
2. What is the place value of the digit 8 in 215.638?
   Solution: thousandths

(6)

Study Exercise One

1. What is the place value of the digit 6 in 21.7065?
2. What is the place value of the digit 2 in 6.8752?
3. How do you read 17.4?
4. What is the place value of each digit in 17.4?

(7)

Reading Decimals

If we wish to read 8.365, we will first decide the place value of each digit.

The place value of 8 is *units*
3 is *tenths*
6 is *hundredths*
5 is *thousandths*

8.365 means *8 and 3 tenths plus 6 hundredths plus 5 thousandths*

(8)

Example 1: What is the meaning of .471?

Solution: .471 means 4 tenths plus 7 hundredths plus 1 thousandth

Example 2: What is the meaning of 2.05?

Solution: 2.05 means 2 and 0 tenths plus 5 hundredths

(Frame 9, contd.)

Example 3: Write the decimal numeral that means 15 and 2 tenths plus 0 hundredths plus 3 thousandths plus 7 ten thousandths

Solution: 15.2037 (9)

Study Exercise Two

Give the meaning of the following decimals

1. 2.32 **2.** .1005 **3.** 1.23134 (10)

Converting Decimals to Fractions

Since .43 = 4 tenths plus 3 hundredths,

Line (a) $.43 = \frac{4}{10} + \frac{3}{100}$

Line (b) $= \frac{4 \times 10}{10 \times 10} + \frac{3}{100}$

Line (c) $= \frac{40}{100} + \frac{3}{100}$

Line (d) $= \frac{43}{100}$ (43 hundredths) (11)

Examples of Converting Decimals to Fractions

Example 1:

Line (a) .71 = 7 tenths plus 1 hundredth

Line (b) $= \frac{7}{10} + \frac{1}{100}$

Line (c) $= \frac{7 \times 10}{10 \times 10} + \frac{1}{100}$

Line (d) $= \frac{70}{100} + \frac{1}{100}$

Line (e) $= \frac{71}{100}$

Example 2:

Line (a) .071 = 0 tenths plus 7 hundredths plus 1 thousandth

Line (b) $= \frac{0}{10} + \frac{7}{100} + \frac{1}{1000}$

Line (c) $= \frac{0 \times 100}{10 \times 100} + \frac{7 \times 10}{100 \times 10} + \frac{1}{1000}$

Line (d) $= \frac{0}{1000} + \frac{70}{1000} + \frac{1}{1000}$

Line (e) $= \frac{71}{1000}$ (12)

Study Exercise Three

1. What is the meaning of .39?
2. Write the decimal numeral that means 3 and 3 tenths plus 0 hundredths plus 7 thousandths.
3. Write the fraction equivalent to .47.
4. Write the fraction equivalent to .039.
5. Write the fraction equivalent to .623.

(13)

Decimals to Fractions Again

Line (a) $.832 = \frac{8}{10} + \frac{3}{100} + \frac{2}{1000}$

Line (b) $= \frac{800}{1000} + \frac{30}{1000} + \frac{2}{1000}$

Line (c) $= \frac{832}{1000}$

Notice that the place value of the last digit determines the denominator.

$.617 = \frac{617}{1000}$ since the place value of the digit 7 is thousandths.

$.29 = \frac{29}{100}$ since the place value of the digit 9 is hundredths.

Shortcut:

1. Write a fraction and in the numerator put the digits that follow the decimal point.
2. In the denominator write the place value of the last digit.

Example 1: $.3177 = \frac{3177}{10{,}000}$

Example 2: $.213 = \frac{213}{1000}$

Example 3: $.77 = \frac{77}{100}$

(15)

Decimal Point

The decimal point plays an important role in any number. For example, 1.33 is *not* equal to 13.3 since $1.33 = 1\frac{33}{100}$ and $13.3 = 13\frac{3}{10}$ and $1\frac{33}{100}$ is *not* equal to $13\frac{3}{10}$.

(16)

When there are no digits after a decimal point, the decimal point is usually omitted.

Example 1: 35. = 35

Example 2: 2. = 2

(17)

The Digit Zero

The digit zero is often used to keep the decimal point in the correct place. Remember that 20 = 20.; the zero is used to keep the digit 2 in tens' place. Placing a zero in front of the digit 2 does not change the number.

Example 1: 020 = 020.

Example 2: 07 = 7

Example 3: .2 = 0.2 (18)

Converting Decimals to Fractions

Examples: Write as fractions:

1. .21

Solution: $\frac{21}{100}$

2. .437

Solution: $\frac{437}{1000}$

3. .0769

Solution: $\frac{0769}{10{,}000} = \frac{769}{10{,}000}$ (19)

Study Exercise Four

1. Give the meaning of .17
2. Give the meaning of 3.07245
3. Write .893 as a fraction
4. Write .01121 as a fraction
5. Write 67 thousandths in decimal notation (20)

REVIEW EXERCISES

A. Multiple Choice—Select the letter of the correct answer.

1. What is the place value of the digit 8 in 61.283?
a. tenths b. hundredths c. hundreds d. thousands

2. What is the place value of the digit 4 in 2.004?
a. tenths b. hundredths c. thousandths d. ten thousandths

3. Twenty-three and 23 hundredths is written
a. 2323 b. 23.023 c. 23.23 d. 232.3

4. The number "107 ten-thousandths" is written
a. .107 b. 107.0010 c. .0107 d. 100.007

5. .43 as a fraction is
a. $\frac{43}{10}$ b. $\frac{43}{100}$ c. $\frac{43}{1000}$ d. $\frac{43}{10,000}$

6. .0081 as a fraction is
a. $\frac{81}{10}$ b. $\frac{81}{100}$ c. $\frac{81}{1000}$ d. $\frac{81}{10,000}$

7. Which is correct?
a. $.09 = \frac{9}{100}$ b. $.009 = \frac{9}{100}$ c. $.0009 = \frac{9}{1,000}$
d. $.00009 = \frac{9}{1,000}$

B. True or False

8. .071 = .71 **9.** $\frac{19}{100} = .19$ **10.** .31 = .3100
11. 0.77 = 77 **12.** 2.0 = 2

(21)

SOLUTIONS TO REVIEW EXERCISES

A. **1.** b **2.** c **3.** c
4. c **5.** b **6.** d
7. a

B. **8.** false **9.** true **10.** true
11. false **12.** true

(22)

SUPPLEMENTARY PROBLEMS

A. Write out in words the following decimals
1. 0.5 **2.** 9.25 **3.** 0.609
4. 3.217 **5.** 0.0013 **6.** 100.01
7. 708.080 **8.** 20.02 **9.** 0.00072

B. Find the place value of the underlined digit
10. $17.\underline{6}23$ **11.** $.11\underline{7}2$ **12.** $.001\underline{1}7$
13. $107.062\underline{3}$ **14.** $0.00\underline{3}4$ **15.** $1.7\underline{0}00$
16. $\underline{2}1.452$ **17.** $0.10170\underline{2}0$ **18.** $4.0\underline{9}9$

C. Write the decimal form of the following numbers written as words
19. forty-eight hundredths **20.** four thousandths
21. five and 93 thousandths **22.** three and 3 tenths

SUPPLEMENTARY PROBLEMS, CONTD.

23. one-hundred-two ten-thousandths
24. two and 50 thousandths

D. Write the following as fractions reduced to lowest terms.

25. 0.25 **26.** .125 **27.** .050

SOLUTIONS TO STUDY EXERCISES

Study Exercise One (Frame 7)

1. thousandths
2. ten-thousandths
3. seventeen and 4 tenths
4. place value of the digit 1 is tens
   place value of the digit 7 is units
   place value of the digit 4 is tenths

Study Exercise Two (Frame 10)

1. 2 and 3 tenths plus 2 hundredths
2. 1 tenth plus 0 hundredths plus 0 thousandths plus 5 ten-thousandths
3. 1 and 2 tenths plus 3 hundredths plus 1 thousandth plus 3 ten-thousandths plus 4 hundred-thousandths

Study Exercise Three (Frame 13)

1. 3 tenths plus 9 hundredths

2. 3.307

3. $$.47 = \frac{4}{10} + \frac{7}{100} = \frac{40}{100} + \frac{7}{100} = \frac{47}{100}$$

4. 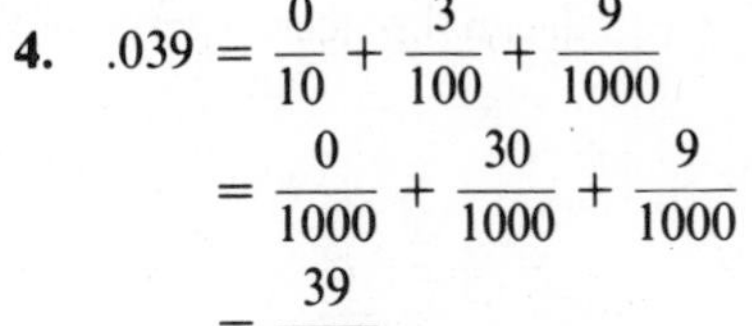

$$.039 = \frac{0}{10} + \frac{3}{100} + \frac{9}{1000} = \frac{0}{1000} + \frac{30}{1000} + \frac{9}{1000} = \frac{39}{1000}$$

5. $$.623 = \frac{6}{10} + \frac{2}{100} + \frac{3}{1000} = \frac{600}{1000} + \frac{20}{1000} + \frac{3}{1000} = \frac{623}{1000}$$

Study Exercise Four (Frame 20)

1. 1 tenth plus 7 hundredths
2. 3 units and 0 tenths plus 7 hundredths plus 2 thousandths plus 4 ten-thousandths plus 5 hundred-thousandths

3. $\frac{893}{1,000}$ **4.** $\frac{1,121}{100,000}$ **5.** .067

unit

19

Converting Decimals to Fractions

Objectives:

By the end of this unit you should:

1. know what a mixed decimal is.
2. be able to convert decimals to fractions. ①

We have seen that:

$.8 = \frac{8}{10}$ (1 decimal place, 1 zero in the denominator)

$.83 = \frac{83}{100}$ (2 decimal places, 2 zeros in the denominator)

$.837 = \frac{837}{1000}$ (3 decimal places, 3 zeros in the denominator)

$.8379 = \frac{8379}{?}$

$.0021 = \frac{21}{?}$ ②

Decimals to Fractions Again

Rule: To change a decimal to a fraction

1. Write the digits after the decimal point in the numerator of the fraction.
2. In the denominator write 1 followed by as many zeros as there are decimal places. ③

Examples:

1. $.237 = \frac{237}{1{,}000}$ ← (digits after the decimal point) ← (1 followed by 3 zeros)

2. $.1133 = \frac{1133}{10{,}000}$ 3. $.207647 = \frac{207647}{1{,}000{,}000}$ ④

Study Exercise One—Oral

A. This part will be given to you orally on the tape.

B. Change the following decimals to fractions.

6. .807 **7.** .0207 **8.** .12347 ⑤

Mixed Decimals

A decimal number which consists of a whole number and fraction is a *mixed decimal.*

Example 1: $9.3 = 9 + .3 = 9 + \frac{3}{10} = 9\frac{3}{10}$

Example 2: $7.07 = 7 + .07 = 7 + \frac{7}{100} = 7\frac{7}{100}$

Example 3: $15.351 = 15 + .351 = 15 + \frac{351}{1000} = 15\frac{351}{1000}$

9.3, 7.07, and 15.351 are mixed decimals

Mixed Decimals Converted to Mixed Numbers

Examples:

1. $2.3 = 2\frac{3}{10}$

2. $14.22 = 14\frac{22}{100}$

$= 14\frac{11}{50}$ (reduce fractions to lowest terms)

3. $2.002 = 2\frac{2}{1000}$

$= 2\frac{1}{500}$

Study Exercise Two

Convert to common fractions or mixed numbers in lowest terms

1. .202	**2.** 3.14	**3.** 103.004
4. 0.75	**5.** 0.40	**6.** 2.125

What happens if zeros are placed after the decimal point? Does the number change?

$$.13 = \frac{13}{100}$$

$$.013 = \frac{13}{1000}$$

$$\text{but } .130 = \frac{130}{1000} = \frac{13}{100} = .13$$

Example 1: .13 is not equal to .013 since $\frac{13}{100}$ is not equal to $\frac{13}{1000}$

Example 2: .13 does equal .130

⑨

Rule: Zeros may be placed on either side of a number without changing the number if the decimal point remains in its original position.

Example 1: 12.2 = 12.20

Example 2: .302 = 0.30200

(10)

Now study the following examples.

1. 123.1 equals 123.10
2. 203.04 equals 203.0400
3. 17.182 does not equal 17.0182
4. .091 equals 0.091
5. 91 equals 0091
6. 20.027 does not equal 20.27

(11)

Simplest Decimal Form

The *simplest decimal form* of a decimal is that form in which the last digit on the right of the decimal point is not the digit 0.
The simplest decimal form of

1. 2.130 is 2.13
2. 4.400 is 4.4
3. 13.23 is 13.23
4. 13.3009 is 13.3009

(12)

Study Exercise Three

True or False

1. 2.007 = 2.700
2. 0.73 = .73
3. The simplest decimal form of 120.070 is 120.07
4. The simplest decimal form of .005 is .5
5. 107 = 107.

(13)

Decimals Ending with a Fraction

Sometimes a decimal ends with a fraction.
Some examples include

$$.12\frac{1}{3},\quad .3\frac{1}{3},\quad .007\frac{1}{7}$$

In converting this type of decimal to a fraction, the fraction at the end is not used to determine place value.

$$.12\frac{1}{3} = \frac{12\frac{1}{3}}{100}$$

$$.3\frac{1}{3} = \frac{3\frac{1}{3}}{10}$$

$$.007\frac{1}{7} = \frac{7\frac{1}{7}}{1000}$$

(14)

In simplifying decimals ending with a fraction to a common fraction, proceed as follows:

Line (a) $.12\frac{1}{3} = \dfrac{12\frac{1}{3}}{100}$

Line (b) $= \dfrac{\frac{37}{3}}{\frac{100}{1}}$

Line (c) $= \dfrac{37}{3} \times \dfrac{1}{100}$

Line (d) $= \dfrac{37}{300}$

Example: Change $.13\frac{1}{3}$ to a fraction in lowest terms.

Solution:

Line (a) $.13\frac{1}{3} = \dfrac{13\frac{1}{3}}{100}$

Line (b) $= \dfrac{\frac{40}{3}}{\frac{100}{1}}$

Line (c) $= \dfrac{40}{3} \times \dfrac{1}{100}$

Line (d) $= \dfrac{\overset{2}{\cancel{40}}}{3} \times \dfrac{1}{\underset{5}{\cancel{100}}}$

Line (e) $= \dfrac{2}{15}$

Study Exercise Four

Change to a fraction in lowest terms:

1. $.6\frac{2}{3}$ **2.** $.33\frac{1}{3}$ **3.** $.16\frac{2}{3}$

4. $.002\frac{1}{4}$ **5.** $1.3\frac{1}{5}$ **6.** $.01\frac{1}{10}$

REVIEW EXERCISES

A. True or False

1. $2.01 = 2.1$

2. $2.01 = 2\frac{1}{10}$

3. $.23\frac{1}{3} = \frac{7}{30}$

4. $.016\frac{2}{3} = \frac{1}{20}$

5. $9.09 = 9\frac{9}{10}$

6. $8.800 = 8.8$

7. $8.07 = 8.70$

B. Multiple Choice: Select the letter of the correct choice.

8. Which is a mixed decimal?
a. .002 b. 1.002 c. .143 d. .5

9. Which is correct?

a. $3\frac{11}{1,000} = 3.11$

b. $5\frac{17}{100} = 5.017$

c. $2\frac{13}{1,000} = 2.013$

d. $\frac{23}{100} = .023$

10. When 7.12 is converted to a mixed number, the result is

a. $7\frac{6}{10}$ b. $7\frac{3}{10}$ c. $7\frac{3}{25}$ d. $7\frac{3}{250}$

11. When $.08\frac{1}{4}$ is converted to a fraction, the result is:

a. $\frac{33}{100}$ b. $\frac{33}{4}$ c. $\frac{33}{100}$ d. $\frac{33}{400}$

SOLUTIONS TO REVIEW EXERCISES

A. **1.** false **2.** false **3.** true **4.** false
5. false **6.** true **7.** false

B. **8.** b **9.** c **10.** c **11.** d

(19)

SUPPLEMENTARY PROBLEMS

A. Write the following as fractions reduced to lowest terms:

1. 0.05 **2.** 0.500 **3.** .008
4. 0.600 **5.** .075 **6.** .2000
7. .45 **8.** .0002 **9.** .0012

B. Convert to a mixed number

10. 2.040 **11.** 12.001 **12.** 2.002
13. 004.02 **14.** 10.01 **15.** 1.001

C. Convert to a fraction in lowest terms:

16. $.4\frac{2}{3}$ **17.** $.04\frac{2}{5}$ **18.** $2.3\frac{1}{3}$

19. $2.33\frac{1}{3}$ **20.** $1.16\frac{2}{3}$ **21.** $3.02\frac{1}{4}$

SOLUTIONS TO STUDY EXERCISES

Study Exercise One (Frame 5)

A. **1.-5.** Answers are given on the tape

B. **6.** $\frac{807}{1{,}000}$ **7.** $\frac{207}{10{,}000}$ **8.** $\frac{12347}{100{,}000}$

Study Exercise Two (Frame 8)

1. $.202 = \frac{202}{1{,}000} = \frac{101}{500}$

2. $3.14 = 3\frac{14}{100} = 3\frac{7}{50}$

3. $103.004 = 103\frac{4}{1{,}000} = 103\frac{1}{250}$

4. $0.75 = \frac{75}{100} = \frac{3}{4}$

5. $0.40 = \frac{40}{100} = \frac{2}{5}$

6. $2.125 = 2\frac{125}{1{,}000} = 2\frac{1}{8}$

Study Exercise Three (Frame 13)

1. false **2.** true **3.** true
4. false **5.** true

13A

Study Exercise Four (Frame 17)

1.
$$.6\frac{2}{3} = \frac{6\frac{2}{3}}{10} = \frac{\frac{20}{3}}{\frac{10}{1}} = \frac{\overset{2}{\cancel{20}}}{3} \times \frac{1}{\underset{1}{\cancel{10}}} = \frac{2}{3}$$

2.
$$.33\frac{1}{3} = \frac{33\frac{1}{3}}{100} = \frac{\frac{100}{3}}{\frac{100}{1}} = \frac{\overset{1}{\cancel{100}}}{3} \times \frac{1}{\underset{1}{\cancel{100}}} = \frac{1}{3}$$

SOLUTIONS TO STUDY EXERCISES, CONTD.

Study Exercise Four (Frame 17, contd.)

3. $.16\frac{2}{3} = \dfrac{16\frac{2}{3}}{100}$

$= \dfrac{\frac{50}{3}}{\frac{100}{1}}$

$= \dfrac{\overset{1}{\cancel{50}}}{3} \times \dfrac{1}{\underset{2}{\cancel{100}}}$

$= \dfrac{1}{6}$

4. $.002\frac{1}{4} = \dfrac{2\frac{1}{4}}{1{,}000}$

$= \dfrac{\frac{9}{4}}{\frac{1{,}000}{1}}$

$= \dfrac{9}{4} \times \dfrac{1}{1{,}000}$

$= \dfrac{9}{4{,}000}$

5. $1.3\frac{1}{5} = 1 + \dfrac{3\frac{1}{5}}{10}$

$= 1 + \dfrac{\frac{16}{5}}{10}$

$= 1 + \dfrac{\frac{16}{5}}{\frac{10}{1}}$

$= 1 + \dfrac{\overset{8}{\cancel{16}}}{5} \times \dfrac{1}{\underset{5}{\cancel{10}}}$

$= 1 + \dfrac{8}{25}$

$= 1\frac{8}{25}$

6. $.01\frac{1}{10} = \dfrac{1\frac{1}{10}}{100}$

$= \dfrac{\frac{11}{10}}{\frac{100}{1}}$

$= \dfrac{11}{10} \times \dfrac{1}{100}$

$= \dfrac{11}{1{,}000}$

unit

20

Addition and Subtraction of Decimals

Objectives:

By the end of this unit you should:

1. learn to add decimals

2. learn subtraction of decimals

Decimals as Fractions

A decimal may be written as a fraction:

(1) $.7 = \frac{7}{10}$ **(2)** $.23 = \frac{23}{100}$ **(3)** $1.5 = 1\frac{5}{10} = 1\frac{1}{2}$

Each digit of a decimal represents a fraction as indicated by its place value.

Line (a) $.23 = \frac{2}{10} + \frac{3}{100}$

Line (b) $= \frac{20}{100} + \frac{3}{100}$

Line (c) $= \frac{23}{100}$

2

We can add decimals by adding the fractions they represent.

Example: Let us add .13 + .21

Line (a) $.13 = \frac{1}{10} + \frac{3}{100}$

Line (b) $.21 = \frac{2}{10} + \frac{1}{100}$

Line (c) Add the tenths: $\frac{1}{10} + \frac{2}{10} = \frac{3}{10}$

Line (d) Add the hundredths: $\frac{3}{100} + \frac{1}{100} = \frac{4}{100}$

Line (e) $.13 + .21 = \frac{3}{10} + \frac{4}{100}$

$= .34$

3

Another Example

Add .33 + .45

Solution:

$$.33 = \frac{3}{10} + \frac{3}{100}$$
$$.45 = \frac{4}{10} + \frac{5}{100}$$

$$.33 + .45 = \frac{7}{10} + \frac{8}{100}$$
$$= .78$$

④

Study Exercise One

Add the following decimals by using fractions.

1. .22 + .46
2. .03 + .21 + .24
3. .44 + .2 + .11
4. 1.2 + 2.3

⑤

A Shorter Method of Adding Decimals

Let us add .23 + .14

Line (a) $.23 + .14 = \frac{2}{10} + \frac{3}{100} + \frac{1}{10} + \frac{4}{100}$

Line (b) $= \frac{2}{10} + \frac{1}{10} + \frac{3}{100} + \frac{4}{100}$

Line (c) $= \frac{3}{10} + \frac{7}{100}$

Line (d) $= \frac{3 \times 10}{10 \times 10} + \frac{7}{100}$

Line (e) $= \frac{30}{100} + \frac{7}{100}$

Line (f) $= \frac{37}{100}$

We will arrange the numbers in a vertical column so the tenths are in the first column and the hundredths in the second column

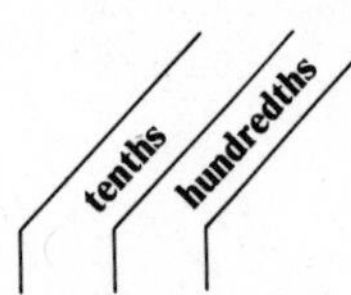

$$\begin{array}{r} .2\ 3 \\ +.1\ 4 \\ \hline .3\ 7 \end{array}$$

place decimal points in the same vertical line.

⑥

Examples:

1. Add .312 + .145

Solution:

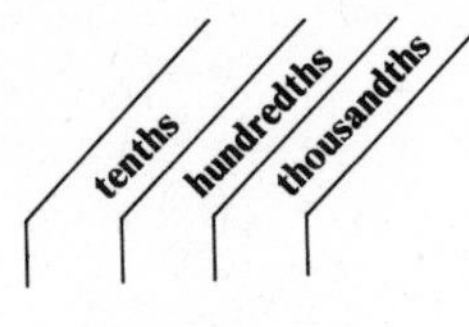

$$\begin{array}{r} .3\ 1\ 2 \\ +\ .1\ 4\ 5 \\ \hline .4\ 5\ 7 \end{array}$$

add columns beginning from the right.

2. Add 1.02 + .143

Solution:

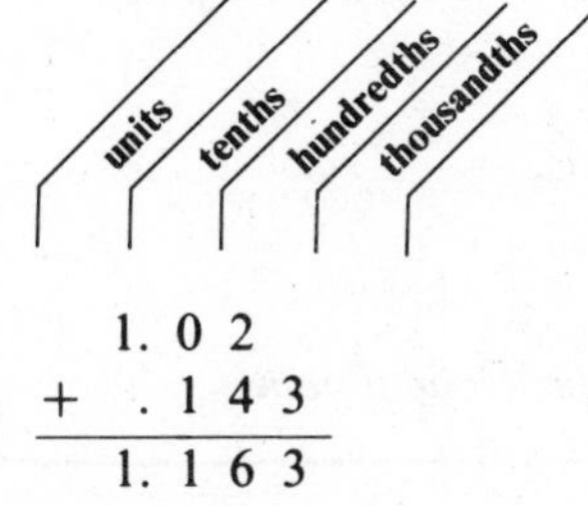

$$\begin{array}{r} 1.\ 0\ 2 \\ +\ \ .\ 1\ 4\ 3 \\ \hline 1.\ 1\ 6\ 3 \end{array}$$

(7)

Remember that 10 hundredths is 1 tenth since $\frac{10}{100} = \frac{1}{10}$ and that 10 thousandths is 1 hundredth since $\frac{10}{1{,}000} = \frac{1}{100}$.

Example: Add .38 + .24

Solution:

$$\begin{array}{r} .38 \\ +.24 \\ \hline \end{array}$$

When we add the hundredths' column, we get 12 hundredths. Twelve hundredths is 10 hundredths plus 2 hundredths; but 10 hundredths is 1 tenth, so we simply carry 1 into the tenths' column.

$$\begin{array}{r} {}^{1} \\ .38 \\ +.24 \\ \hline .62 \end{array}$$

1 ← space for carrying

.62 *answer*

(8)

We can use carrying in adding since each place value is ten times the next place value.

Example: Add 41.874 + 28.257

Solution:

$$\begin{array}{r} {}^{1\,1\ \,1\,1} \\ 41.874 \\ +28.257 \\ \hline 70.131 \end{array}$$

11 11 ← space for carrying

70.131 *answer*

Carrying

Most of the time, carrying is done mentally. In adding decimals remember always to line up the decimal points.

Examples:

1. Add 32.85 + 1.46

Solution:

$$\begin{array}{r} \scriptstyle 1\ 1 \\ 32.85 \\ +1.46 \\ \hline 34.31 \end{array}$$

2. Add 2.04 + .762 + .3

Solution:

$$\begin{array}{r} \scriptstyle 1\ 1 \\ 2.04 \\ .762 \\ +.3 \\ \hline 3.102 \end{array}$$

In the previous example where we added 2.04 + .762 + .3, zeros may be added if they do not change the value of the number.

2.04	write as	2.040
.762	write as	0.762
+.3	write as	+0.300

↑ *remember to line up the decimal point*

Study Exercise Two

Find:

1. 1.34 + 2.41

2. 3.74 + 2.94

3. 6.273 + .198 + .223

4. 23.59 + 16.81 + 1.03

5. 1.07 + 31.2 + .0767

6. 2.6 + 5.983 + 11.75

7. 7.56 + 1.96 + .21

8. 31.8579 + 11.264 + 32.79

9. 7.456 + .923 + 1.04 + 7.3

Subtraction of Decimals

As an example of subtraction, let us subtract 2.78 − 1.42

$$\begin{array}{r} 2.78 \\ -1.42 \\ \hline 1.36 \end{array}$$

Place decimal points in a vertical line.

Arrange vertically and subtract from right to left.

2.856 ← *minuend*
− 1.233 ← *subtrahend*
1.623 ← *difference*

Our check is: **difference + subtrahend = minuend**

$$\begin{array}{r} 1.623 \\ +1.233 \\ \hline 2.856 \end{array}$$

it checks!

(14)

Study the examples of subtraction below.

Example 1: Subtract 4.38 − 1.14

Solution:

```
  4.38
 −1.14
 -----
  3.24
```

Check:

```
  3.24
 +1.14
 -----
  4.38
```

Example 2: Subtract 13.87 − 2.6

Solution:

```
  13.87
 − 2.6
 ------
  11.27
```

Check:

```
  11.27
 + 2.6
 ------
  13.87
```

Example 3: Subtract 3.4897 − 1.1635

Solution:

```
  3.4897
 −1.1635
 -------
  2.3262
```

Check:

```
  2.3262
 +1.1635
 -------
  3.4897
```

(15)

Study Exercise Three

Perform the subtractions and check your answers.

1. 83.58 − 61.24

2. 17.77 − 6.5

3. 101.02 − 1.01

4.

```
  2.7615
 −1.4102
 -------
```

(16)

Borrowing

Subtraction may require borrowing.

Example: Subtract .84 − .46

Solution:

```
  .84     regroup .84
 −.46     1 tenth is 10 hundredths
 ----
          .84 = 8 tenths plus 4 hundredths
              = 7 tenths plus 14 hundredths
```

```
then   .84   can be written   7 tenths + 14 hundredths
      −.46                    4 tenths +  6 hundredths
      ----                    ------------------------
                              3 tenths +  8 hundredths   or   .38
```

In short form

```
   7
  .8̸¹4
 −.4 6
 -----
  .3 8     answer
```

(17)

In a similar fashion we may borrow from any column since our place value system is such that each place value is 10 times the value on its right.

Example: 4.347 − 1.478

Solution:

```
     3            1             1 1
               2 3          3  2 3
 4.3 4¹7       4.3 4¹7      4. 3 4¹7
−1.4 7 8     −1.4 7 8     −1. 4 7 8
 ? ?? 9        ? ? 6 9      2. 8 6 9    answer
```

Check:

```
  2.869
+1.478
  4.347    it checks
```

Example: Let us subtract 4 − 1.86

Solution: First write 4 as 4.00

```
  4.00      3.¹00      3.9¹0
−1.86     −1. 86     −1.8 6
                       2.1 4   answer
```

Check:

```
  2.14
+1.86
  4.00
```

Study Exercise Four

Perform the indicated operations and check your answers.

1. 5.17 − 2.79

2. 4.681 − 3.022

3. 16.84 − 2.55

4. 4.2 − 1.892

5. 3 − 1.784

6.
```
  4.8
−1.9736
```

7.
```
  4.03
−2.1856
```

Using Decimals in Problems

Joe went to the market for his mother and bought items which cost 69¢, 45¢, $1.93, 70¢, and 17¢. Excluding tax, how much did he spend?

Solution:

```
$ .69
   .45
  1.93
   .70
   .17
 $3.94
```

Joe spent $3.94

Uses of Subtraction

Subtraction is used when we wish to find:

(1) the difference between 2 quantities,
(2) how much greater the larger quantity is than the smaller, or
(3) how much is left when a quantity is taken away.

Example: Jim averaged 34 points in ten basketball games.
Bob averaged 18 points. Jim's average is how much higher than Bob's?

Solution:

$$\begin{array}{r} {}^{2}\!\not{3}{}^{1}4 \\ -18 \\ \hline 16 \end{array}$$

Jim's average is 16 more than Bob's.

Study Exercise Five

1. Shipments of 1,745.8 lb, 2,348.72 lb, and 1,694.2 lb of coal are placed in a bin. Find the total number of pounds of coal in the bin.
2. Harry's deposit in the bank was \$64.79 before depositing \$62.34, \$17.65, and \$8.28. Find his balance after he made the deposits.
3. The odometer on Joe's car registered 3,567.3 miles before he left on a trip. Upon his return, it registered 4020.9 miles. How many miles did he travel on his trip?
4. A carpenter agrees to build a cabinet for \$25.00. The lumber and hardware cost him \$10.48. What will be his profit?
5. Mrs. Davis had 106.4 yards of ribbon and used 78.8 yards. How many yards had she left?

REVIEW EXERCISES

Perform the indicated operations.

1. $.6 + 1.8 + 3.24$
2. $24.16 + 90.75$
3. $$\begin{array}{r} 18.16 \\ 2.95 \\ +\ .91 \\ \hline \end{array}$$
4. $$\begin{array}{r} .83 \\ 2.9 \\ .7 \\ +1.0 \\ \hline \end{array}$$
5. $35.18 - .72$ (check your answer)
6. $3.03 - .72$ (check your answer)
7. $$\begin{array}{r} 3.191 \\ -1.063 \\ \hline \end{array}$$
8. $$\begin{array}{r} 1.2 \\ -\ .874 \\ \hline \end{array}$$
9. Add by changing to fractions:
   $.23 + .31$
10. If a triangle has sides 24.05 inches, 18.63 inches, and 9.54 inches, what is the sum of the three sides?
11. Art had \$4.27 and spent \$2.39 for gasoline. How much does he have remaining?

SOLUTIONS TO REVIEW EXERCISES

1. $$\begin{array}{r} .60 \\ 1.80 \\ +3.24 \\ \hline 5.64 \end{array}$$
2. $$\begin{array}{r} 24.16 \\ +90.75 \\ \hline 114.91 \end{array}$$
3. $$\begin{array}{r} 18.16 \\ 2.95 \\ +\ .91 \\ \hline 22.02 \end{array}$$
4. $$\begin{array}{r} .83 \\ 2.90 \\ .70 \\ +1.00 \\ \hline 5.43 \end{array}$$
5. $$\begin{array}{r} 35.18 \\ -\ .72 \\ \hline 34.46 \end{array}$$ **Check:** $$\begin{array}{r} 34.46 \\ +\ .72 \\ \hline 35.18 \end{array}$$
6. $$\begin{array}{r} 3.03 \\ -\ .72 \\ \hline 2.31 \end{array}$$ **Check:** $$\begin{array}{r} 2.31 \\ +\ .72 \\ \hline 3.03 \end{array}$$
7. $$\begin{array}{r} 3.191 \\ -1.063 \\ \hline 2.128 \end{array}$$
8. $$\begin{array}{r} 1.200 \\ -\ .874 \\ \hline .326 \end{array}$$
9. $$\begin{array}{rcl} .23 & = & \frac{2}{10} + \frac{3}{100} \\ +\ .31 & = & \frac{3}{10} + \frac{1}{100} \\ \hline .23 + .31 & = & \frac{5}{10} + \frac{4}{100} \\ & = & .54 \end{array}$$
10. $$\begin{array}{r} 24.05 \\ 18.63 \\ +\ 9.54 \\ \hline 52.22 \end{array}$$ The sum of the 3 sides is 52.22 inches
11. $$\begin{array}{r} \$4.27 \\ -2.39 \\ \hline \$1.88 \end{array}$$ Art has \$1.88 remaining

SUPPLEMENTARY PROBLEMS

A. Add

1. 6.38 + 1.97
2. 4.79 + 6.829
3. 6.09 + .077
4. 7.56 + 3.2 + 8.579
5. 3.61 + 2.6 + .658
6. 3.071 + 2.0968 + .75
7. 74.382 + 9.76 + 11.5498 + 21.7
8. 32.071 + 1.009 + 23.6 + 1.5876

B. Subtract and check

9. 4.769 − 2.143
10. 6.529 − 3.417
11. 11.383 − 5.496
12. 5.0403 − 3.8767
13. 5 − 1.617
14. 6.2007 − 4.9568
15. 11.92 − 11.9
16. 32.214 − 17.839

C. Miscellaneous

17. Subtract 4.23 from the sum of 2.17 and 4.85.
18. Mike had $70 and then he spent $10.75 for shoes, $4.25 for a shirt and $3.50 for a tie. How much money does he have left?
19. How much change should you receive from a $5 bill if you owe $3.62?
20. If you buy an article that regularly sells for $7.49 at a reduction of $1.98, how much will you pay?
21. The outside diameter of a copper pipe is 2.375 inches and its wall thickness is .083 inches. What is the inside diameter?
22. During a certain month, the weather bureau recorded rainfall of 1.02 inches, 2 inches, .58 inches and 0.4 inches. What was the total rainfall for that month?
23. Study the following problems:

$$\begin{array}{r} 27.2 \\ +1.3 \\ \hline 28.5 \end{array} \qquad \begin{array}{r} 2.04 \\ -.11 \\ \hline 1.93 \end{array}$$

Which of the above numbers is

a) a sum
b) an addend
c) a difference
d) a subtrahend
e) a minuend
f) a number less than one

SOLUTIONS TO STUDY EXERCISES

Study Exercise One (Frame 5)

1.

$$.22 = \frac{2}{10} + \frac{2}{100}$$
$$.46 = \frac{4}{10} + \frac{6}{100}$$
$$.22 + .46 = \frac{6}{10} + \frac{8}{100}$$
$$= \frac{68}{100}$$
$$= .68$$

2.

$$.03 = \frac{0}{10} + \frac{3}{100}$$
$$.21 = \frac{2}{10} + \frac{1}{100}$$
$$.24 = \frac{2}{10} + \frac{4}{100}$$
$$.03 + .21 + .24 = \frac{4}{10} + \frac{8}{100}$$
$$= .48$$

SOLUTIONS TO STUDY EXERCISES, CONTD.

Study Exercise One (Frame 5, contd.)

3.
$$.44 = \frac{4}{10} + \frac{4}{100}$$
$$.2 = \frac{2}{10}$$
$$.11 = \frac{1}{10} + \frac{1}{100}$$
$$.44 + .2 + .11 = \frac{7}{10} + \frac{5}{100}$$
$$= .75$$

4.
$$1.2 = 1 + \frac{2}{10}$$
$$2.3 = 2 + \frac{3}{10}$$
$$1.2 + 2.3 = 3 + \frac{5}{10}$$
$$= 3.5$$

5A

Study Exercise Two (Frame 12)

1. 3.75 **2.** 6.68 **3.** 6.694
4. 41.43 **5.** 32.3467 **6.** 20.333
7. 9.73 **8.** 75.9119 **9.** 16.719

12A

Study Exercise Three (Frame 16)

1. $\begin{array}{r} 83.58 \\ -61.24 \\ \hline 22.34 \end{array}$ **Check:** $\begin{array}{r} 22.34 \\ +61.24 \\ \hline 83.58 \end{array}$

2. $\begin{array}{r} 17.77 \\ -\ \ 6.5\ \ \\ \hline 11.27 \end{array}$ **Check:** $\begin{array}{r} 11.27 \\ +\ \ 6.5\ \ \\ \hline 17.77 \end{array}$

3. $\begin{array}{r} 101.02 \\ -\ \ \ 1.01 \\ \hline 100.01 \end{array}$ **Check:** $\begin{array}{r} 100.01 \\ +\ \ \ 1.01 \\ \hline 101.02 \end{array}$

4. $\begin{array}{r} 2.7615 \\ -1.4102 \\ \hline 1.3513 \end{array}$ **Check:** $\begin{array}{r} 1.3513 \\ +1.4102 \\ \hline 2.7615 \end{array}$

16

Study Exercise Four (Frame 20)

1. $\begin{array}{r} 5.17 \\ -2.79 \\ \hline 2.38 \end{array}$ **Check:** $\begin{array}{r} 2.38 \\ +2.79 \\ \hline 5.17 \end{array}$

2. $\begin{array}{r} 4.681 \\ -3.022 \\ \hline 1.659 \end{array}$ **Check:** $\begin{array}{r} 1.659 \\ +3.022 \\ \hline 4.681 \end{array}$

3. $\begin{array}{r} 16.84 \\ -\ \ 2.55 \\ \hline 14.29 \end{array}$ **Check:** $\begin{array}{r} 14.29 \\ +\ \ 2.55 \\ \hline 16.84 \end{array}$

4. $\begin{array}{r} 4.200 \\ -1.892 \\ \hline 2.308 \end{array}$ **Check:** $\begin{array}{r} 2.308 \\ +1.892 \\ \hline 4.200 \end{array}$

SOLUTIONS TO STUDY EXERCISES, CONTD.

Study Exercise Four (Frame 20, contd.)

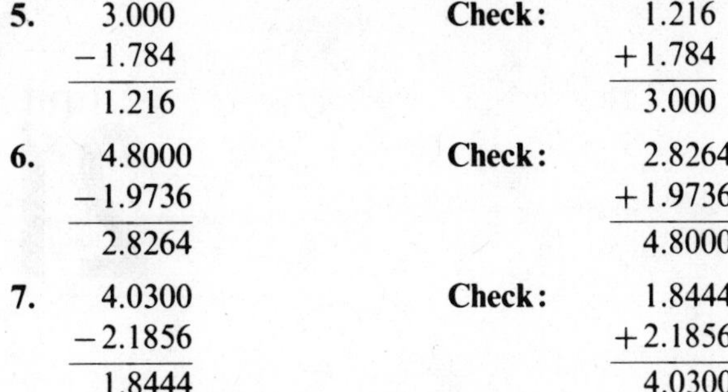

5. 3.000 − 1.784 = 1.216 **Check:** 1.216 + 1.784 = 3.000

6. 4.8000 − 1.9736 = 2.8264 **Check:** 2.8264 + 1.9736 = 4.8000

7. 4.0300 − 2.1856 = 1.8444 **Check:** 1.8444 + 2.1856 = 4.0300

20A

Study Exercise Five (Frame 23)

1. 1,745.8 + 2,348.72 + 1,694.2 = 5,788.72 There are 5,788.72 pounds of coal

2. $62.34 + 17.65 + 8.28 = $88.27; $ 64.79 + 88.27 = $153.06 Total balance is $153.06

3. 4,020.9 − 3,567.3 = 453.6 Joe traveled 453.6 miles

4. $25.00 − 10.48 = $14.52 His profit will be $14.52

5. 106.4 − 78.8 = 27.6 There were 27.6 yards left.

23A

---

unit

21

Multiplying Decimals

Objectives:

By the end of this unit you should be able to:

1. change a decimal to a fraction
2. change fractions with denominators of powers of ten to decimals
3. multiply decimals
4. multiply by powers of ten
5. simplify complex decimals

(1)

Changing Decimals to Fractions

In the previous unit we learned to change decimals to fractions. For example,

(1) $.43 = \frac{43}{100}$

(2) $.177 = \frac{177}{1{,}000}$

Remember that the number of zeros in the denominator is the same as the number of decimal places.

(2)

Fractions to Decimals

How would $\frac{63}{100}$ be written as a decimal?

Solution: The two zeros in the denominator tell us that there will be two decimal places. Thus, $\frac{63}{100} = .63$

What is $\frac{13}{1{,}000}$ as a decimal?

Solution: The three zeros in the denominator indicate three decimal places in the decimal. Thus, $\frac{13}{1{,}000} = .013$

(3)

Study Exercise One

Change the following fractions to decimal form:

1. $\frac{26}{100}$ **2.** $\frac{17}{1{,}000}$ **3.** $\frac{28}{10{,}000}$

(4)

Mixed Decimals to Fractions

A mixed decimal is written as a fraction by the same method as previously explained. For example,

Line (a) $3.13 = 3\frac{13}{100}$

Line (b) $= 3 + \frac{13}{100}$

Line (c) $= \frac{300}{100} + \frac{13}{100}$

Line (d) $= \frac{313}{100}$

Further Examples

1. *Line (a)* $5.3 = 5\frac{3}{10}$

Line (b) $= 5 + \frac{3}{10}$

Line (c) $= \frac{50}{10} + \frac{3}{10}$

Line (d) $= \frac{53}{10}$

2. *Line (a)* $121.07 = 121\frac{7}{100}$

Line (b) $= 121 + \frac{7}{100}$

Line (c) $= \frac{121 \times 100}{100} + \frac{7}{100}$

Line (d) $= \frac{12100}{100} + \frac{7}{100}$

Line (e) $= \frac{12107}{100}$

Observe that the numerator consists of all the digits in the original number.

We thus may skip some steps and write

(1) $5.4 = \frac{54}{10}$

(2) $13.24 = \frac{1324}{100}$

(3) $6.01 = \frac{601}{100}$

Note: **1.** The numerator consists of all the digits in the original number.
2. The number of zeros in the denominator is the same as the number of the digits to the right of the decimal point.

We can reverse this procedure to change back to decimals.

Examples:

1. $\frac{391}{100} = 3.91$ (2 zeros and 2 decimal places)

2. $\frac{1{,}503}{10} = 150.3$ (1 zero and 1 decimal place)

3. $\frac{47{,}728}{10{,}000} = 4.7728$ (4 zeros and 4 decimal places)

Study Exercise Two

A. Convert to improper fractions.

1. 7.9 **2.** 12.21

B. Convert to a decimal.

3. $\frac{1707}{100}$ **4.** $\frac{2001}{1000}$

9

Multiplication

We will first multiply two decimals by changing them to fractions.

Line (a) $2.3 \times .7 = \frac{23}{10} \times \frac{7}{10}$

Line (b) $= \frac{161}{100}$

Line (c) $= 1.61$

Another example of multiplication by changing to fractions:

Line (a) $.421 \times 1.3 = \frac{421}{1,000} \times \frac{13}{10}$

Line (b) $= \frac{5,473}{10,000}$

Line (c) $= .5473$

Rule: In multiplying two decimals, the number of decimal places in the product is always the sum total of the decimal places of the original two numbers.

Study Exercise Three

Multiply by changing to fractions.

1. 2.23 × .4 **2.** 12.23 × 1.3 **3.** .58 × .21

Multiplication of Decimals

Multiply 2.231 × .32

```
    2.231 ←—— Multiplicand has 3 decimal places
  ×   .32 ←—— Multiplier has 2 decimal places
  -------
     4462
    6693
  -------
   .71392 ←—— Product has 5 decimal places
```

Study Exercise Four

Multiply

1. .612 × .14 **2.** 1.33 × 1.2

14

Now let us multiply .06 × .007

By changing to fractions, we get

$$.06 \times .007 = \frac{6}{100} \times \frac{7}{1000}$$
$$= \frac{42}{100{,}000}$$
$$= .00042$$

However, it is quicker to multiply 6 × 7 = 42 and place the decimal point by counting 5 decimal places. Observe that we must write three zeros on the left side of 42 to locate the decimal point correctly.

Another example: Multiply 213 × 1.28

Solution:

```
   213 ←—— 0 decimal places
   1.28 ←—— 2 decimal places
  -----
  17 04
  42 6
 213
 ------
 272.64 ←—— 2 decimal places
```

When the word "of" is used in mathematics, it implies multiplication. Thus 213 of 1.28 implies 213 × 1.28.

Study Exercise Five

Find:

1. 1.97 × .018 **2.** 2.48 × 1.25 **3.** .025 of .08

4. 1.002 of 2.3 **5.** .1423 × 2.5

(17)

Multiplying Several Numbers Together

When several decimal numbers are multiplied together, the number of decimal places in the answer is found by counting the total number of decimal places in all the numbers to be multiplied.

Example: 2.71 × .14 × 2.1

Solution:

```
    2.71 ←—— 2 decimal places
 ×   .14 ←—— 2 decimal places
  ------
    1084
    271
  ------
    3794
 ×   2.1 ←—— 1 decimal place
  ------
    3794
   7588
  ------
  .79674 ←—— 5 decimal places
```

Study Exercise Six

Multiply:

1. .32 × 1.2 × 3.4 **2.** .172 × .012 × 12

(19)

Multiplication by Powers of Ten

First let us review multiplication of whole numbers by 10 and by 100.
Let us multiply 235 × 10 and 235 × 100

```
  235        235
 × 10      × 100
 ----      -----
  000        000
 235        000
 ----      235
 2350      -----
           23500
```

Thus 235 × 10 = 2,350
235 × 100 = 23,500

Rule: To multiply a whole number by 10, 100, 1,000, 10,000, etc., simply transfer the zeros to the end of the whole number.

Example 1: 14 × 100 = 1,400

Example 2: 2,351 × 1,000 = 2,351,000

Now let us multiply a decimal by 100.
For example, 7.271 × 100
First multiply 7271 × 100

$$7271 \times 100 = 727100$$

Now the answer must have 3 decimal places

$$7.271 \times 100 = 727.100$$

The zeros at the end may be dropped

$$7.271 \times 100 = 727.1$$

Examples:

1. 3.318 × 10 = 33.180 or 33.18
2. 3.318 × 100 = 331.800 or 331.8
3. 3.318 × 1,000 = 3318.000 or 3318

Notice the answer is simply found by moving the decimal point in the original number.

1. 3.318 × 10 = 33.18 (move decimal point 1 place to the right)
2. 3.318 × 100 = 331.8 (move decimal point 2 places to the right)
3. 3.318 × 1,000 = 3318 (move decimal point 3 places to the right)

Multiplying Decimal Numerals by Powers of Ten

Rule: To multiply by 10, move the decimal point 1 place to the *right*.
To multiply by 100, move the decimal point 2 places to the *right*.
To multiply by 1,000, move the decimal point 3 places to the *right*.
To multiply by 10,000, move the decimal point 4 places to the *right*.

Remember if there is no decimal point, it is at the end of the number. For example, 37 = 37. When multiplying by 10, 100, 1,000, etc., and moving the decimal point to the right, it may be necessary to add zeros.

For example, $10 \times 37 = 10 \times 37.0$
$= 370$

(25)

Study these examples:

1. $4.721 \times 10 = 47.21$
2. $.0327 \times 100 = 03.27$ or 3.27
3. $2.238 \times 1{,}000 = 2238$
4. $12.2 \times 1{,}000 = 12200$

Study Exercise Seven

1. $.00572 \times 100$
2. $3.51 \times 1{,}000$
3. $.007 \times 10$
4. $5.072 \times 10{,}000$
5. $.0027 \times 10$
6. $100 \times .0453$
7. $1{,}000 \times 1.5$

Complex Decimal Fractions

An example of a complex decimal fraction is $\frac{.23}{.69}$.

We will simplify such a fraction by multiplying numerator and denominator by 100 (in this case) to get rid of the decimals.

$$\frac{.23}{.69} = \frac{.23 \times 100}{.69 \times 100} = \frac{23}{69} = \frac{1}{3}$$

To simplify $\frac{.16}{.08}$ we will again multiply numerator and denominator by 100.

$$\frac{.16}{.08} = \frac{.16 \times 100}{.08 \times 100} = \frac{16}{8} = 2$$

(Frame 29, contd.)

To simplify $\frac{.12}{.018}$, we first write each decimal with the same number of decimal places and then multiply by 1,000 in numerator and denominator.

$$\frac{.12}{.018} = \frac{.120}{.018}$$
$$= \frac{.120 \times 1{,}000}{.018 \times 1{,}000}$$
$$= \frac{120}{18}$$
$$= \frac{20}{3} \text{ or } 6\frac{2}{3}$$

29

A more complicated example is $\frac{.2 \times .07}{.06 \times .4}$

Solution: First multiply the numbers in the numerator and then multiply the numbers in the denominator.

$$\frac{.2 \times .07}{.06 \times .4} = \frac{.014}{.024}$$

Then multiply numerator and denominator by 1,000

$$\frac{.014}{.024} = \frac{.014 \times 1{,}000}{.024 \times 1{,}000}$$
$$= \frac{14}{24}$$
$$= \frac{7}{12}$$

Study Exercise Eight

Simplify:

1. $\frac{.08}{.36}$ **2.** $\frac{.16}{.004}$ **3.** $\frac{3.6}{.09}$

4. $\frac{.2 \times .06}{.3 \times .4}$ **5.** $\frac{2}{.4}$

REVIEW EXERCISES

A. **1.** Change $\frac{46}{100}$ to a decimal.

2. Change 18.2 to an improper fraction.

3. Change $\frac{4001}{100}$ to a decimal

4. Multiply .68 × 1.4

5. Multiply 2.432 × .41

6. Find 2.02 of 1.1

7. Multiply .11 × .22 × .33

8. Simplify $\frac{.12}{.4}$

B. Multiply and select the one with the correct answer. Mark a, b, or c on your paper.

9. a. 100 × 2.603 = 26.3
b. 10 × .814 = 81.4
c. 1,000 × .026 = 26

10. a. .0041 × 100 = .041
b. .0716 × 1,000 = 71.6
c. .325 × 10 = 32.5

11. a. 10 × 1.34 = 13.4
b. 100 × .0217 = .217
c. 36.5 × 1,000 = 3650

SOLUTIONS TO REVIEW EXERCISES

A. **1.** .46

2. $\frac{182}{10}$ or $\frac{91}{5}$

3. 40.01

4.
$$\begin{array}{r} .68 \\ 1.4 \\ \hline 272 \\ 68 \\ \hline .952 \end{array}$$

5.
$$\begin{array}{r} 2.432 \\ .41 \\ \hline 2432 \\ 9728 \\ \hline .99712 \end{array}$$

6.
$$\begin{array}{r} 2.02 \\ 1.1 \\ \hline 202 \\ 202 \\ \hline 2.222 \end{array}$$

7.
$$\begin{array}{r} .11 \\ \times .22 \\ \hline 22 \\ 22 \\ \hline 242 \\ \times .33 \\ \hline 726 \\ 726 \\ \hline .007986 \end{array}$$

8.
$$\frac{.12 \times 100}{.4 \times 100} = \frac{12}{40}$$
$$= \frac{3}{10}$$

B. **9.** c

10. b

11. a

33

SUPPLEMENTARY PROBLEMS

A. Change to improper fractions:

1. 8.3 **2.** 10.1 **3.** 23.11 **4.** 2.022

B. Change to a decimal:

5. $\frac{1605}{10}$ **6.** $\frac{1605}{100}$ **7.** $\frac{201}{100}$

8. $\frac{302}{1,000}$ **9.** $\frac{4078}{10,000}$ **10.** $\frac{1724}{10}$

C. Multiply

11. $6 \times .3$ **12.** $.7 \times .06$
13. $.05 \times .12$ **14.** $.09 \times .09$
15. $.003 \times .2$ **16.** $.025 \times 8$
17. $.016 \times .205$ **18.** 1.03×2.3
19. 3.14×3.5 **20.** 10×2.073
21. 100×33.4 **22.** $1,000 \times .036$
23. $.0816 \times 100$ **24.** $.0237 \times 10$
25. $.0273 \times 10,000$ **26.** 4.1427×100
27. $100 \times .5176$ **28.** $20.1 \times 1,000$
29. $23.2 \times 1.3 \times 2.1$ **30.** $12.22 \times 1.1 \times 3.2$

D. Simplify:

31. $\frac{.02}{.12}$ **32.** $\frac{.40}{.2}$ **33.** $\frac{1.8}{.9}$

34. $\frac{.6 \times .2}{1.2}$ **35.** $\frac{.4}{2}$ **36.** $\frac{.2}{4}$

37. $\frac{.121}{.11}$ **38.** $\frac{.002}{.1}$

SOLUTIONS TO STUDY EXERCISES

Study Exercise One (Frame 4)

1. .26 **2.** .017 **3.** .0028

4A

Study Exercise Two (Frame 9)

A. 1. $7\frac{9}{10} = 7 + \frac{9}{10}$

$= \frac{70}{10} + \frac{9}{10}$

$= \frac{79}{10}$

2. $12\frac{21}{100} = 12 + \frac{21}{100}$

$= \frac{1200}{100} + \frac{21}{100}$

$= \frac{1221}{100}$

B. 3. 17.07 **4.** 2.001

Study Exercise Three (Frame 12)

1. $\frac{223}{100} \times \frac{4}{10} = \frac{892}{1,000}$
$= .892$

2. $\frac{1223}{100} \times \frac{13}{10} = \frac{15899}{1,000}$
$= 15.899$

3. $\frac{58}{100} \times \frac{21}{100} = \frac{1218}{10,000}$
$= .1218$

12A

SOLUTIONS TO STUDY EXERCISES, CONTD.

Study Exercise Four (Frame 14)

1.
```
   .612 ←—— 3 decimal places
    .14 ←—— 2 decimal places
   ----
   2448
   612
 ------
 .08568 ←—— 5 decimal places
```

2.
```
  1.33 ←—— 2 decimal places
   1.2 ←—— 1 decimal place
  ----
   266
  133
 -----
 1.596 ←—— 3 decimal places
```

14A

Study Exercise Five (Frame 17)

1.
```
   1.97
   .018
  -----
   1576
   197
  000
 ------
 .03546
```

2.
```
   2.48
   1.25
   ----
   1240
   496
  248
 ------
 3.1000 or 3.1
```

3.
```
    .025
     .08
    ----
     200
    000
  ------
  .00200 or .002
```

4.
```
  1.002
    2.3
  -----
   3006
  2004
 ------
 2.3046
```

5.
```
  .1423
    2.5
  -----
   7115
  2846
 ------
 .35575
```

17A

Study Exercise Six (Frame 19)

1.
```
    .32
  × 1.2
  -----
     64
    32
  -----
    384
  × 3.4
  -----
   1536
  1152
 ------
 1.3056
```

2.
```
     .172
   × .012
   ------
      344
     172
    000
   ------
    02064
   ×   12
   ------
    04128
   02064
  -------
  .024768
```

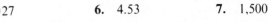

Study Exercise Seven (Frame 27)

1. .572 **2.** 3,510 **3.** .07 **4.** 50,720

5. .027 **6.** 4.53 **7.** 1,500

27A

SOLUTIONS TO STUDY EXERCISES, CONTD.

Study Exercise Eight (Frame 31)

1. $\dfrac{.08 \times 100}{.36 \times 100} = \dfrac{8}{36}$

$= \dfrac{2}{9}$

2. $\dfrac{.16 \times 1{,}000}{.004 \times 1{,}000} = \dfrac{160}{4}$

$= 40$

3. $\dfrac{3.6 \times 100}{.09 \times 100} = \dfrac{360}{9}$

$= 40$

4. $\dfrac{.2 \times .06}{.3 \times .4} = \dfrac{.012}{.12}$

$= \dfrac{.012 \times 1{,}000}{.12 \times 1{,}000}$

$= \dfrac{12}{120}$

$= \dfrac{1}{10}$

5. $\dfrac{2. \times 10}{.4 \times 10} = \dfrac{20}{4}$

$= 5$

31A

unit

22

Rounding Decimals

Objectives:

By the end of this unit you should be able to:

1. round decimals to any desired accuracy
2. expand from one to many
3. compare decimals

①

Some numbers when expressed as decimals have many decimal places. In order to make practical use of such numbers, we will cut them off and use only a limited number of decimal places.

②

In our previous study of decimals we learned that for the number 8.3579,

3 is the tenths' digit
5 is the hundredths' digit
7 is the thousandths' digit
9 is the ten-thousandths' digit

③

To round a decimal to the nearest tenth, we will keep only the digits up to and including the tenths' place. But we want to keep the nearest tenth.

④

Rules For Rounding Decimals.

1. Draw a box around the last digit to be kept.
2. Look at the next digit after the box you drew.
3. a) If that next digit is 0, 1, 2, 3, or 4, replace all digits after the box with zeros.
   b) If that next digit is 5, 6, 7, 8, or 9, add one to the digit in the box and then replace all digits after the box with zeros.

(Frame 5, contd.)

Example: Round 12.8368 to the nearest hundredth

Solution:

Last digit to be kept is 3

$12.8\boxed{3}68$

Next digit after 3 is 6

Therefore, add 1 to 3 and replace digits after the box with zeros.

$12.8\boxed{4}00$ or 12.84

Thus, 12.8368 rounds to 12.84 to the nearest hundredth. (5)

Examples:

1. Round 4.2348 to the nearest tenth
   Solution: $4.\boxed{2}348$ gives 4.2
2. Round 31.4651 to the nearest thousandth
   Solution: $31.46\boxed{5}1$ gives 31.465
3. Round 7.876 to the nearest hundredth
   Solution: $7.8\boxed{7}6$ gives 7.88 (6)

Study Exercise One

Round as indicated:
1. 4.4578 to the nearest hundredth
2. 7.462 to the nearest hundredth
3. 4.4578 to the nearest tenth
4. 19.507 to the nearest hundredth
5. 6.39918 to the nearest thousandth
6. 6.39918 to the nearest ten-thousandth
7. 6.39918 to the nearest hundredth (7)

Sometimes in rounding, the number of decimals is specified. For example, to round off 32.4178 to 2 decimals,

$32.4\boxed{1}78$ gives 32.42 (8)

Other times the required accuracy is given by specifying the number of digits to be kept. For example, to round 4.2852 to 3 digits,

$4.2\boxed{8}52$ gives 4.29 (9)

The Significant Digits of a Number

Zeros at the beginning of a number are not counted as significant digits since they are merely used to position the decimal point.

Examples:

1. .7623 has 4 significant digits
2. .07623 has 4 significant digits
3. .0306 has 3 significant digits
4. .0012 has 2 significant digits (10)

Rounding can be specified in several ways. Study the examples below.

Example 1: Round 14.2765 to 3 decimals
Solution: 14.27[6]5 gives 14.277

Example 2: Round 14.2765 to 3 significant digits
Solution: 14.[2]765 gives 14.3

Example 3: Round .00481 to 1 significant digit.

Solution: .00[4]81 gives .005

Study Exercise Two

1. Round 7.63148 to
   a) the nearest hundredth
   b) 4 decimals
   c) 4 significant digits

2. Round .007891 to
   a) 1 significant digit
   b) 2 decimals
   c) the nearest thousandth

Multiplication

Example: Multiply .0234 × 47.1 and round the result to 4 significant digits.
Solution:

```
  .0234 ←——— 4 decimal places
   47.1 ←——— 1 decimal place
  -----
   0234
  1638
 0936
 -------
 1.10214 ←——— 5 decimal places
```

1.10[2]14 rounded to 4 significant digits is 1.102

Study Exercise Three

1. Multiply 2.31 × 2.31 and round the result to the nearest hundredth.
2. Multiply 42.8 × 2.2 and round the result to 3 significant digits.
3. Multiply .088 × .088 and round the result to 1 significant digit.

(14)

Expansion

Expansion means *changing from one to several.*

Example: If 1 dozen items weigh 1.7 lb., find the weight of 2.6 dozen items.
Solution: First write the known fact
1 dozen items weigh 1.7 lb.
Now expand to 2.6 dozen by multiplying by 1.7
2.6 dozen items weigh 2.6 × 1.7 lb. = 4.42 lbs.

Example: If 1 lb. costs 6.2¢, find the cost of 8.3 lbs.
Solution: 1 lb. costs 6.2¢
8.3 lbs. cost 8.3 × 6.2¢ = 51.46¢

Study Exercise Four

1. If 1 lb. costs 8.1¢, find the cost of 6.1 lbs.
2. If 1 lb. costs 16.2¢, find the cost of 4.21 lbs.
3. Find the time for a missile to travel 16.8 miles if it travels 1 mile in 1.9 seconds.

Finding the Number When a Decimal Part is Known

Example: Find the number if .1 of the number is 18

Solution: We will first change the decimal part to a fraction

$$.1 = \frac{1}{10}$$

Thus, $\frac{1}{10}$ of the number is 18

To find the number, expand to $\frac{10}{10}$.

The number is 10 × 18 or 180

Example: Find the number if .01 of the number is .2371

Solution: $.01 = \frac{1}{100}$

Thus, $\frac{1}{100}$ of the number is .2371

To find the number, expand to $\frac{100}{100}$.

The number is 100 × .2371 = 23.71

Study Exercise Five

1. Find the number if .1 of the number is 23.45
2. Find the number if .01 of the number is .0172
3. Find the number if .5 of the number is 2.761
4. Find the number if .05 of the number is .87

(20)

Comparing Decimals

Which do you think is larger, .03 or .2?
To compare decimals:

1. Change the decimals so each has the same number of decimal places; then take the larger number as the larger decimal.
2. A mixed decimal is larger than a decimal fraction.

In the problem above,

$$.03 = .03$$
$$.2 = .20$$

20 is larger than 3; therefore, .2 is larger than .03

Example 1: Which is larger, .154 or .12?

Solution:

.154 = .154
.12 = .120

154 is larger than 120, therefore, .154 is larger.

Example 2: Which is larger, .14 or 1.02?

Solution: 1.02 is larger since 1.02 is a mixed decimal and .14 is a decimal.

(22)

Study Exercise Six

Which is larger:

1. .47 or .278?
2. .04 or .004?
3. 2.9 or .89?
4. .2 or .21?
5. .0051 or .006?

(23)

REVIEW EXERCISES

Part I. Select the correct statement and mark a, b, or c.

1. 42.5462 = (a) 42.6 to the nearest tenth
(b) 42.55 to the nearest hundredth
(c) 42.547 to the nearest thousandth

2. .097836 = (a) .098 to the nearest hundredth
(b) .0978 to the nearest thousandth
(c) .0978 to the nearest ten-thousandth

3. .015908 = (a) .02 to 3 decimals
(b) .0160 to 4 decimals
(c) .02 to 2 decimals

4. .00817 = (a) .01 to 1 significant digit
(b) .0 to 1 significant digit
(c) .008 to 1 significant digit

Part II.

5. Multiply 2.87 × 4.6 and round the result to the nearest hundredth.
6. If 1 lb. costs 17.8¢, find the cost of 7.3 lbs.
7. Find the number if .01 of the number is 8.7694
8. Which is larger, .82 or .089?
9. Which is larger, 3.01 or .899?

SOLUTIONS TO REVIEW EXERCISES

Part I.

1. b **2.** c **3.** c **4.** c

Part II.

5.
$$\begin{array}{r} 2.87 \\ 4.6 \\ \hline 1722 \\ 1148 \\ \hline 13.202 \end{array}$$
Rounds to 13.20 to the nearest hundredth

6.
$$\begin{array}{r} 17.8 \\ 7.3 \\ \hline 534 \\ 1246 \\ \hline 129.94 \end{array}$$
cost is 129.94¢

7. $\frac{1}{100}$ of the number is 8.7694

$\frac{100}{100}$ of the number is 100 × 8.7694 or 876.94

8. .82 = .820 820 is larger than 89
.089 = .089

Therefore, .82 is larger

9. 3.01 is larger since 3.01 is a mixed decimal and .899 is a decimal. (25)

SUPPLEMENTARY PROBLEMS

1. Round 34.785
(a) to the nearest tenth
(b) to the nearest hundredth
(c) to the nearest thousandth

2. Round 6.39948
(a) to the nearest tenth
(b) to the nearest hundredth
(c) to the nearest thousandth

3. Round .0285
(a) to 1 decimal
(b) to 2 decimals
(c) to 3 decimals

4. Round 6.4079
(a) to 1 decimal
(b) to 2 decimals
(c) to 3 decimals

5. Round 1.2052
(a) to 2 significant digits
(b) to 3 significant digits
(c) to 4 significant digits

6. Round .070707
(a) to 1 significant digit
(b) to 2 significant digits
(c) to 3 significant digits

7. Round 1.234567 to the nearest ten-thousandth.

8. Multiply .0716 × 48.2 and round the result to 4 significant digits.

9. Multiply 8.001 × 1.08 and round the result to the nearest thousandth.

10. If 1 lb. of candy costs 3.8¢, find the cost of 6.4 lbs. of candy.

11. If a car travels 1 mile in 1.8 minutes, how long will it take it to travel 6.4 miles?

12. If a factory produces 1 item in .46 hours, how long will it take it to produce 13 items?

13. Find the number if .1 of the number is .173.

14. Find the number if .5 of the number is 4.67.

15. Find the number if .04 of the number is 1.071.

16. Which is larger .638 or .5376?

17. Which is larger .50 or .05?

18. Which is larger 1.91 or .998?

19. Arrange in order of size (largest first).
.38, 1.5, .457, and .0605

20. Arrange in order of size (largest first).
.01, .001, .1, 1.1, and .0001

SOLUTIONS TO STUDY EXERCISES

Study Exercise One (Frame 7)

1. 4.46 **2.** 7.46 **3.** 4.5 **4.** 19.51
5. 6.399 **6.** 6.3992 **7.** 6.40

7A

Study Exercise Two (Frame 12)

1. a) 7.63 b) 7.6315 c) 7.631
2. a) .008 b) .01 c) .008

SOLUTIONS TO STUDY EXERCISES, CONTD.

Study Exercise Three (Frame 14)

1.
```
    2.31
    2.31
   -----
     231
    693
   462
  ------
  5.3361     Rounds to 5.34
```

2.
```
   42.8
    2.2
   ----
    856
   856
  -----
  94.16     Rounds to 94.2
```

3.
```
     .088
     .088
     ----
      704
     704
    000
  -------
  .007744     Rounds to .008
```

14A

Study Exercise Four (Frame 17)

1. 1 lb. costs 8.1¢
6.1 lb. cost 6.1 × 8.1¢
Answer: 49.41¢

```
   8.1
   6.1
   ---
    81
   486
  -----
  49.41
```

2. 1 lb. costs 16.2¢
4.21 lbs. cost 4.21 × 16.2¢
Answer: 68.202¢

```
    16.2
    4.21
    ----
     162
    324
   648
  ------
  68.202
```

3. 1 mile takes 1.9 seconds
16.8 miles takes 16.8 × 1.9 seconds
Answer: 31.92 seconds

```
   16.8
    1.9
   ----
   1512
   168
  -----
  31.92
```

17

Study Exercise Five (Frame 20)

1. $\frac{1}{10}$ of the number is 23.45

$\frac{10}{10}$ of the number is 10 × 23.45 or 234.5

2. $\frac{1}{100}$ of the number is .0172

$\frac{100}{100}$ of the number is 100 × .0172 or 1.72

3. $\frac{5}{10}$ or $\frac{1}{2}$ of the number is 2.761

$\frac{2}{2}$ of the number is 2 × 2.761 or 5.522

SOLUTIONS TO STUDY EXERCISES, CONTD.

Study Exercise Five (Frame 20, contd.)

4. $\frac{5}{100}$ or $\frac{1}{20}$ of the number is .87

$\frac{20}{20}$ of the number is $20 \times .87$ or 17.4

20A

Study Exercise Six (Frame 23)

1. .47 = .470 — 470 is larger than 278
.278 = .278 — .47 is larger

2. .04 = .040 — 40 is larger than 4
.004 = .004 — .04 is larger

3. 2.9 is larger since 2.9 is a mixed decimal and .89 is a decimal

4. .2 = .20 — 21 is larger than 20
.21 = .21 — .21 is larger

5. .0051 = .0051 — 60 is larger than 51
.006 = .0060 — .006 is larger

23A

---

unit

23

Division of Decimal Numerals

Objectives:

By the end of this unit you should be able to:

1. divide by decimals.
2. round off the quotient in a division problem.
3. convert fractions to decimals.
4. divide by multiples of ten by the short method. (1)

Division of decimals is similar to division of whole numbers.

A division problem may be written in three ways. For example, six divided by 3 equals 2, may be written

(1) $6 \div 3 = 2$ or

(2) $\frac{6}{3} = 2$ or

(3) $3\overline{)6}$ with quotient 2 (2)

When we divide by decimals, we first will write them as fractions. Study the examples below.

1. $2.12 \div 3.14$ will first be written $\frac{2.12}{3.14}$
2. $14.28 \div .28$ will first be written $\frac{14.28}{.28}$ (3)

After writing the problem as a fraction we will change the denominator to a whole number. Study the example below:

$$2.75 \div .5 = \frac{2.75}{.5}$$

$$= \frac{2.75 \times 10}{.5 \times 10}$$

$$= \frac{27.5}{5}$$

It will now be easy to divide by the whole number, 5. (4)

We will look at two more examples of changing a problem of division by a decimal to division by a whole number.

Example 1: $.8614 \div .14 = \dfrac{.8614}{.14}$

$= \dfrac{.8614 \times 100}{.14 \times 100}$

$= \dfrac{86.14}{14}$

Example 2: $23.1 \div .112 = \dfrac{23.1}{.112}$

$= \dfrac{23.1 \times 1000}{.112 \times 1000}$

$= \dfrac{23100}{112}$

(5)

Notice in Example 1 of the preceding frame, we could accomplish the same thing by moving both decimal points 2 places to the right.

That is, $\dfrac{.8614}{.14} = \dfrac{86.14}{14}$

In Example 2, both decimal points are moved 3 places to the right.

That is, $\dfrac{23.1}{.112} = \dfrac{23100}{112}$

Important: Both decimal points must be moved the same number of places to the right.

Study Exercise One

Change the following to division by a whole number.

1. $23.7 \div .2$ **2.** $1.434 \div 21.34$

3. $.147 \div .32$ **4.** $6.8 \div 2.183$

(7)

The problem $6.8 \div 2.183$ was simplified in the preceding exercise to $6800 \div 2183$. As a long division problem, $6.8 \div 2.183$ is written as follows:

divisor → 2.183)6.8 ← *dividend*

or in simplified form as:

divisor → 2183)6800 ← *dividend*

(8)

The usual practice is to show the position of the decimal points as follows:

2.183)6.8 was simplified by moving the decimal points 3 places to the right.

2.183.)6.800. We mark the new position of the decimal points with ⤻

(9)

Examples:

1. $1.4\overline{)28.24}$ will be written $1.4.\overline{)28.2.4}$

2. $2.21\overline{).281}$ will be written $2.21.\overline{).28.1}$

3. $.131\overline{)42.3}$ will be written $.131.\overline{)42.300.}$ (10)

Study Exercise Two

Rewrite so the divisor is a whole number.

1. $2.42 \div .21$

2. $1.7 \div .34$

3. $1.701\overline{)28.3456}$

4. $4.07\overline{)7.083}$ (11)

Division of Decimals by Whole Numbers

Problem: Divide .68 by 2

Solution: Since .68 = 6 tenths plus 8 hundredths, $.68 \div 2$ gives 3 tenths plus 4 hundredths or .34

.34 ← *quotient*

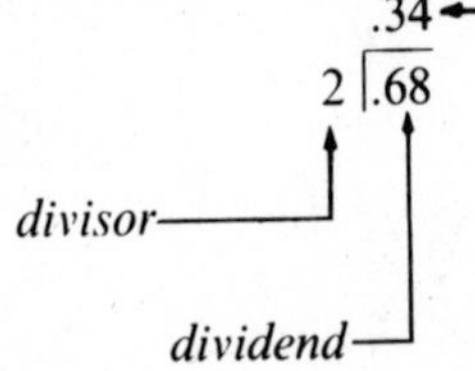

(Decimal point in the quotient must line up with the decimal point in the dividend.) (12)

Example 1: Find $.84 \div 4$

Solution:

$$\begin{array}{r} .21 \\ 4\overline{).84} \end{array}$$

So we find $.84 \div 4 = .21$

Example 2: Find $2.684 \div 2$

Solution:

$$\begin{array}{r} 1.342 \\ 2\overline{)2.684} \end{array}$$

So, $2.684 \div 2 = 1.342$

Example 3: Find $6.003 \div 3$

Solution:

$$\begin{array}{r} 2.001 \\ 3\overline{)6.003} \end{array}$$

So, $6.003 \div 3 = 2.001$ (13)

Study Exercise Three

1. Find $3.069 \div 3$

2. Find $4.804 \div 4$

3. Find $18.006 \div 2$

4. Find $8.26402 \div 2$ (14)

Division by a Decimal

Problem: Divide .048 by .2

Solution:

Step (1) First arrange in long division form

.2 ⟌.048

Step (2) Next, the divisor must be made a whole number

.2. ⟌.0.48

Step (3) Then line up the decimal point in the quotient

.2. ⟌.0.48

Step (4) Finally, divide

```
      .24
.2. ⟌.0.48        The quotient is .24
```

(15)

Example 1: .216 ÷ .03

Solution:

```
.03. ⟌.21.6
           7.2      The quotient is 7.2
.03. ⟌.21.6
```

Example 2: 4.601 ÷ 4.3

Solution:

```
          1.07      The quotient is 1.07
4.3. ⟌4.6.01
       4 3
         3 0
           0
         3 01
         3 01
```

(16)

Study Exercise Four

1. .36 ÷ .6 **2.** .048 ÷ .2 **3.** .076 ÷ .04

4. 16.308 ÷ .36 **5.** .7881 ÷ 3.7 **6.** 9.128 ÷ .028

(17)

In a division problem like 9 ÷ 4, even though no decimal points are shown, it will be necessary to write 9 as 9.00.
Remember we can place as many zeros as we wish after the decimal point.

$4\overline{)9}$ will be written $4\overline{)9.00}$

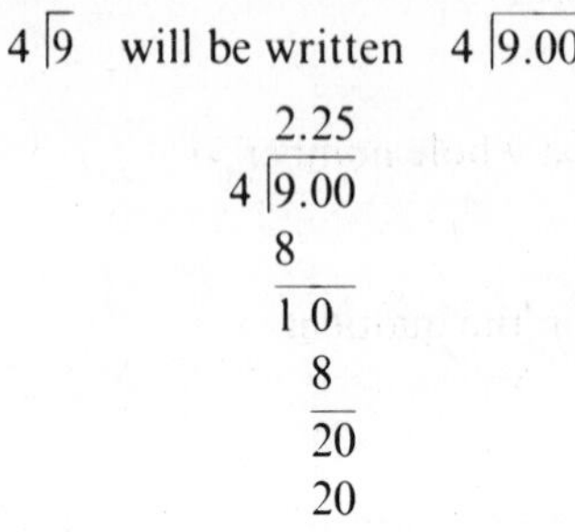

(18)

Division problems can be checked by multiplication. Let us divide 3 by .6.

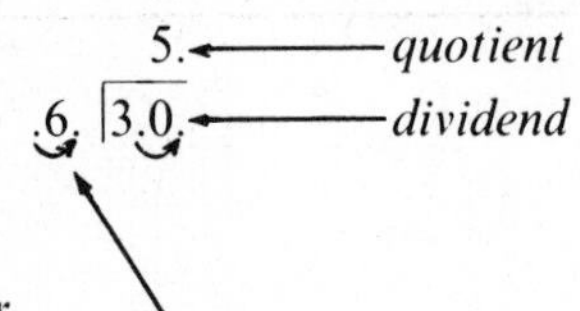

divisor

To check, **divisor × quotient equals dividend.**

Check: .6 × 5 = 3 (19)

Study Exercise Five

Divide and check:

1. .0006 ÷ .012 **2.** 15 ÷ .625 **3.** 7 ÷ 1.75 (20)

Changing a Fraction to a Decimal

A fraction may be changed to a decimal by long division.

For example, $\frac{3}{4}$ is converted to a decimal as follows:

```
    .75
 4 |3.0
   2 8
   ---
     20
     20
     --
      0
```

The decimal equivalent of $\frac{3}{4}$ is .75

(21)

Study Exercise Six

Find the decimal equivalent of the following fractions.

1. $\frac{1}{4}$ **2.** $\frac{7}{10}$ **3.** $\frac{3}{5}$ **4.** $\frac{4}{5}$

5. $\frac{2}{5}$ **6.** $\frac{1}{16}$ **7.** $\frac{5}{8}$ **8.** $\frac{3}{8}$ (22)

When a division problem ends with a zero remainder, the division is said to *terminate*. Sometimes the division does not terminate.

Example:

```
     .1444...
  9 |1.3000...
     9
     40
     36
      40
      36
       40
       36
        4
```

.1444... is called a *repeating decimal.*

23

If we wish to terminate a quotient which is a repeating decimal, we will express the remainder as a fraction.

```
     .1           .14           .144
  9 |1.3      9 |1.30      9 |1.300
     9            9            9
     4            40           40
                  36           36
                   4            40
                                36
                                 4
```

Quotient is $.1\frac{4}{9}$

Quotient is $.14\frac{4}{9}$

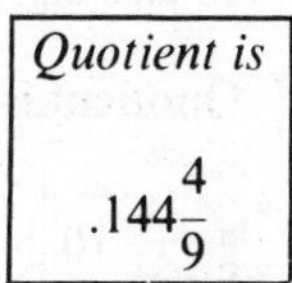

Quotient is $.144\frac{4}{9}$

24

Study Exercise Seven

Divide as far as two decimal places and terminate the answer with a fraction.

1. $2 \div 3$ **2.** $7 \div 16$ **3.** $.732 \div 3.4$

Many times we will want to round off our answer to a specified accuracy.

Example: Find $1 \div 7$ and round off to two decimals.

Solution:

$$7\overline{)1.00}\quad .14$$

```
   .14
7 |1.00
   7
   --
   30
   28
   --
    2
```

Quotient is $.14\frac{2}{7}$

If the fraction in the quotient is less than $\frac{1}{2}$ we discard it.

Since $\frac{2}{7}$ is less than $\frac{1}{2}$, $1 \div 7 = .14$ to two decimals.

If the fraction in the quotient is $\frac{1}{2}$ or more, discard the fraction and add 1 to the last digit to be kept.

Example: Find $2 \div 3$ and round off to two decimals.

Solution:

```
   .66
3 |2.00
  1 8
  ---
   20
   18
   --
    2
```

Quotient is $.66\frac{2}{3}$

Since $\frac{2}{3}$ is more than $\frac{1}{2}$, we will add 1 to the last 6.

$2 \div 3 = .67$ to two decimals.

Example 1: Find $.0718 \div 1.2$ to two significant digits.

Solution:

```
        .059
1.2, |.0,718
        60
       ---
       118
       108
       ---
        10
```

Quotient is $.059\frac{10}{12}$

Since $\frac{10}{12}$ is more than $\frac{1}{2}$, we discard the fraction and add 1 to the last digit to be kept.

Thus, $.0718 \div 1.2 = .060$ to two significant digits.

Another method of rounding off a division problem is to carry the division to one extra figure and then round back by the rules of rounding.

For example, let us find $.27163 \div .05$ to 1 decimal.

```
        5.43
.05, |.27,163
      25
      --
       2 1
       2 0
       ---
         16
         15
         --
          13
```

Now round 5.43 to 5.4

The quotient is 5.4 to 1 decimal

The Two Methods of Terminating a Division Problem

1. Terminate the answer with a fraction and then round off by considering the size of the fraction.

2. Carry the division to one extra digit and round off by the rules of rounding.

Study Exercise Eight

A. Divide and round off correctly to two decimals by terminating with a fraction.

1. .75 ÷ .178 **2.** .081 ÷ .14

B. Divide and round off correctly to two decimals by continuing one extra place and rounding back.

3. .638 ÷ 2.6 **4.** 1.37 ÷ .76

Division by Powers of Ten

You may remember the short method we have for multiplying by 10, 100, 1,000, etc. Now we want a short method of dividing by 10, 100, 1,000, etc.

Study the examples below:

```
14.3 ÷ 10        14.3 ÷ 100        14.3 ÷ 1,000

      1.43              .143               .0143
10. |14.30       100 |14.300       1000 |14.3000
     10               10 0               10 00
     ---              ----               -----
      4 3              4 30               4 300
      4 0              4 00               4 000
      ---              ----               -----
       30               300                3000
       30               300                3000
       --               ---                ----
```

Rules: To divide by 10, move the decimal point one place to the *left*.
To divide by 100, move the decimal point two places to the *left*.
To divide by 1,000, move the decimal point three places to the *left*.

Example 1: 142.3 ÷ 100 = 1.423

Example 2: 17.21 ÷ 10 = 1.721

Example 3: .012 ÷ 1,000 = .000012

Study Exercise Nine

Divide as indicated by the short method:

1. 7.56 ÷ 10 **2.** .304 ÷ 100 **3.** .701 ÷ 1,000 **4.** 3.11 ÷ 100

5. .051 ÷ 10 **6.** 689 ÷ 1000 **7.** 12 ÷ 100 **8.** .0101 ÷ 10

(34)

Division by Multiples of Ten

Many division problems where the divisor is a multiple of 10 can be worked by a short method.

Example: $29.9 \div 1300$

Solution: $29.9 \div 1300 = \dfrac{29.9}{1300}$

Remove the zeros from the denominator by dividing numerator and denominator by 100. This is done by moving the decimal points two places to the left.

$$\frac{29.9}{1300} = \frac{.299}{13}$$
$$= .023$$

35

Example 1: Divide $.0524 \div 40$ by the short method.

Solution:

Line (a) $.0524 \div 40 = \dfrac{.0524}{40}$

Line (b) $\dfrac{.0524}{40} = \dfrac{.00524}{4}$

Line (c) $\dfrac{.00524}{4} = .00131$

Example 2: Find $20.6 \div 5{,}000$ by the short method.

Solution:

Line (a) $20.6 \div 5{,}000 = \dfrac{20.6}{5000}$

Line (b) $\dfrac{20.6}{5000} = \dfrac{.0206}{5}$

Line (c) $\dfrac{.0206}{5} = .00412$

36

Study Exercise Ten

Divide by the short method.

1. $21.8 \div 200$ **2.** $.620 \div 50$

3. $8.1 \div 9{,}000$ **4.** $92{,}070 \div 300$

37

REVIEW EXERCISES

A. Change to division by a whole number.

1. 1.89 ÷ .3 **2.** .0417 ÷ .032

B. Divide as indicated.

3. .018 ÷ .02 **4.** .24 ÷ .6

5. 16.308 ÷ .36

C. Divide and check.

6. 3 ÷ .04

D. Find the decimal equivalent.

7. $\frac{5}{4}$ **8.** $\frac{5}{8}$ **9.** $\frac{3}{4}$ **10.** $\frac{2}{5}$

E. Divide and round off to 2 decimals.

11. 13 ÷ 17 **12.** 1 ÷ 6

F. Divide by the short method.

13. 8.017 ÷ 100 **14.** .008 ÷ 10

15. 2.15 ÷ 500 **16.** 17.24 ÷ 20

SOLUTIONS TO REVIEW EXERCISES

A. **1.** 18.9 ÷ 3 **2.** 41.7 ÷ 32

B. **3.**

```
        .9
.02.)|.01.8
```

4.

```
      .4
.6.)|2.4
```

5.

```
          45.3
.36.)|16.30.8
      14 4
       1 90
       1 80
         10 8
         10 8
```

C. **6.**

```
       75.
.04.)|3.00.
```

Check:

```
   75
 × .04
  3.00
```

D. **7.**

```
    1.25
4 |5.00
   4
   1 0
     8
     20
     20
```

8.

```
    .625
8 |5.00
   4 8
     20
     16
      40
      40
```

9.

```
    .75
4 |3.00
   2 8
     20
     20
```

10.

```
    .4
5 |2.0
   2 0
```

SOLUTIONS TO REVIEW EXERCISES, CONTD.

(Frame 39, contd.)

E. **11.**

```
       .764     Quotient: .76
17 |13.00
    11 9
    1 10
    1 02
      80
      68
      12
```

12.

```
      .166     Quotient: .17
6 |1.00
     6
    40
    36
    40
    36
     4
```

F. **13.** $8.017 \div 100 = .08017$

14. $.008 \div 10 = .0008$

15. $\frac{2.15}{500} = \frac{.0215}{5}$
$= .0043$

16. $\frac{17.24}{20} = \frac{1.724}{2}$
$= .862$

39

SUPPLEMENTARY PROBLEMS

A. Divide as indicated.

1. $35.6 \div .4$ **2.** $.0016 \div .8$
3. $72 \div .4$ **4.** $40 \div 12.8$
5. $.00012 \div .03$ **6.** $18.7572 \div 5.39$
7. $670.8 \div .78$ **8.** $1.634 \div .043$
9. $972 \div .108$ **10.** $.63 \div .0007$

B. Find the decimal equivalent of the following fractions.

11. $\frac{6}{10}$ **12.** $\frac{3}{8}$ **13.** $\frac{9}{16}$ **14.** $\frac{1}{20}$
15. $\frac{5}{8}$ **16.** $\frac{4}{5}$ **17.** $\frac{7}{20}$

C. Divide and check.

18. $.12 \div 3.2$ **19.** $10.85 \div .0775$
20. $.1875 \div .25$

D. Divide and round off the answer to two decimal places.

21. $3.06 \div .11$ **22.** $24.1 \div 6.001$
23. $.527 \div 1.37$ **24.** $731.8 \div 14.6$
25. $51.7 \div .292$ **26.** $12.2 \div .098$

E. Divide by the short method.

27. $68.2 \div 100$ **28.** $86.4 \div 200$
29. $.143 \div 10$ **30.** $76.4 \div 4000$
31. $2.824 \div 1{,}000$

SOLUTIONS TO STUDY EXERCISES

Study Exercise One (Frame 7)

1. $\frac{23.7}{.2} = \frac{237}{2}$

2. $\frac{1.434}{21.34} = \frac{143.4}{2134}$

3. $\frac{.147}{.32} = \frac{14.7}{32}$

4. $\frac{6.8}{2.183} = \frac{6800}{2183}$

(7A)

Study Exercise Two (Frame 11)

1. .21 ⟌2.42 = .21‸ ⟌2.42‸

2. .34 ⟌1.7 = .34‸ ⟌1.70‸

3. 1.701 ⟌28.3456 = 1.701‸ ⟌28.345‸6

4. 4.07 ⟌7.083 = 4.07‸ ⟌7.08‸3

Study Exercise Three (Frame 14)

1. 1.023 2. 1.201 3. 9.003 4. 4.13201

(14A)

Study Exercise Four (Frame 17)

1. .6
.6‸ ⟌.3‸6

2. .24
.2‸ ⟌.0‸48

3. 1.9
.04‸ ⟌.07‸6

4. 45.3
.36‸ ⟌16.30‸8

5. .213
3.7‸ ⟌.7‸881

6. 326.
.028‸ ⟌9.128‸

(17A)

Study Exercise Five (Frame 20)

1.
```
          .05
.012‸ ⟌.000‸60
             60
```
Check:
```
 .012
  .05
.00060
```

2.
```
           24.
.625‸ ⟌15.000‸
        12 50
         2 500
         2 500
```
Check:
```
  .625
    24
  2500
 1250
15.000
```

3.
```
          4.
1.75‸ ⟌7.00‸
        7 00
```
Check:
```
1.75
   4
7.00
```

(20A)

SOLUTIONS TO STUDY EXERCISES, CONTD.

Study Exercise Six (Frame 22)

1. .25	**2.** .7	**3.** .6	**4.** .8
5. .4	**6.** .0625	**7.** .625	**8.** .375

22A

Study Exercise Seven (Frame 25)

1.

$$\begin{array}{r} .66 \\ 3\overline{)2.00} \\ 1\,8 \\ \hline 20 \\ 18 \\ \hline 2 \end{array}$$

Quotient: $.66 + \frac{2}{3} = .66\frac{2}{3}$

2.

$$\begin{array}{r} .43 \\ 16\overline{)7.00} \\ 6\,4 \\ \hline 60 \\ 48 \\ \hline 12 \end{array}$$

Quotient: $.43 + \frac{12}{16} = .43 + \frac{3}{4}$

$= .43\frac{3}{4}$

3.

$$\begin{array}{r} .21 \\ 3.4.\overline{).7.32} \\ \\ 6\,8 \\ \hline 52 \\ 34 \\ \hline 18 \end{array}$$

Quotient: $.21 + \frac{18}{34} = .21 + \frac{9}{17}$

$= .21\frac{9}{17}$

25A

Study Exercise Eight (Frame 31)

A. 1.

$$\begin{array}{r} 4.21\frac{62}{178} \\ .178.\overline{).750.00} \\ \\ 712 \\ \hline 38\,0 \\ 35\,6 \\ \hline 2\,40 \\ 1\,78 \\ \hline 62 \end{array}$$

Since $\frac{62}{178}$ is less than $\frac{1}{2}$, the quotient is 4.21

2.

$$\begin{array}{r} .57\frac{12}{14} \\ .14.\overline{).08.10} \\ \\ 7\,0 \\ \hline 1\,10 \\ 98 \\ \hline 12 \end{array}$$

Since $\frac{12}{14}$ is more than $\frac{1}{2}$, the quotient is .58

SOLUTIONS TO STUDY EXERCISES, CONTD.

Study Exercise Eight (Frame 31, contd.)

B. **3.**

```
        .245
2.6.)  .6.380
        5 2
        1 18
        1 04
         140
         130
          10
```

Since the third decimal place is 5, we will round up.
The quotient is .25

4.

```
        1.802
.76.) 1.37.000
        76
        61 0
        60 8
          200
          152
           48
```

Since the third decimal place is a 2, we will drop the 2.
The quotient is 1.80

31A

Study Exercise Nine (Frame 34)

1. .756 **2.** .00304 **3.** .000701

4. .0311 **5.** .0051 **6.** .689

7. .12 **8.** .00101

Study Exercise Ten (Frame 37)

1. $\frac{21.8}{200} = \frac{.218}{2}$

$= .109$

2. $\frac{.620}{50} = \frac{.0620}{5}$

$= .0124$

3. $\frac{8.1}{9000} = \frac{.0081}{9}$

$= .0009$

4. $\frac{92{,}070}{300} = \frac{920.7}{3}$

$= 306.9$

unit

24

Working With Decimals

Objectives:

By the end of this unit you should be able to:

1. work reduction problems.
2. find a number when a decimal part is known.
3. find what decimal one number is of another.
4. find powers of decimals. (1)

Expansion

In an earlier unit we studied expansion. By expansion we mean changing from one to several. For example, if one pencil costs 4¢, how much do 8 pencils cost? We get the answer by multiplying 4¢ × 8. Thus, 8 pencils cost 32¢.

Reduction

Reduction means *changing from several back to one.*

Example: If 9 pencils cost 54¢, find the cost of 1 pencil.

Solution: 9 pencils cost 54¢
1 pencil costs 54¢ ÷ 9 = 6¢
Notice that we reduce to one by dividing.

Example: If 2.4 articles weigh 10.8 lbs., find the weight of one article.

Solution: First write the known fact.
2.4 articles weigh 10.8 lbs.
Next reduce to 1 article by dividing.
1 article weighs 10.8 lbs. ÷ 2.4 = 4.5 lbs.

Study Exercise One

1. If a car uses .6 gallon of gas to go 8.4 miles, how far does it travel on 1 gallon of gas?
2. If 3.8 lbs. cost 9.5¢, find the cost of 5.4 lbs.
3. If 3.6¢ buys .972 lb, how many lbs. can you buy for 15¢?
4. If a factory produces 2.4 articles in 9.6 hours, how many articles are produced in $16\frac{1}{2}$ hours?

Finding a Number When a Decimal Part is Known

Finding a number when a decimal part is known can be done by *division*.

Question 1: 3 times what number is 12?

Solution: Divide 3 into 12 and get 4.

Question 2: .3 times what number is .12?

Solution: Divide .3 into .12 and get .4

(6)

Example: If .3 of a number is 15, find the number.

Solution: We will let a box with a question mark inside, $\boxed{?}$, represent the number we want.

Step (1) The problem can be written

$$.3 \times \boxed{?} = 15$$

Step (2) Divide by .3

$$\frac{.3 \times \boxed{?}}{.3} = \frac{15}{.3}$$

Step (3) Cancel

$$\frac{\overset{1}{\cancel{.3}} \times \boxed{?}}{\underset{1}{\cancel{.3}}} = \frac{15}{.3}$$

$$\boxed{?} = \frac{15}{.3}$$

Step (4) Divide

$$\boxed{?} = 50$$

$$\begin{array}{r} 5\,0. \\ .3.\overline{)15.0.} \\ 15 \\ \hline 0 \\ 0 \\ \hline \end{array}$$

The number is 50

Example: If .25 of a number is 32, find the number.

Solution:

Line (a) $.25 \times \boxed{?} = 32$

Line (b) $\dfrac{.25 \times \boxed{?}}{.25} = \dfrac{32}{.25}$

Line (c) $\dfrac{\overset{1}{\cancel{.25}} \times \boxed{?}}{\underset{1}{\cancel{.25}}} = \dfrac{32}{.25}$

Line (d) $\boxed{?} = 128$

$$\begin{array}{r} 1\,28. \\ .25.\overline{)\,32.00.} \\ 25 \\ \hline 7\,0 \\ 5\,0 \\ \hline 2\,00 \\ 2\,00 \\ \hline \end{array}$$

The number is 128

(8)

Study Exercise Two

1. If .4 of a number is 12, find the number.
2. If .12 of a number is 3, find the number.
3. If .02 of a number is .2, find the number.

Instead of a box, it is easier to let the letter *n* stand for the number.

Example: If .12 of a number is .72, find the number.

Solution:

Line (a) $.12 \times n = .72$

Line (b) $\dfrac{.12 \times n}{.12} = \dfrac{.72}{.12}$

Line (c) $\dfrac{\overset{1}{\cancel{.12}} \times n}{\underset{1}{\cancel{.12}}} = \dfrac{.72}{.12}$

Line (d) $n = 6$

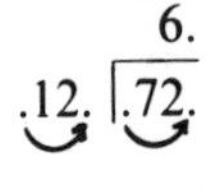

$$\begin{array}{r} 6. \\ .12.\overline{)\,.72.} \\ 72 \\ \hline \end{array}$$

The number is 6.

Study Exercise Three

1. If .8 of a number is 7.2, find the number.
2. If .25 of a number is 2.12, find the number.
3. Find n if $.50 \times n = .75$
4. Find n if $.12 \times n = 2.4$
5. Find n if $.3 \times n = 1.5$
6. Find n if $.025 \times n = 4.2$
7. Find n if $14.2 \times n = 115.02$

(11)

Finding What Decimal Part One Number is of Another

Many times the problem of finding what decimal part one number is of another can be done easily.

For example, what decimal part of 2 is 1? It is easy to see the answer is $\frac{1}{2}$ or .5. To check our answer we multiply .5 by 2. $(.5 \times 2 = 1)$. (12)

We want a method that would work with more complicated problems such as what decimal part of .27 is .18?

Rule: To find what decimal part one number is *of another*, make a fraction of the form:

$$\frac{\textbf{“is” number}}{\textbf{“of” number}}$$

Example: What decimal part of 6 is 2?

Solution: $\dfrac{\text{“is” number}}{\text{“of” number}}$

$$\frac{2}{6} = \frac{1}{3} = .33\frac{1}{3}$$

(13)

Example: What decimal part of .27 is .18?

Solution: $\dfrac{\text{“is” number}}{\text{“of” number}}$

$$\frac{.18}{.27} = .18 \div .27$$

$$.66\frac{18}{27} = .66\frac{2}{3}$$

```
         .66 18/27
.27.) .18.0
       16 2
        1 80
        1 62
          18
```

Check: $.66\frac{2}{3} \times .27 = \frac{2}{3} \times .27$

$= .18$

(14)

Study Exercise Four

1. What decimal part of 24 is 6?
2. What decimal part of .48 is .16?
3. What decimal part of 1.04 is .2496?
4. What decimal part of .12 is .4?
5. What decimal part of .51 is .34?
6. What decimal part of 2.8 is .08? 

Powers of Decimals

We will review exponents before proceeding to powers of decimals.

$4 \times 4 \times 4$ may be written 4^3

4^3 ← *exponent* (the 3), ← *base* (the 4)

An *exponent* is a number that tells how many times the base appears as a factor in multiplication.

Thus, 5^4 means $5 \times 5 \times 5 \times 5$

We will find powers of decimals in the same way.

Example 1: Find $(.3)^2$

Solution: $(.3)^2 = .3 \times .3$
$= .09$

Example 2: Find $(.02)^3$

Solution: $(.02)^3 = .02 \times .02 \times .02$
$= .000008$

Example 3: Find $(1.06)^2$

Solution: $(1.06)^2 = 1.06 \times 1.06$
$= 1.1236$

Study Exercise Five

Find the following powers:

1. $(.4)^2$ **2.** $(.01)^3$ **3.** $(.12)^2$

4. $(1.2)^2$ **5.** $(.2)^4$ **6.** $(.03)^3$

7. $(12.1)^2$ **8.** $(.86)^2$

18

REVIEW EXERCISES

1. If four new dollar bills are .016 inches in thickness, how thick is a stack of 100 new dollar bills?
2. At the rate of 3 for 10¢, find the cost of a dozen candy bars.
3. If 4 gallons of fuel cost 115.6¢, find the cost of 10 gallons of fuel.
4. Find the number if .05 of the number is 12.1.
5. What is the number if 1.1 of the number is .154?
6. What decimal part of .48 is .8?
7. What decimal part of 2.6 is 1.3?
8. Find $(.06)^2$
9. Find $(1.02)^2$
10. Find $(1.1)^3$
11. Find n if $1.8 \times n = .54$

SOLUTIONS TO REVIEW EXERCISES

1. 4 bills are .016 inches thick.
   1 bill is $.016 \div 4 = .004$ inches thick.
   100 bills are $100 \times .004 = .4$ inch thick.
2. 3 bars cost 10¢.

   1 bar costs $10¢ \div 3 = \frac{10}{3}¢$.

   12 bars cost $12 \times \frac{10}{3} = 40¢$.
3. 4 gallons cost 115.6¢.
   1 gallon costs $115.6¢ \div 4 = 28.9¢$.
   10 gallons cost $10 \times 28.9¢ = 289¢$ or \$2.89.
4. $.05 \times n = 12.1$

   $$\frac{.05 \times n}{.05} = \frac{12.1}{.05}$$

   $$n = \frac{12.1}{.05}$$

   $$n = 242$$

   $$\begin{array}{r} 2\,42. \\ .05.\overline{)12.10.} \\ 10 \\ \hline 2\,1 \\ 2\,0 \\ \hline 10 \\ 10 \\ \hline \end{array}$$

   The number is 242
5. $1.1 \times n = .154$

   $$\begin{array}{r} .14 \\ 1.1.\overline{).1.54} \\ 1\,1 \\ \hline 44 \\ 44 \\ \hline \end{array}$$

   $$\frac{1.1 \times n}{1.1} = \frac{.154}{1.1}$$

   $$n = .14$$

   The number is .14

SOLUTIONS TO REVIEW EXERCISES, CONTD.

(Frame 20, contd.)

6. $\frac{.8}{.48} = \frac{.8 \times 100}{.48 \times 100}$

$= \frac{80}{48}$

$= \frac{10}{6}$

$= \frac{5}{3}$

$= 1\frac{2}{3}$

$= 1.66\frac{2}{3}$

7. $\frac{1.3}{2.6} = \frac{1.3 \times 10}{2.6 \times 10}$

$= \frac{13}{26}$

$= \frac{1}{2}$

$= .5$

8. $.06 \times .06 = .0036$

9. $1.02 \times 1.02 = 1.0404$

```
  1.02
  1.02
  ----
   204
  000
 102
------
1.0404
```

10. $n = (1.1)^3$

$= 1.1 \times 1.1 \times 1.1$

$= 1.331$

```
 1.1      1.21
 1.1       1.1
 ---      ----
  11       121
 11       121
----     -----
1.21     1.331
```

11. $1.8 \times n = .54$

$\frac{\overset{1}{\cancel{1.8}} \times n}{\underset{1}{\cancel{1.8}}} = \frac{.54}{1.8}$

$n = \frac{.54}{1.8}$

$n = .3$

```
        .3
1.8 ) .5.4
       5 4
       ---
```

SUPPLEMENTARY PROBLEMS

1. If 2 lbs. cost 42.6¢, find the cost of 6.8 lbs.
2. Find the weight of 1 cubic inch if 3.25 cubic inches weigh 6.8 lbs. (Answer to the nearest hundredth.)
3. If a car travels 6.25 miles in 11.26 minutes, how far will it travel in 3.2 minutes? (Answer to the nearest hundredth.)
4. If meat sells for 3 lbs. for $1.41, how many pounds can you buy for $1? (Answer to the nearest hundredth.)
5. If a factory produces 3.4 articles in 21.08 hours, how many will it produce in 12.2 hours? (Answer to the nearest tenth.)

SUPPLEMENTARY PROBLEMS, CONTD.

6. Find the number if .001 of the number is 8.

7. Find the number if .0754 of the number is .3. (Answer to the nearest hundredth.)

8. What decimal part of .2 is .002?

9. What decimal part of .55 is 1.705?

10. What decimal part of 3.6 is 18?

11. Find n if $.6 \times n = .84$.

12. Find n if $.25 \times n = .36$.

13. Find n if $.035 \times n = 2.1$.

14. Find n if $2.6 \times n = 2.34$.

15. What decimal part of .56 is .14?

16. What decimal part of .48 is .03?

17. What decimal part of .36 is .6? (Answer to the nearest hundredth.)

18. What decimal part of 1.06 is .25? (Answer to the nearest tenth.)

19. Find $(.5)^2$

20. Find $(.08)^3$

21. Find $(1.3)^2$

22. Find $(2.13)^2$

23. Find $(.09)^3$

24. Find $(.01)^4$

25. Find $(.08)^2$

SOLUTIONS TO STUDY EXERCISES

Study Exercise One (Frame 5)

1. .6 gallon gives 8.4 miles

1 gallon gives $8.4 \div .6 = 14$ miles

```
     1 4.
 .6.|8.4.
     6
     2 4
     2 4
```

2. 3.8 lbs. cost 9.5¢

1 lb. costs $9.5¢ \div 3.8 = 2.5¢$

5.4 lb. cost $2.5 \times 5.4 = 13.5¢$

```
        2.5
 3.8.|9.5.
       7 6
       1 9 0
       1 9 0
```

3. 3.6¢ buys .972 lb.

1¢ buys .972 lb. $\div$ 3.6 = .27 lb.

15¢ buys 15¢ $\times$ .27 lbs. = 4.05 lbs.

```
        .27
 3.6.|.9.72
       7 2
       2 52
       2 52
```

4. 9.6 hours to produce 2.4 articles

1 hour to produce 2.4 articles $\div$ 9.6 = .25 article

In $16\frac{1}{2}$ hours, $.25 \times 16\frac{1}{2} = 4.125$ articles

```
         .25
 9.6.|2.4.00
       1 9 2
         4 80
         4 80
```

5A

SOLUTIONS TO STUDY EXERCISES, CONTD.

Study Exercise Two (Frame 9)

1. $.4 \times \boxed{?} = 12$

$$\frac{.4 \times \boxed{?}}{.4} = \frac{12}{.4}$$

$$\boxed{?} = 30$$

The number is 30

$$.4_{\wedge}\overline{)12.0_{\wedge}} = 30.$$

$$\begin{array}{r} 12 \\ \hline 0 \\ 0 \\ \hline \end{array}$$

2. $.12 \times \boxed{?} = 3$

$$\frac{.12 \times \boxed{?}}{.12} = \frac{3}{.12}$$

$$\boxed{?} = 25$$

The number is 25

$$.12_{\wedge}\overline{)3.00_{\wedge}} = 25.$$

$$\begin{array}{r} 24 \\ \hline 60 \\ 60 \\ \hline \end{array}$$

3. $.02 \times \boxed{?} = .2$

$$\frac{.02 \times \boxed{?}}{.02} = \frac{.2}{.02}$$

$$\boxed{?} = 10$$

The number is 10

$$.02_{\wedge}\overline{).20_{\wedge}} = 10.$$

$$\begin{array}{r} 2 \\ \hline 0 \\ 0 \\ \hline \end{array}$$

9A

Study Exercise Three (Frame 11)

1. $.8 \times n = 7.2$

$$\frac{\overset{1}{\cancel{.8}} \times n}{\underset{1}{\cancel{.8}}} = \frac{7.2}{.8}$$

$$n = 9$$

The number is 9

2. $25 \times n = 2.12$

$$\frac{\overset{1}{\cancel{.25}} \times n}{\underset{1}{\cancel{.25}}} = \frac{2.12}{.25}$$

$$n = 8.48$$

The number is 8.48

3. $.50 \times n = .75$

$$\frac{\overset{1}{\cancel{.50}} \times n}{\underset{1}{\cancel{.50}}} = \frac{.75}{.50}$$

$$n = 1.5$$

The number is 1.5

4. $.12 \times n = 2.4$

$$\frac{\overset{1}{\cancel{.12}} \times n}{\underset{1}{\cancel{.12}}} = \frac{2.4}{.12}$$

$$n = 20$$

The number is 20

5. $.3 \times n = 1.5$

$$\frac{\overset{1}{\cancel{.3}} \times n}{\underset{1}{\cancel{.3}}} = \frac{1.5}{.3}$$

$$n = 5$$

The number is 5

6. $.025 \times n = 4.2$

$$\frac{\overset{1}{\cancel{.025}} \times n}{\underset{1}{\cancel{.025}}} = \frac{4.2}{.025}$$

$$n = 168$$

The number is 168

SOLUTIONS TO STUDY EXERCISES, CONTD.

Study Exercise Three (Frame 11, contd.)

7. $14.2 \times n = 115.02$

$$\frac{\overset{1}{\cancel{14.2}} \times n}{\underset{1}{\cancel{14.2}}} = \frac{115.02}{14.2}$$

$$n = 8.1$$

The number is 8.1

(11A)

Study Exercise Four (Frame 15)

1. $\frac{6}{24} = \frac{1}{4}$

$= .25$

2. $\frac{.16}{.48} = \frac{.16 \times 100}{.48 \times 100}$

$= \frac{16}{48}$

$= \frac{1}{3}$

$= .33\frac{1}{3}$

3. $\frac{.2496}{1.04} = .24$

$$\begin{array}{r} .24 \\ 1.04.\overline{).24.96} \\ 20\ 8 \\ \hline 4\ 16 \\ 4\ 16 \\ \hline \end{array}$$

4. $\frac{.4}{.12} = 3.33\frac{1}{3}$

$$\begin{array}{r} 3.33\frac{1}{3} \\ .12.\overline{).40.} \\ 36 \\ \hline 4\ 0 \\ 3\ 6 \\ \hline 4 \end{array}$$

5. $\frac{.34}{.51} = .66\frac{2}{3}$

$$\begin{array}{r} .66\frac{34}{51} = .66\frac{2}{3} \\ .51.\overline{).34.0} \\ 30\ 6 \\ \hline 3\ 40 \\ 3\ 06 \\ \hline 34 \end{array}$$

6. $\frac{.08}{2.8} = .02\frac{6}{7}$

$$\begin{array}{r} .02\frac{6}{7} \\ 2.8.\overline{).0.80} \\ 56 \\ \hline 24 \end{array}$$

(15A)

SOLUTIONS TO STUDY EXERCISES, CONTD.

Study Exercise Five (Frame 18)

1. $(.4)^2 = .4 \times .4$
$= .16$

2. $(.01)^3 = .01 \times .01 \times .01$
$= .000001$

3. $(.12)^2 = .12 \times .12$
$= .0144$

4. $(1.2)^2 = 1.2 \times 1.2$
$= 1.44$

5. $(.2)^4 = .2 \times .2 \times .2 \times .2$
$= .0016$

6. $(.03)^3 = .03 \times .03 \times .03$
$= .000027$

7. $(12.1)^2 = 12.1 \times 12.1$
$= 146.41$

8. $(.86)^2 = .86 \times .86$
$= .7396$

unit

25

Square Root Algorithm

Objectives:

By the end of this unit you should be able to find the square root of a number by the square root algorithm. ①

Square Root

One of two equal factors of a number is called a *square root* of the number.
For example, since $4 \times 4 = 16$

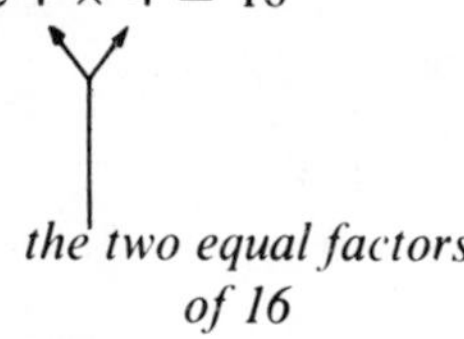

4 is a square root of 16. ②

Since $7 \times 7 = 49$, then 7 is a square root of 49.

To indicate square root, we use the symbol $\sqrt{\ }$ which is called a *radical.*

Thus, the square root of 49 is indicated by $\sqrt{49}$.

$$\sqrt{49} = 7$$

③

Square roots of certain numbers can be found by simply guessing, but the square root of every number cannot be found by this method. Numbers that are perfect squares will have square roots which can be found by guessing.
The perfect squares are:

$1^2 = 1$	$5^2 = 25$	$9^2 = 81$
$2^2 = 4$	$6^2 = 36$	$10^2 = 100$
$3^2 = 9$	$7^2 = 49$	$11^2 = 121$
$4^2 = 16$	$8^2 = 64$	$12^2 = 144$, etc.

1, 4, 9, 16, 25, 36, 49, 64, 81, 100, 121, 144, ... are perfect squares. ④

Examples:

1. $\sqrt{64} = 8$ since $8 \times 8 = 64$
2. $\sqrt{9} = 3$ since $3 \times 3 = 9$
3. $\sqrt{900} = 30$ since $30 \times 30 = 900$
4. $\sqrt{289} = 17$ since $17 \times 17 = 289$
5. $\sqrt{.16} = .4$ since $.4 \times .4 = .16$
6. $\sqrt{.0009} = .03$ since $.03 \times .03 = .0009$ (5)

Study Exercise One

The following numbers are all perfect squares. Find the square root and check by multiplying.

1. $\sqrt{81}$
2. $\sqrt{100}$
3. $\sqrt{400}$
4. $\sqrt{256}$
5. $\sqrt{1225}$
6. $\sqrt{.01}$
7. $\sqrt{225}$
8. $\sqrt{196}$
9. $\sqrt{.0036}$
10. $\sqrt{.04}$
11. $\sqrt{1.44}$ (6)

If a number is not a perfect square, we will discuss in later frames a method to approximate the square root. For example, 7 is not a perfect square and $\sqrt{7}$ can only be approximated. It will help when thinking about square roots that *the larger a number, the larger its square root*. For example, $\sqrt{2}$ is larger than $\sqrt{1}$ and $\sqrt{3}$ is larger than $\sqrt{2}$. Thus, $\sqrt{2}$ is larger than $\sqrt{1}$ and smaller than $\sqrt{3}$. (7)

Perfect squares:

1, 4, 9, 16, 25, 36, 49, 64, 81, ... etc.

Problem: Between what two perfect squares is the number 13?

Answer: 13 is between 9 and 16.

Therefore, $\sqrt{13}$ is between $\sqrt{9}$ and $\sqrt{16}$ and since $\sqrt{9} = 3$ and $\sqrt{16} = 4$, $\sqrt{13}$ is between 3 and 4. (8)

Study Exercise Two

Between which two perfect squares is each of the following numbers found?

1. 50
2. 18
3. 37
4. 23
5. 6
6. 90 (9)

Square Root Algorithm

By the *square root algorithm* we will mean a rule or method that will find the square root of a number.

The method is in many ways similar to a long division problem.

Review below: 20 ÷ 16

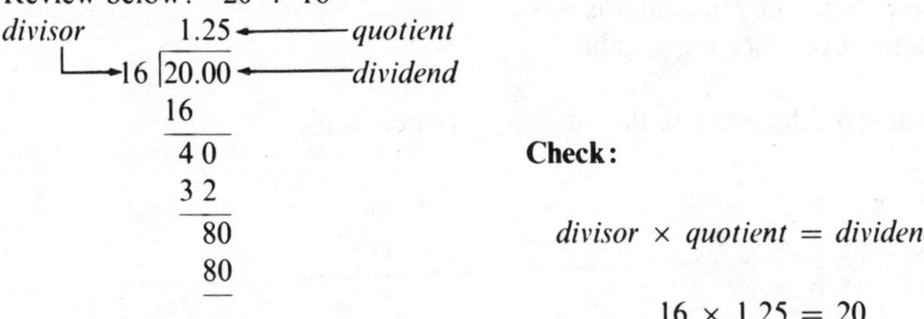

Check:

divisor × *quotient* = *dividend*

$$16 \times 1.25 = 20$$

(10)

Square Roots of Perfect Squares

Problem: Let us find $\sqrt{36}$.

$$\begin{array}{r} ?. \leftarrow \textit{quotient} \\ \textit{divisor} \rightarrow ?\sqrt{36.} \leftarrow \textit{dividend} \end{array}$$

Since divisor × quotient = dividend, and the quotient will be the square root of 36, quotient and divisor must be equal (a square root is one of two equal factors of a number). Since we have no divisor, we will make one by placing the *same* digit in the divisor and quotient, in this case, 6.

$$\begin{array}{r} 6. \\ 6\sqrt{36.} \\ 36 \\ \hline 0 \end{array} \qquad 6 \times 6 = 36$$

(11)

The Square Root Algorithm

This process will be developed slowly in the next few frames.

Problem: Find $\sqrt{20}$

Step 1:

$$\begin{array}{r} ?. \\ ?\sqrt{20.} \end{array}$$

We must place the same number in divisor and quotient. Between which two perfect squares is 20? The number 20 is found between 16 and 25. We will use the lower number 16 and take its square root and use as a divisor and quotient.

$$\begin{array}{r} 4. \\ 4\sqrt{20.} \end{array}$$

(12)

Step 2:

$$\begin{array}{r} 4. \\ 4\overline{\smash{\big)}\,20.} \\ 16 \\ \hline 4 \end{array}$$

Multiply as in a division problem and subtract. The remainder is 4.
Before continuing with the process, you must practice steps 1 and 2.

Note: Sometimes the remainder will come out larger than the "divisor." For example:

$$\begin{array}{r} 3 \\ 3\overline{\smash{\big)}\,15} \\ 9 \\ \hline 6 \end{array}$$

(13)

Study Exercise Three

Use steps 1 and 2 and find the quotient and remainder.

1. $\sqrt{29}$ **2.** $\sqrt{40}$ **3.** $\sqrt{85}$ **4.** $\sqrt{8}$

5. $\sqrt{110}$ **6.** $\sqrt{63}$ **7.** $\sqrt{73}$ **8.** $\sqrt{35}$

(14)

Square Root Algorithm, contd.

The first step in the process is to divide the number into pairs moving from the decimal point to the *left* and then again moving to the *right* in pairs from the decimal point. The digit 0 may be placed to make a pair, if needed.

Example 1: $\sqrt{23.42}$ will be written in pairs

$\sqrt{\underset{\smile}{23}.\underset{\smile}{42}}$

Example 2: $\sqrt{143}$ will be written in pairs

$\sqrt{\underset{\smile}{01}\ \underset{\smile}{43}.}$

Example 3: $\sqrt{127.6}$ will be written in pairs

$\sqrt{\underset{\smile}{01}\ \underset{\smile}{27}.\underset{\smile}{60}}$

Remember, we always begin at the decimal point and move left and then start again from the decimal point and move right.

Study Exercise Four

Write as pairs:

1. $\sqrt{47.22}$ **2.** $\sqrt{130.01}$ **3.** $\sqrt{83.3}$

4. $\sqrt{1,237}$ **5.** $\sqrt{100.762}$ **6.** $\sqrt{8.1462}$

7. $\sqrt{765.00125}$ **8.** $\sqrt{15321.68}$

Square Root Algorithm, contd.

Now let us find $\sqrt{324}$

Step (1) Write as pairs:

$$\sqrt{03\ 24.}$$

Step (2) Consider only the first pair:

$$\begin{array}{r} ?\ \ . \\ ?\ \sqrt{03\ 24.} \end{array}$$

Step (3) 3 is between the perfect squares 1 and 4. So we choose 1 as the first digit in the quotient.

$$\begin{array}{r} 1\ \ \ . \\ 1\ \sqrt{03\ 24.} \\ \\ 1 \\ \hline 2 \end{array}$$

Multiply, subtract, and bring down the next *pair*.

In using the square root algorithm we always bring down pairs.

The previous problem in frame 17 can now be written:

$$\begin{array}{r} 1\ \ \ . \\ 1\sqrt{03\ 24.} \\ \\ 1 \\ \hline \rfloor\ 2\ 24 \end{array}$$

At each step along the problem we must make a new divisor. This new divisor is always *double* the quotient obtained up to that point.

$$\begin{array}{r} 1\ \ \ . \\ 1\sqrt{03\ 24.} \\ \\ 1 \\ \hline 2?\rfloor\ 2\ 24 \end{array}$$

new divisor
2 × 1 = 2

(Frame 18, contd.)

We place the 2 in the divisor but leave a space after 2 because we must again place the same digit here and in the quotient. We must guess this digit so we come as close as possible to 224 or exactly to 224.

```
   1 ? .
1√03 24.

   1
2?| 2 24
```

2? × ? = 224 or a number close to 224 without going over.

```
   1  8.
1√03 24.
```

Guess 7

27 × 7 = 189

```
    1
28| 2 24
    2 24
```

Guess 8

28 × 8 = 224

8 is the digit we want

18

Example: Find $\sqrt{529}$

Solution:

Step (1) Divide in pairs $\sqrt{05\ 29.}$

Step (2) 5 is between the perfect squares 4 and 9.
Thus we place a 2 in quotient and divisor.

```
   2
2√05 29

   4
   1 29
```

Step (3) Multiply, subtract, and bring down a pair.

Step (4) Double the quotient (2 × 2 = 4) and place 4 as the new divisor.

```
   2 ?
2√05 29.

    4
4?| 1 29
```

Step (5) By trial, find the next digit, so that 4? × ? = 129.
The digit is 3.

```
   2  3.
2√05 29.

    4
43| 1 29
    1 29
       0
```

So, $\sqrt{529} = 23$.

Check: 23 × 23 = 529

Study Exercise Five

By the square root algorithm, find the square root of the numbers below and check your answers.

1. $\sqrt{256}$ **2.** $\sqrt{361}$

3. $\sqrt{1156}$ **4.** $\sqrt{676}$

Square Roots of Decimals

Example: $\sqrt{21.16}$

Solution: $\sqrt{\underset{\smile}{21}.\underset{\smile}{16}}$ pair from the decimal point.

$$\begin{array}{r} 4 \\ 4\sqrt{21.16} \\ 16 \\ \hline 5\ 16 \end{array}$$

place 4 in quotient and divisor, multiply, subtract, bring down

$$\begin{array}{r} 4.\ 6 \\ 4\sqrt{21.16} \end{array}$$

double quotient ($2 \times 4 = 8$) and place 8? as new divisor.

$$\begin{array}{r} 16 \\ 86 \,|\, 5\ 16 \\ 5\ 16 \\ \hline \end{array}$$

by trial, find the next digit 6

So $\sqrt{21.16} = 4.6$

Check: $4.6 \times 4.6 = 21.16$

Study Exercise Six

Find:

1. $\sqrt{11.56}$ **2.** $\sqrt{.0729}$

More Difficult Square Root Problems

We will now consider a problem where a remainder occurs.

Example: Find $\sqrt{28.30\ 24}$

Solution:

```
        5  3  2
    5 √28.30 24      pair from the decimal point
       25
 10? |
 103 | 3 30          bring down the pair 30
       3 09          double the quotient (2 × 5 = 10) and place 10? as a new
                     divisor; by trial find the next digit 3.
 106? |
 1062 | 21 24        bring down the next pair 24; double the quotient (2 × 53 =
        21 24        106) and place 106? as a new divisor; by trial find the
                     next digit 2.
```

Now set the decimal point directly above the decimal point in 28.3024.

The square root is 5.32

Check: $5.32 \times 5.32 = 28.3024$

23

Study Exercise Seven

Find the following square roots and check your results.

1. $\sqrt{6.8121}$ **2.** $\sqrt{.0676}$

24

Now let us find $\sqrt{.000441}$

Solution:

```
        0  2  1      First group is 00, so we write 0 in the quotient and move
   2 √.00 04 41      on to the next pair as the first group.

           4
   41 |  0 41
           41
```

Now set the decimal point.

Thus, the square root is .021

Check: $.021 \times .021 = .000441$

25

If a square root does not come out with a zero remainder, it may be calculated to any desired accuracy.

To find a square root accurate to one decimal place, we find the answer to one extra decimal and round off the answer as required.

Example: Find $\sqrt{2.24}$ accurate to 1 decimal.

Solution:

$$\begin{array}{r} 1\quad 4\quad 9 \\ 1\sqrt{\overline{\underset{\smile}{02}.\underset{\smile}{24}\ \underset{\smile}{00}}} \end{array}$$

$$\begin{array}{lcr|r}
 & & & 1 \\
\left(\begin{matrix}2\times 1 = 2\\ \overline{2?}\end{matrix}\right) & \longrightarrow & \underline{24} & 1\ 24 \\
 & & & 96 \\
\left(\begin{matrix}(2\times 14) = 28\\ 28?\end{matrix}\right) & \longrightarrow & \underline{289} & 28\ 00 \\
 & & & 26\ 01 \\
 & & & \overline{1\ 99}
\end{array}$$

1.49 rounds to 1.5 to one decimal.
Thus the square root is 1.5 to 1 decimal.

Example: Find $\sqrt{3}$ accurate to 2 decimals.

Solution:

$$\begin{array}{r} 1\quad 7\quad 3\quad 2 \\ 1\sqrt{\overline{\underset{\smile}{03}.\underset{\smile}{00}\ \underset{\smile}{00}\ \underset{\smile}{00}}} \end{array}$$

$$\begin{array}{lcr|r}
 & & & 1 \\
\left(\begin{matrix}2\times 1 = 2\\ 2?\end{matrix}\right) & \longrightarrow & \underline{27} & 2\ 00 \\
 & & & 1\ 89 \\
\left(\begin{matrix}2\times 17 = 34\\ 34?\end{matrix}\right) & \longrightarrow & \underline{343} & 11\ 00 \\
 & & & 10\ 29 \\
\left(\begin{matrix}2\times 173 = 346\\ 346?\end{matrix}\right) & \longrightarrow & \underline{3462} & 71\ 00 \\
 & & & 69\ 24 \\
 & & & \overline{1\ 76}
\end{array}$$

$\sqrt{3}$ is 1.73 accurate to 2 decimals.

Study Exercise Eight

Find the following square roots accurate to 2 decimals.

1. $\sqrt{.814}$ **2.** $\sqrt{5.2176}$ **3.** $\sqrt{2}$

REVIEW EXERCISES

A. True or False: Determine if the following square roots are correct by multiplication.

Example: $\sqrt{9} = 3$; true, since $3 \times 3 = 9$

1. $\sqrt{.16} = .4$ **2.** $\sqrt{.4} = .2$

3. $\sqrt{.0121} = .11$ **4.** $\sqrt{.256} = 1.6$

B. Between what two perfect squares are the following numbers found.

5. 12 **6.** 66

C. Using the square root algorithm, find the indicated square roots.

7. $\sqrt{1024}$ **8.** $\sqrt{.0729}$

D. Round off the answer to 2 decimals.

9. $\sqrt{5}$ **10.** $\sqrt{.813}$ 30

SOLUTIONS TO REVIEW EXERCISES

A. **1.** true, since $.4 \times .4 = .16$
2. false, since $.2 \times .2 = .04$
3. true, since $.11 \times .11 = .0121$
4. false, since $1.6 \times 1.6 = 2.56$

B. **5.** 12 is between 9 and 16
6. 66 is between 64 and 81

C. **7.**

```
                         3  2.
                       3√10 24.
                          9
(2 × 3 = 6) ——————→ 62 | 1 24
     6?                  1 24
```

8.

```
                         . 2  7
                       2√.07 29
                            4
(2 × 2 = 4) ———→ 47 | 3 29
     4?               3 29
```

D. **9.**

```
                               2  2  3  6
                             2√05.00 00 00
                               4
(2 × 2 = 4) ——————————→ 42 | 1 00
     4?
                               84
(2 × 22 = 44) ———→ 443 | 16 00
     44?                 13 29
(2 × 223 = 446) ——→ 4466 | 2 71 00
     446?                  2 67 96
                              3 04
```

The square root is 2.24 to 2 decimals.

10.

```
                                   9  0  1
                                 9√.81 30 00
                                   81
(2 × 9 = 18) ———→ 180 |  30
     18?                  0
(2 × 90 = 180) ———→ 1801 | 30 00
     180?                  18 01
                           11 99
```

The square root is .90 to 2 decimals. 31

SUPPLEMENTARY PROBLEMS

A. Between which two perfect squares is each of the following numbers found.

1. 8 **2.** 3 **3.** 110

4. 58 **5.** 13 **6.** 83

B. Write as pairs. Example: $\sqrt{1.43} = \sqrt{\underbrace{01}.\underbrace{43}}$

7. $\sqrt{3.141}$ **8.** $\sqrt{4,232}$

9. $\sqrt{23.007}$ **10.** $\sqrt{176.21783}$

C. The following numbers below are all perfect squares. Find the square root and check your answer by multiplying. Use the square root algorithm.

11. $\sqrt{225}$ **12.** $\sqrt{1681}$ **13.** $\sqrt{3.24}$

14. $\sqrt{2809}$ **15.** $\sqrt{.0289}$ **16.** $\sqrt{361}$

17. $\sqrt{21.2521}$ **18.** $\sqrt{9.3636}$

D. Find the square root accurate to two decimal places. Use the square root algorithm.

19. $\sqrt{18.3}$ **20.** $\sqrt{3}$ **21.** $\sqrt{.0762}$

22. $\sqrt{162.4}$ **23.** $\sqrt{2,104}$ **24.** $\sqrt{130}$

25. $\sqrt{3.7}$ **26.** $\sqrt{27}$

SOLUTIONS TO STUDY EXERCISES

Study Exercise One (Frame 6)

1. 9; $9 \times 9 = 81$ **2.** 10; $10 \times 10 = 100$

3. 20; $20 \times 20 = 400$ **4.** 16; $16 \times 16 = 256$

5. 35; $35 \times 35 = 1225$ **6.** .1; $.1 \times .1 = .01$

7. 15; $15 \times 15 = 225$ **8.** 14; $14 \times 14 = 196$

9. .06; $.06 \times .06 = .0036$ **10.** .2; $.2 \times .2 = .04$

11. 1.2; $1.2 \times 1.2 = 1.44$

6A

Study Exercise Two (Frame 9)

1. Between 49 and 64 **2.** Between 16 and 25

3. Between 36 and 49 **4.** Between 16 and 25

5. Between 4 and 9 **6.** Between 81 and 100

Study Exercise Three (Frame 14)

1.
```
   5
5√29
  25
  --
   4
```

2.
```
   6
6√40
  36
  --
   4
```

3.
```
   9
9√85
  81
  --
   4
```

SOLUTIONS TO STUDY EXERCISES, CONTD.

Study Exercise Three (Frame 14, contd.)

4. $\begin{array}{r} 2 \\ 2\sqrt{8} \\ 4 \\ \hline 4 \end{array}$

5. $\begin{array}{r} 10 \\ 10\sqrt{110} \\ 100 \\ \hline 10 \end{array}$

6. $\begin{array}{r} 7 \\ 7\sqrt{63} \\ 49 \\ \hline 14 \end{array}$

7. $\begin{array}{r} 8 \\ 8\sqrt{73} \\ 64 \\ \hline 9 \end{array}$

8. $\begin{array}{r} 5 \\ 5\sqrt{35} \\ 25 \\ \hline 10 \end{array}$

14A

Study Exercise Four (Frame 16)

1. $\sqrt{47.22}$

2. $\sqrt{01\ 30.01}$

3. $\sqrt{83.30}$

4. $\sqrt{1.2\ 37.}$

5. $\sqrt{01\ 00.76\ 20}$

6. $\sqrt{08.14\ 62}$

7. $\sqrt{07\ 65.00\ 12\ 50}$

8. $\sqrt{01\ 53\ 21.68}$

16A

Study Exercise Five (Frame 20)

1.

$\begin{array}{r} 1\ \ 6. \\ 1\sqrt{02\ 56.} \\ 1 \\ 26 \,|\, 1\ 56 \\ 1\ 56 \\ \hline \end{array}$ $\left(\begin{array}{c} 2 \times 1 = 2 \\ 2? \end{array}\right) \rightarrow 26$

Check:

$16 \times 16 = 256$

2.

$\begin{array}{r} 1\ \ 9 \\ 1\sqrt{03\ 61.} \\ 1 \\ 29 \,|\, 2\ 61 \\ 2\ 61 \\ \hline \end{array}$ $\left(\begin{array}{c} 2 \times 1 = 2 \\ 2? \end{array}\right) \rightarrow 29$

Check:

$19 \times 19 = 361$

3.

$\begin{array}{r} 3\ \ 4. \\ 3\sqrt{11\ 56.} \\ 9 \\ 64 \,|\, 2\ 56 \\ 2\ 56 \\ \hline \end{array}$ $\left(\begin{array}{c} 2 \times 3 = 6 \\ 6? \end{array}\right) \rightarrow 64$

Check:

$34 \times 34 = 1156$

4.

$\begin{array}{r} 2\ \ 6. \\ 2\sqrt{06\ 76.} \\ 4 \\ 46 \,|\, 2\ 76 \\ 2\ 76 \\ \hline \end{array}$ $\left(\begin{array}{c} 2 \times 2 = 4 \\ 4? \end{array}\right) \rightarrow 46$

Check:

$26 \times 26 = 676$

20A

SOLUTIONS TO STUDY EXERCISES, CONTD.

Study Exercise Six (Frame 22)

1.

$$3\sqrt{11.56} = 3\ 4$$

$(2 \times 3 = 6;\ 6?) \rightarrow 64 \overline{)\ 2\ 56}$, quotient 9; $2\ 56$

The square root is 3.4.

2.

$$2\sqrt{.07\ 29} = 2\ 7$$

$(2 \times 2 = 4;\ 4?) \rightarrow 47 \overline{)\ 3\ 29}$, quotient 4; $3\ 29$

The square root is .27.

22A

Study Exercise Seven (Frame 24)

1.

$$2\sqrt{06.81\ 21} = 2\ 6\ 1$$

$(2 \times 2 = 4;\ 4?) \rightarrow 46 \overline{)\ 2\ 81}$, quotient 4; $2\ 76$

$(2 \times 26 = 52;\ 52?) \rightarrow 521 \overline{)\ 5\ 21}$; $5\ 21$

The square root is 2.61.

Check:

$2.61 \times 2.61 = 6.8121$

2.

$$2\sqrt{.06\ 76} = 2\ 6$$

$(2 \times 2 = 4;\ 4?) \rightarrow 46 \overline{)\ 2\ 76}$, quotient 4; $2\ 76$

The square root is .26.

Check:

$.26 \times .26 = .0676$

24A

Study Exercise Eight (Frame 29)

1.

$$9\sqrt{.81\ 40\ 00} = 9\ 0\ 2$$

81

$(2 \times 9 = 18;\ 18?) \rightarrow 180 \overline{)\ 40}$; 0

$(2 \times 90 = 180;\ 180?) \rightarrow 1802 \overline{)\ 40\ 00}$; $36\ 04$; remainder $3\ 96$

The square root of .814 accurate to two decimals is .90.

SOLUTIONS TO STUDY EXERCISES, CONTD.

Study Exercise Eight (Frame 29, contd.)

2.

```
                                2  2  8  4
                              2√05.21 76 00

                                 4
(2 × 2 = 4)  ──────────► 42 | 1 21
    4?                          84
(2 × 22 = 44) ─────────► 448 |37 76
    44?                        35 84
(2 × 228 = 456) ───────► 4564 | 1 92 00
    456?                         1 82 56
                                    9 44
```

The square root of 5.2176 accurate to two decimals is 2.28.

3.

```
                                1  4  1  4
                              1√02.00 00 00

                                 1
(2 × 1 = 2)  ──────────► 24 | 1 00
    2?                          96
(2 × 14 = 28) ─────────► 281 | 4 00
    28?                         2 81
(2 × 141 = 282) ───────► 2824 |1 19 00
    282?                        1 12 96
                                   6 04
```

The square root of 2 accurate to two decimals is 1.41.

29A

Practice Test—Decimal Numerals—Units 18–25

1. What is the place value of 3 in 1.203?
2. Add: $2.034 + .2 + 1.76 + .03$
3. Subtract: $7.86 - 2.9871$
4. Multiply: $2.107 \times .23$
5. Divide: $11.076 \div 1.2$
6. Arrange in order of size from the smallest to the largest:
   2.78, .0278, 0.278, 27.8
7. Divide and give the quotient correct to the nearest hundredth:
   $582 \div 39.37$
8. Change to a fraction or mixed number:
   a) $.66\frac{2}{3}$ b) 2.06
9. Change to decimals:
   a) $\frac{5}{8}$ b) $\frac{9}{2,000}$
10. Round to the nearest hundredth: 3.4892
11. Write as a decimal: Six hundred and twenty-five thousandths.
12. Square 1.01
13. Round to two significant digits: .00385
14. Multiply by the short method: $1,000 \times 48.21$
15. Divide by the short method: $37.2 \div 10$
16. Change $\frac{3}{7}$ to a decimal correct to 3 decimal places.
17. What part of 3.2 is 15.12?
18. Find the square root of 3.61.
19. If 3.6 lb. cost 72¢, find the cost of 1 lb.
20. If 4.2 lb. cost 2.4¢, how many lb. can be purchased for 3.5¢?
21. Find $\sqrt{108.2}$ correct to one decimal place.
22. Simplify $\frac{.63}{.09}$
23. Find n if $.2 \times n = 1.2$
24. Find $.03 \times 6 \times .4$
25. What is the number if $.03\frac{1}{2}$ of the number is 2.1?

Answers—Practice Test—Decimal Numerals

1. thousandths

2. 4.024

3. 4.8729

4. .48461

5. 9.23

6. .0278, 0.278, 2.78, 27.8

7. 14.78

8. (a) $\frac{2}{3}$ (b) $2\frac{3}{50}$

9. (a) .625 (b) .0045

10. 3.49

11. 600.025

12. 1.0201

13. .0039

14. 48,210

15. 3.72

16. .429

17. 4.725

18. 1.9

19. 20¢

20. 6.125 lb.

21. 10.4

22. 7

23. 6

24. .072

25. 60

unit

26

Changing Per Cents to Fractions

Objectives:

By the end of this unit you should be able to:

1. explain the meaning of *per cent*.
2. change a per cent to a common fraction.
3. change a per cent to a decimal fraction.

Per Cent

A *per cent* is a fraction in which the numeral preceding the % symbol is the numerator and the % symbol implies a denominator of 100.

Example 1: $5\% = \frac{5}{100}$

Example 2: $286\% = \frac{286}{100}$

Example 3: $6\frac{1}{4}\% = \frac{6\frac{1}{4}}{100}$

Example 4: $4.25\% = \frac{4.25}{100}$

Example 5: $100\% = \frac{100}{100}$

Per cent is not a new kind of number. It is merely a more convenient way of expressing hundredths.

Example 1: "Our bank pays $5\frac{3}{4}\%$ interest on savings accounts."

Example 2: "The carrying charge on this refrigerator is 12%."

Example 3: "The salesman's commission amounts to $33\frac{1}{3}\%$ of the selling price."

Changing Per Cent to Common Fractions

The % symbol is convenient to write and to say but we must *convert* to either a common or decimal fraction in order to use it.

Example 1: $10\% = \frac{10}{100} = \frac{10 \div 10}{100 \div 10} = \frac{1}{10}$

Example 2: $25\% = \frac{25}{100} = \frac{25 \div 25}{100 \div 25} = \frac{1}{4}$

Example 3: $52\% = \frac{52}{100} = \frac{52 \div 4}{100 \div 4} = \frac{13}{25}$

Example 4: $19\% = \frac{19}{100}$ ④

Study Exercise One

Express each of the following per cents as a numeral naming a common fraction in lowest terms:

1. 50% **2.** 75% **3.** 20% **4.** 68% **5.** 59% ⑤

Per Cents Involving Common Fractions

Example 1: $12\frac{1}{2}\% = \frac{12\frac{1}{2}}{100}$

Line (*a*) $= 12\frac{1}{2} \div 100$

Line (*b*) $= \frac{25}{2} \div \frac{100}{1}$

Line (*c*) $= \frac{25}{2} \times \frac{1}{100}$

Line (*d*) $= \frac{\cancel{25}^{1}}{2} \times \frac{1}{\cancel{100}_{4}}$

$= \frac{1}{8}$

Example 2: $33\frac{1}{3}\% = \frac{33\frac{1}{3}}{100}$

Line (*a*) $= 33\frac{1}{3} \div 100$

Line (*b*) $= \frac{100}{3} \div \frac{100}{1}$

Line (*c*) $= \frac{100}{3} \times \frac{1}{100}$

Line (*d*) $= \frac{\cancel{100}^{1}}{3} \times \frac{1}{\cancel{100}_{1}}$

$= \frac{1}{3}$ ⑥

Study Exercise Two

Express each of the following per cents as a numeral naming a common fraction in lowest terms:

1. $16\frac{2}{3}\%$ **2.** $87\frac{1}{2}\%$ **3.** $6\frac{1}{4}\%$ ⑦

Per Cents Greater Than One Hundred

Example 1: $500\% = \frac{500}{100} = 5$

Example 2: $425\% = \frac{425}{100}$

Line (a) $= 4\frac{25}{100}$

Line (b) $= 4\frac{1}{4}$

Example 3: $116\frac{2}{3}\% = \dfrac{116\frac{2}{3}}{100}$

Line (a) $= \dfrac{\frac{350}{3}}{100}$

Line (b) $= \frac{350}{3} \div 100$

Line (c) $= \frac{350}{3} \times \frac{1}{100}$

Line (d) $= \frac{\cancel{350}^{\cancel{35}\,7}}{3} \times \frac{1}{\cancel{100}_{\cancel{10}\,2}} = \frac{7}{6}$

Line (e) $= \frac{7}{6} = 1\frac{1}{6}$

(8)

Study Exercise Three

Express each of the following per cents as mixed numerals or whole numbers:

1. 100%
2. 280%
3. $133\frac{1}{3}\%$
4. $387\frac{1}{2}\%$
5. 150%
6. 800%

(9)

Changing Per Cents to Decimal Fractions

Example 1: $14\% = \frac{14}{100} = .14$

Example 2: $115\% = \frac{115}{100} = 1.15$

Example 3: $215.8\% = \frac{215.8}{100} = 2.158$

Example 4: $3.4\frac{1}{3}\% = \dfrac{3.4\frac{1}{3}}{100} = .034\frac{1}{3}$

(10)

A Short Cut

The % symbol represents *hundredths* or 2 decimal places.
Therefore, to change a per cent to a decimal fraction:
First: *Move the decimal 2 places to the left.*
Then: *Remove the % symbol.*

Example 1: $215.3\% = 2.153$

Example 2: $100\% = 1.00 = 1$

Example 3: $5\% = .05$

Example 4: $3\frac{1}{4}\% = .03\frac{1}{4}$

11

Per Cents Less Than 1 %

Example 1: $\frac{1}{2}\% = \frac{\frac{1}{2}}{100} = .00\frac{1}{2}$

Example 2: $\frac{1}{2}\% = .5\% = .005$

Therefore $.00\frac{1}{2} = .005 = \frac{1}{2}\%$

Example 3: $\frac{1}{4}\% = \frac{\frac{1}{4}}{100} = .00\frac{1}{4}$

Example 4: $\frac{1}{4}\% = .25\% = .0025$

Therefore $.00\frac{1}{4} = .0025 = \frac{1}{4}\%$

12

Study Exercise Four

Express the following per cents as decimal numerals:

1. 24.7% **2.** 4.7% **3.** 148.63% **4.** $.3\%$

5. $.475\%$ **6.** 102% **7.** $4\frac{3}{4}\%$ **8.** $37\frac{1}{2}\%$

9. $\frac{1}{8}\%$ **10.** $\frac{1}{3}\%$

REVIEW EXERCISES

A. **1.** Explain the meaning of *per cent*.

B. **2.** The % symbol acts like a denominator of ____________.

C. Change the following per cents to numerals naming a common fraction in lowest terms:

3. 30% **4.** $87\frac{1}{2}\%$ **5.** $37\frac{1}{2}\%$ **6.** 80%

D. Change the following per cents to mixed numerals or whole numbers:

7. 100% **8.** 225% **9.** $166\frac{2}{3}\%$ **10.** 500%

E. Change the following per cents to decimal numerals:

11. 23.6% **12.** $\frac{1}{7}\%$ **13.** $5\frac{3}{4}\%$ **14.** 25%

(14)

SOLUTIONS TO REVIEW EXERCISES

A. **1.** See Frame 2

B. **2.** 100

C. **3.** $30\% = \frac{30 \div 10}{100 \div 10} = \frac{3}{10}$

4. $87\frac{1}{2}\% = \frac{87\frac{1}{2}}{100} = \frac{\frac{175}{2}}{100} = \frac{175}{2} \div 100 = \frac{\overset{7}{\cancel{175}}}{2} \times \frac{1}{\underset{4}{\cancel{100}}} = \frac{7}{8}$

5. $37\frac{1}{2}\% = \frac{37\frac{1}{2}}{100} = \frac{\frac{75}{2}}{100} = \frac{75}{2} \div 100 = \frac{\overset{3}{\cancel{75}}}{2} \times \frac{1}{\underset{4}{\cancel{100}}} = \frac{3}{8}$

6. $80\% = \frac{80 \div 10}{100 \div 10} = \frac{8 \div 2}{10 \div 2} = \frac{4}{5}$

D. **7.** $100\% = \frac{100}{100} = 1$

8. $225\% = \frac{225}{100} = 2\frac{25}{100} = 2\frac{25 \div 25}{100 \div 25} = 2\frac{1}{4}$

9. $166\frac{2}{3}\% = \frac{166\frac{2}{3}}{100} = \frac{\frac{500}{3}}{100} = \frac{500}{3} \div 100 = \frac{\overset{5}{\cancel{500}}}{3} \times \frac{1}{\underset{1}{\cancel{100}}} = \frac{5}{3} = 1\frac{2}{3}$

10. $500\% = \frac{500}{100} = 5$

E. **11.** $23.6\% = .236$

12. $\frac{1}{7}\% = .00\frac{1}{7}$

13. $5\frac{3}{4}\% = .05\frac{3}{4}$

14. $25\% = .25$

(15)

SUPPLEMENTARY PROBLEMS

A. Change the following per cents to decimal fractions:

1. 6%	**2.** $83\frac{1}{3}\%$	**3.** $5\frac{3}{4}\%$
4. 225%	**5.** 120%	**6.** $.00\frac{1}{8}\%$
7. 175%	**8.** $12\frac{1}{2}\%$	**9.** $6\frac{1}{4}\%$
10. 3.6%	**11.** 22%	**12.** $16\frac{2}{3}\%$
13. 56%	**14.** .5%	**15.** .25%

B. Change the per cents in **A** above to common fractions.

C. When changing a per cent to a fraction, the numeral preceding the % symbol becomes the ____________ while the % symbol becomes the ____________, 100.

SOLUTIONS TO STUDY EXERCISES

Study Exercise One (Frame 5)

1. $50\% = \dfrac{50 \div 50}{100 \div 50} = \dfrac{1}{2}$

2. $75\% = \dfrac{75 \div 25}{100 \div 25} = \dfrac{3}{4}$

3. $20\% = \dfrac{20 \div 20}{100 \div 20} = \dfrac{1}{5}$

4. $68\% = \dfrac{68 \div 4}{100 \div 4} = \dfrac{17}{25}$

5. $59\% = \dfrac{59}{100}$

Study Exercise Two (Frame 7)

1. $16\frac{2}{3}\% = \dfrac{16\frac{2}{3}}{100} = \dfrac{\frac{50}{3}}{100} = \dfrac{50}{3} \div 100 = \dfrac{\overset{1}{\cancel{50}}}{3} \times \dfrac{1}{\underset{2}{\cancel{100}}} = \dfrac{1}{6}$

2. $87\frac{1}{2}\% = \dfrac{87\frac{1}{2}}{100} = \dfrac{\frac{175}{2}}{100} = \dfrac{175}{2} \div 100 = \dfrac{\overset{\overset{7}{\cancel{35}}}{\cancel{175}}}{2} \times \dfrac{1}{\underset{\underset{4}{\cancel{20}}}{\cancel{100}}} = \dfrac{7}{8}$

3. $6\frac{1}{4}\% = \dfrac{6\frac{1}{4}}{100} = \dfrac{\frac{25}{4}}{100} = \dfrac{25}{4} \div 100 = \dfrac{\overset{1}{\cancel{25}}}{4} \times \dfrac{1}{\underset{4}{\cancel{100}}} = \dfrac{1}{16}$

SOLUTIONS TO STUDY EXERCISES, CONTD.

Study Exercise Three (Frame 9)

1. $100\% = \dfrac{100}{100} = 1$

2. $280\% = \dfrac{280}{100} = 2\dfrac{80 \div 10}{100 \div 10} = 2\dfrac{8 \div 2}{10 \div 2} = 2\dfrac{4}{5}$

3. $133\frac{1}{3}\% = \dfrac{133\frac{1}{3}}{100} = \dfrac{\frac{400}{3}}{100} = \dfrac{400}{3} \div 100 = \dfrac{\overset{4}{\cancel{400}}}{3} \times \dfrac{1}{\underset{1}{\cancel{100}}} = \dfrac{4}{3} = 1\dfrac{1}{3}$

4. $387\frac{1}{2}\% = \dfrac{387\frac{1}{2}}{100} = \dfrac{\frac{775}{2}}{100} = \dfrac{775}{2} \div 100 = \dfrac{\overset{\overset{31}{\cancel{155}}}{\cancel{775}}}{2} \times \dfrac{1}{\underset{\underset{4}{\cancel{20}}}{\cancel{100}}} = \dfrac{31}{8} = 3\dfrac{7}{8}$

5. $150\% = \dfrac{150 \div 10}{100 \div 10} = \dfrac{15 \div 5}{10 \div 5} = \dfrac{3}{2} = 1\dfrac{1}{2}$

6. $800\% = \dfrac{800}{100} = 8$

9A

Study Exercise Four (Frame 13)

1. .247 **2.** .047 **3.** 1.4863 **4.** .003

5. .00475 **6.** 1.02 **7.** $.04\frac{3}{4}$ **8.** $.37\frac{1}{2}$

9. $.00\frac{1}{8}$ **10.** $.00\frac{1}{3}$

13A

---

unit

27

Changing Fractions to Per Cents

Objectives:

By the end of this unit you should be able to:

1. change a decimal fraction to a per cent.
2. change a common fraction to a per cent.
3. convert some fractions to per cents and some per cents to fractions from memory.

(1)

Converting Decimal Fractions to Per Cents

Since we know the % symbol represents *hundredths* (two decimal places) we can use the % symbol instead of two decimal places. Therefore, to express a decimal fraction as a per cent do the following:

First: Move the decimal 2 places to the *right*.
Then: Write the % symbol.

Example 1: .63 = 63%

Example 2: 4.2 = 420%

Example 3: .0003 = .03%

Example 4: .2 = 20%

Example 5: 8 = 800%

(2)

Study Exercise One

Express each of the following decimals as a per cent:

1. .35

2. .03

3. 5

4. $.33\frac{1}{3}$

5. $3.16\frac{2}{3}$

6. $.00\frac{1}{2}$

(3)

Common Fractions as Per Cents

The % symbol represents the denominator 100. Therefore, to change a fraction with a denominator of 100 to a per cent, remove the denominator 100 and write the % symbol.

Example 1: $\frac{25}{100} = 25\%$

Example 2: $\frac{42}{100} = 42\%$

Example 3: $\frac{387}{100} = 387\%$

Example 4: $\frac{100}{100} = 100\%$

(4)

Some denominators can easily be changed to 100.

Example 1: $\frac{3}{20} = \frac{3 \times 5}{20 \times 5} = \frac{15}{100} = 15\%$

Example 2: $\frac{7}{10} = \frac{7 \times 10}{10 \times 10} = \frac{70}{100} = 70\%$

Example 3: $\frac{8}{25} = \frac{8 \times 4}{25 \times 4} = \frac{32}{100} = 32\%$

Example 4: $\frac{15}{300} = \frac{15 \div 3}{300 \div 3} = \frac{5}{100} = 5\%$

Example 5: $\frac{3}{200} = \frac{3 \div 2}{200 \div 2} = \frac{1\frac{1}{2}}{100} = 1\frac{1}{2}\%$

(5)

When the denominator can not be changed easily to 100, we convert the common fraction to a two-place decimal fraction and proceed as follows:

Example 1: $\frac{3}{8}$ may be changed to a two-place decimal.

$$8\,\overline{)3.00} = .37\frac{1}{2}$$

Line (a) $8\,\overline{)3.00}$ (quotient $.37\frac{1}{2}$)

Line (b) $.37\frac{1}{2} = 37\frac{1}{2}\%$

Example 2: $\frac{43}{7}$ may be changed to a two-place decimal.

Line (a) $7\,\overline{)43.00}$ (quotient $6.14\frac{2}{7}$)

Line (b) $6.14\frac{2}{7} = 614\frac{2}{7}\%$

(6)

Study Exercise Two

Change the following common fractions to per cents:

1. $\frac{7}{100}$ **2.** $\frac{82}{100}$ **3.** $\frac{13}{20}$ **4.** $\frac{124}{200}$

5. $\frac{17}{300}$ **6.** $\frac{5}{6}$ **7.** $2\frac{1}{8}$

Useful Equivalents

Per Cent	Common Fraction	Decimal Fraction
10%	$\frac{1}{10}$	.1
20%	$\frac{1}{5}$	.2
30%	$\frac{3}{10}$	.3
40%	$\frac{2}{5}$	.4
50%	$\frac{1}{2}$	.5
60%	$\frac{3}{5}$	.6
70%	$\frac{7}{10}$	.7
80%	$\frac{4}{5}$	.8
90%	$\frac{9}{10}$	.9
100%	1	1.0
25%	$\frac{1}{4}$	.25
75%	$\frac{3}{4}$	.75
$33\frac{1}{3}\%$	$\frac{1}{3}$	$.33\frac{1}{3}$
$66\frac{2}{3}\%$	$\frac{2}{3}$	$.66\frac{2}{3}$

(8)

Study Exercise Three

This exercise will be given to you on the tape.

(9)

REVIEW EXERCISES

A. Express each of the following decimals as a per cent:

1. .57 **2.** .08 **3.** 6 **4.** $.16\frac{2}{3}$ **5.** $.00\frac{3}{4}$

B. Express each of the following common fractions as a per cent:

6. $\frac{9}{100}$ **7.** $\frac{17}{25}$ **8.** $\frac{14}{300}$ **9.** $\frac{6}{7}$ **10.** $2\frac{1}{2}$

C. Complete the following chart from memory:

Per Cent	Common Fraction	Decimal Fraction
25%		
	$\frac{3}{4}$	
		$.33\frac{1}{3}$
	$\frac{2}{5}$	
$66\frac{2}{3}\%$		
	$\frac{7}{10}$	
30%		
	$\frac{1}{2}$	

(10)

SOLUTIONS TO REVIEW EXERCISES

A. **1.** $.57 = 57\%$ **2.** $.08 = 8\%$

3. $6 = 600\%$ **4.** $.16\frac{2}{3} = 16\frac{2}{3}\%$

5. $.00\frac{3}{4} = \frac{3}{4}\%$

B. **6.** $\frac{9}{100} = 9\%$ **7.** $\frac{17}{25} = \frac{17 \times 4}{25 \times 4} = \frac{68}{100} = 68\%$

8. $\frac{14}{300} = \frac{14 \div 3}{300 \div 3} = \frac{4\frac{2}{3}}{100} = 4\frac{2}{3}\%$ **9.** $\frac{6}{7} = 7\overline{)6.00}$ with quotient $.85\frac{5}{7}$ $= 85\frac{5}{7}\%$

10. $2\frac{1}{2} = \frac{5}{2} = 2\overline{)5.00}$ with quotient 2.50 $= 250\%$

SOLUTIONS TO REVIEW EXERCISES, CONTD.

(Frame 11, contd.)

C.

Per Cent	Common Fraction	Decimal Fraction
25%	$\frac{1}{4}$	.25
75%	$\frac{3}{4}$	.75
$33\frac{1}{3}\%$	$\frac{1}{3}$	$.33\frac{1}{3}$
40%	$\frac{2}{5}$	.4
$66\frac{2}{3}\%$	$\frac{2}{3}$	$.66\frac{2}{3}$
70%	$\frac{7}{10}$	.7
30%	$\frac{3}{10}$	.3
50%	$\frac{1}{2}$	.5

(11)

SUPPLEMENTARY PROBLEMS

A. Change the following numerals to per cents:

1. $\frac{2}{5}$ **2.** $\frac{3}{16}$ **3.** .8327 **4.** $\frac{1}{8}$ **5.** 2.25

6. $.004\frac{1}{4}$ **7.** $\frac{19}{300}$ **8.** $\frac{1}{16}$ **9.** 4.75 **10.** $.04\frac{3}{4}$

11. $\frac{17}{400}$ **12.** $\frac{14}{20}$ **13.** $\frac{6}{25}$ **14.** $\frac{8.5}{10}$ **15.** $8\frac{1}{3}$

SUPPLEMENTARY PROBLEMS, CONTD.

(Frame 12, contd.)

B. Complete the following chart from memory:

	Per Cent	Common Fraction
16.	20%	
17.	25%	
18.	30%	
19.	$33\frac{1}{3}\%$	
20.	10%	
21.	75%	
22.	40%	
23.	$66\frac{2}{3}\%$	
24.	80%	
25.	50%	
26.	70%	
27.	90%	

(12)

SOLUTIONS TO STUDY EXERCISES

Study Exercise One (Frame 3)

1. $.35 = 35\%$ **2.** $.03 = 3\%$ **3.** $5 = 500\%$

4. $.33\frac{1}{3} = 33\frac{1}{3}\%$ **5.** $3.16\frac{2}{3} = 316\frac{2}{3}\%$ **6.** $.00\frac{1}{2} = \frac{1}{2}\%$ (3A)

Study Exercise Two (Frame 7)

1. $\frac{7}{100} = 7\%$ **2.** $\frac{82}{100} = 82\%$

3. $\frac{13}{20} = \frac{13 \times 5}{20 \times 5} = \frac{65}{100} = 65\%$ **4.** $\frac{124}{200} = \frac{124 \div 2}{200 \div 2} = \frac{62}{100} = 62\%$

5. $\frac{17}{300} = \frac{17 \div 3}{300 \div 3} = \frac{5\frac{2}{3}}{100} = 5\frac{2}{3}\%$ **6.** $\frac{5}{6} = 6\overline{)5.00}$ with quotient $.83\frac{2}{6}$; $= 83\frac{1}{3}\%$

7. $2\frac{1}{8} = \frac{17}{8} = 8\overline{)17.00}$ with quotient $2.12\frac{4}{8}$; $= 2.12\frac{1}{2} = 212\frac{1}{2}\%$ (7A)

Study Exercise Three (Frame 9)

Answers Given on Tape (9A)

unit

28

The Three Types of Per Cent Problems

Objectives:

By the end of this unit you should be able to find:

1. a per cent of a number.
2. what per cent one number is of another.
3. the number when a per cent of it is known.

(1)

Finding a Per Cent of a Number

Example 1: Find 14% of 83:

Line (a) 14% of $83 = n$
Line (b) $14\% = .14$
Line (c) $.14 \times 83 = n$
Line (d) $11.62 = n$
Line (e) $n = 11.62$ *answer:* 11.62

(2)

Example 2: Find 75% of 792:

Line (a) 75% of $792 = n$
Line (b) $75\% = \frac{3}{4}$
Line (c) $\frac{3}{4} \times 792 = n$
Line (d) $594 = n$
Line (e) $n = 594$ *answer:* 594

(3)

Study Exercise One

Find:

1. 42% of 57	**2.** 4% of 30	**3.** 10% of 40
4. 1% of 3672	**5.** 7% of 6.3	**6.** $5\frac{3}{4}\%$ of 896
7. 6.2% of 50	**8.** 120% of 16	**9.** $116\frac{2}{3}\%$ of 80

(4)

Finding a Per Cent of an Amount of Money

Example 1: Find 3% of \$7.28

Line (a) $.03 \times \$7.28 = n$

Line (b) $\$.2184 = n$

Line (c) $n = \$.2184$ *answer:* \$.22

Example 2: Find $33\frac{1}{3}\%$ of \$142

Line (a) $\frac{1}{3} \times \$142 = n$

Line (b) $\$47.33\frac{1}{3} = n$

Line (c) $n = \$47.33\frac{1}{3}$ *answer:* \$47.33

(5)

Study Exercise Two

Find each of the following to the nearest cent:

1. 36% of \$70 **2.** 1% of \$6 **3.** 6% of \$98.42

4. 125% of \$600 **5.** $66\frac{2}{3}\%$ of \$70 **6.** $\frac{1}{4}\%$ of \$9,267

(6)

Finding What Per Cent One Number is of Another

Example 1: What fractional part of 8 is 6?

Line (a) $\frac{6}{8} = \frac{6 \div 2}{8 \div 2} = \frac{3}{4}$ *answer:* $\frac{3}{4}$

Example 2: What decimal part of 8 is 6?

Line (a) $\frac{6}{8} = \frac{3}{4} = 4\overline{)3.00}$ (quotient .75) *answer:* .75

Example 3: What per cent of 8 is 6?

Line (a) $\frac{3}{4} = 75\%$

Line (b) $.75 = 75\%$ *answer:* 75%

(7)

To find what per cent one number is of another, we find what common fractional part one number is of the other expressed in lowest terms. Then we change this fraction to a per cent, if we know it; otherwise, we change the fraction to a two-place decimal, then to a per cent.

(8)

Common Fraction to a Per Cent

Example 1: What per cent of 99 is 66?

Line (a) $\frac{66}{99} = \frac{66 \div 11}{99 \div 11}$

Line (b) $\frac{6}{9} = \frac{6 \div 3}{9 \div 3}$

Line (c) $\frac{2}{3} = 66\frac{2}{3}\%$ *answer:* $66\frac{2}{3}\%$

Decimal Fraction to a Per Cent

Example 1: What per cent of 86 is 16?

Line (a) $\frac{16}{86} = \frac{16 \div 2}{86 \div 2}$

Line (b) $= \frac{8}{43}$

Line (c)

$$\begin{array}{r} .18\frac{26}{43} \\ 43\overline{)8.00} \\ 4\ 3 \\ \hline 3\ 70 \\ 3\ 44 \\ \hline 26 \end{array}$$

Line (d) $.18\frac{26}{43} = 18\frac{26}{43}\%$ *answer:* $18\frac{26}{43}\%$

Some Denominators That Are Easily Changed to 100

Example 1: What per cent of 25 is 7?

Line (a) $\frac{7}{25} = \frac{7 \times 4}{25 \times 4} = \frac{28}{100} = 28\%$

Example 2: What per cent of 5 is 3?

Line (a) $\frac{3}{5} = \frac{3 \times 20}{5 \times 20} = \frac{60}{100} = 60\%$

Example 3: What per cent of 300 is 27?

Line (a) $\frac{27}{300} = \frac{27 \div 3}{300 \div 3} = \frac{9}{100} = 9\%$

An Improper Fraction is Equal to 100% or More

Example 1: What per cent of 5 is 9? (Using Common Fractions)

Line (a) $\frac{9}{5} = 1\frac{4}{5}$

Line (b) $1 + \frac{4}{5}$

Line (c) 100% + 80%

Line (d) 180%

Example 2: What per cent of 5 is 9? (Using a Decimal Fraction)

Line (a) $\frac{9}{5} = 5\overline{)9.00}$ = 1.80

Line (b) 1.80 = 180%

(12)

Study Exercise Three

1. What per cent of 6 is 2?
2. What per cent of 100 is 41?
3. What per cent of 20 is 17?
4. What per cent of 49 is 13?
5. What per cent of 25 is 500?
6. What per cent of 500 is 25?
7. What per cent of 144 is 96?
8. What per cent of 14 is 30?

(13)

Expressing a Per Cent as an Approximate Decimal

Example 1: What per cent of 7 is 1?

Line (a) $\frac{1}{7} = 7\overline{)1}$

Line (b) $7\overline{)1.0000000000000}$ = .142857142857...

When Do We Stop?

Rounding Per Cents

Division	Per Cent	Nearest Whole Per Cent	Nearest Tenth Per Cent	Nearest Hundredth Per Cent	Nearest Thousandth Per Cent
$\overset{.142}{7\overline{)1.000}}$	14.2%	14%			
$\overset{.1428}{7\overline{)1.0000}}$	14.28%		14.3%		
$\overset{.14285}{7\overline{)1.00000}}$	14.285%			14.29%	
$\overset{.142857}{7\overline{)1.000000}}$	14.2857%				14.286%

Study Exercise Four

A. Round to the nearest whole per cent:
1. What per cent of 3 is 1?
2. What per cent of 28 is 16?

B. Round to the nearest tenth of a per cent:
3. What per cent of 17 is 4?
4. What per cent of 46 is 2?

C. Round to the nearest hundredth of a per cent:
5. What per cent of 46 is 2?

D. Round to the nearest thousandth of a per cent:
6. What per cent of 46 is 2?

Find the Number When a Per Cent of It Is Known

Example 1: 75% of what number is 6?

Line (a) 75% of $n = 6$

Line (b) $\frac{3}{4} \times n = 6$

Line (c) $n = 6 \div \frac{3}{4}$

Line (d) $n = 6 \times \frac{4}{3}$

Line (e) $n = 8$ *Answer:* 8

Check: Does 75% of 8 = 6?

$\frac{3}{\cancel{4}_{1}} \times \overset{2}{\cancel{8}} = 6$ Yes

Example 1: 42% of what number is 12?

Line (a) 42% of $n = 12$
Line (b) $.42 \times n = 12$
Line (c) $n = 12 \div .42$
Line (d) $n = 28\frac{4}{7}$ *Answer:* $28\frac{4}{7}$

(18)

Study Exercise Five

Find each of the following:

1. $16\frac{2}{3}\%$ of what number is 38?
2. 32% of what number is 1.24? (To nearest tenth)
3. 125% of what number is 15?
4. 70% of what number is 49?
5. $5\frac{3}{4}\%$ of what number is \$649? (To nearest dollar)

REVIEW EXERCISES

A. Find:

1. 36% of 20
2. 75% of 128
3. 9% of $384
4. 16% of 36
5. 92% of 1,842
6. 15% of $34.25 (To nearest cent)
7. 23% of 1.5
8. 20% of 35

B. Find:

9. What per cent of 6 is 1?
10. What per cent of 200 is 46?
11. What per cent of 15 is 30?
12. What per cent of 60 is 20?

C. Find each of the following:

13. $12\frac{1}{2}\%$ of what number is 32?
14. 10% of what number is 430?
15. 25% of what number is 60?
16. $137\frac{1}{2}\%$ of what number is 22?
17. $\frac{1}{2}\%$ of what number is $5?
18. What per cent of 41 is 9? (To the nearest hundredth of a per cent) 20

SOLUTIONS TO REVIEW EXERCISES

A.

1. 36% of 20 = .36 × 20 = 7.2

2. 75% of 128 = $\frac{3}{\cancel{4}_{1}} \times \cancel{128}^{32} = 96$

3. 9% of $384 = .09 × $384 = $34.56

4. 16% of 36 = .16 × 36 = 5.76

5. 92% of 1842 = .92 × 1842 = 1694.64

6. 15% of $34.25 = .15 × $34.25 = $5.14

7. 23% of 1.5 = .23 × 1.5 = .345

8. 20% of 35 = $\frac{1}{\cancel{5}_{1}} \times \cancel{35}^{7} = 7$

B.

9. $\frac{1}{6} = 6\overline{)1.00}$ (quotient $.16\frac{2}{3}$) $= 16\frac{2}{3}\%$

10. $\frac{46}{200} = \frac{46 \div 2}{200 \div 2} = \frac{23}{100} = 23\%$

11. $\frac{30}{15} = 2 = 200\%$

12. $\frac{20}{60} = \frac{20 \div 20}{60 \div 20} = \frac{1}{3} = 33\frac{1}{3}\%$

SOLUTIONS TO REVIEW EXERCISES, CONTD.

(Frame 21, contd.)

C. **13.** $12\frac{1}{2}\%$ of $n = 32$

$$\frac{12\frac{1}{2}}{100} \times n = 32$$

$$\frac{\frac{25}{2}}{100} \times n = 32$$

$$\frac{\overset{1}{\cancel{25}}}{2} \times \frac{1}{\underset{4}{\cancel{100}}} \times n = 32$$

$$\frac{1}{8} \times n = 32$$

$$n = 32 \div \frac{1}{8} = 32 \times \frac{8}{1} = 256$$

14. 10% of $n = 430$

$$\frac{1}{10} \times n = 430$$

$$n = 430 \div \frac{1}{10} = 430 \times \frac{10}{1} = 4300$$

15. 25% of $n = 60$

$$\frac{1}{4} \times n = 60$$

$$n = 60 \div \frac{1}{4} = 60 \times \frac{4}{1} = 240$$

16. $$137\frac{1}{2}\% = \frac{137\frac{1}{2}}{100} = \frac{\frac{275}{2}}{100} = \frac{275}{2} \div 100 = \frac{\overset{11}{\cancel{275}}}{2} \times \frac{1}{\underset{4}{\cancel{100}}} = \frac{11}{8}$$

$$\frac{11}{8} \times n = 22$$

$$n = 22 \div \frac{11}{8} = \overset{2}{\cancel{22}} \times \frac{8}{\underset{1}{\cancel{11}}} = 16$$

17. $\frac{1}{2}\% = .5\% = .005$

$$.005 \times n = \$5$$

$$n = \$5 \div .005 = 5\overline{)\$5000} = \$1000$$

18. $$\frac{9}{41} = 41\overline{)9.00000} = 21.95\%$$ (quotient $.21951$)

(21)

SUPPLEMENTARY PROBLEMS

A. Find:

1. 10% of 60	**2.** $66\frac{2}{3}\%$ of 27	**3.** 120% of 160
4. 13% of 16	**5.** $112\frac{1}{2}\%$ of 48	**6.** $206\frac{1}{4}\%$ of 16
7. 90% of .63	**8.** 150% of 1.8	**9.** 275% of 1.28

SUPPLEMENTARY PROBLEMS, CONTD.

(Frame 22, contd.)

B. Find:

10. What per cent of 12 is 8?

11. What per cent of 50 is 60?

12. What per cent of 300 is 45?

13. What per cent of 45 is 300?

14. What per cent of 200 is 38?

15. What per cent of .27 is 8.1?

C. Find each of the following:

16. 8% of what number is 32?

17. 3.6% of what number is 40?

18. 300% of what number is 150?

19. 1% of what number is 68?

20. $33\frac{1}{3}\%$ of what number is 4.3?

21. 6.75% of what number is $480?

22. What per cent of 23 is 3? (To nearest hundredth of a per cent)

SOLUTIONS TO STUDY EXERCISES

Study Exercise One (Frame 4)

1. 42% of $57 = .42 \times 57 = 23.94$

2. 4% of $30 = .04 \times 30 = 1.2$

3. 10% of $40 = \frac{1}{10} \times 40 = 4$

4. 1% of $3672 = .01 \times 3672 = 36.72$

5. 7% of $6.3 = .07 \times 6.3 = .441$

6. $5\frac{3}{4}\%$ of $896 = 5.75\% \times 896 = .0575 \times 896 = 51.52$

7. 6.2% of $50 = .062 \times 50 = 3.1$

8. 120% of $16 = 1.2 \times 16 = 19.2$

9. $116\frac{2}{3}\%$ of $80 = \frac{7}{6} \times 80 = \frac{7}{\cancel{6}_{3}} \times \cancel{80}^{40} = \frac{280}{3} = 93\frac{1}{3}$

4A

Study Exercise Two (Frame 6)

1. 36% of $\$70 = .36 \times \$70 = \$25.20$

2. 1% of $\$6 = .01 \times \$6 = \$.06$

3. 6% of $\$98.42 = .06 \times \$98.42 = \$5.91$

4. 125% of $\$600 = \frac{5}{4} \times \$600 = \frac{5}{\cancel{4}_{1}} \times \cancel{\$600}^{150} = \$750$

5. $66\frac{2}{3}\%$ of $\$70 = \frac{2}{3} \times \$70 = \frac{\$140}{3} = \$46.67$

6. $\frac{1}{4}\%$ of $\$9,267 = .0025 \times \$9,267 = \$23.17$

6A

SOLUTIONS TO STUDY EXERCISES, CONTD.

Study Exercise Three (Frame 13)

1. $\frac{2}{6} = \frac{1}{3} = 33\frac{1}{3}\%$

2. $\frac{41}{100} = 41\%$

3. $\frac{17}{20} = \frac{17 \times 5}{20 \times 5} = \frac{85}{100} = 85\%$

4. $\frac{13}{49} = 49\,\overline{)13.00}$ with quotient $.26\frac{26}{49}$ $= 26\frac{26}{49}\%$

5. $\frac{500}{25} = 20 = 2000\%$

6. $\frac{25}{500} = \frac{25 \div 5}{500 \div 5} = \frac{5}{100} = 5\%$

7. $\frac{96}{144} = \frac{96 \div 12}{144 \div 12} = \frac{8 \div 4}{12 \div 4} = \frac{2}{3} = 66\frac{2}{3}\%$

8. $\frac{30}{14} = \frac{30 \div 2}{14 \div 2} = \frac{15}{7} = 7\,\overline{)15.00}$ with quotient $2.14\frac{2}{7}$ $= 214\frac{2}{7}\%$

13A

Study Exercise Four (Frame 16)

A. 1. $\frac{1}{3} = 3\,\overline{)1.000}$ with quotient $.333$ $\approx 33\%$

2. $\frac{16}{28} = \frac{16 \div 4}{28 \div 4} = \frac{4}{7} = 7\,\overline{)4.000}$ with quotient $.571$ $\approx 57\%$

B. 3. $\frac{4}{17} = 17\,\overline{)4.0000}$ with quotient $.2352$ $\approx 23.5\%$

4. $\frac{2}{46} = \frac{2 \div 2}{46 \div 2} = \frac{1}{23} = 23\,\overline{)1.0000}$ with quotient $.0434$ $\approx 4.3\%$

C. 5. $\frac{2}{46} = \frac{1}{23} = 23\,\overline{)1.00000}$ with quotient $.04347$ $\approx 4.35\%$

D. 6. $\frac{2}{46} = \frac{1}{23} = 23\,\overline{)1.000000}$ with quotient $.043478$ $\approx 4.348\%$

16A

Study Exercise Five (Frame 19)

1. $16\frac{2}{3}\% = \frac{16\frac{2}{3}}{100} = \frac{\frac{50}{3}}{100} = \frac{50}{3} \div 100 = \frac{\overset{1}{\cancel{50}}}{3} \times \frac{1}{\underset{2}{\cancel{100}}} = \frac{1}{6}$

$\frac{1}{6} \times n = 38$

$n = 38 \div \frac{1}{6} = 38 \times \frac{6}{1} = 228$

2. 32% of $n = 1.24$

$.32 \times n = 1.24$

$n = 1.24 \div .32 = 32\,\overline{)124.00}$ with quotient 3.87 ≈ 3.9

SOLUTIONS TO STUDY EXERCISES, CONTD.

Study Exercise Five (Frame 19, contd.)

3. $125\% = 100\% + 25\% = 1 + \frac{1}{4} = \frac{5}{4}$

$\frac{5}{4} \times n = 15$

$n = 15 \div \frac{5}{4} = \overset{3}{\cancel{15}} \times \frac{4}{\underset{1}{\cancel{5}}} = 12$

4. 70% of $n = 49$

$\frac{7}{10} \times n = 49$

$n = 49 \div \frac{7}{10} = \overset{7}{\cancel{49}} \times \frac{10}{\underset{1}{\cancel{7}}} = 70$

5. $5\frac{3}{4}\% = 5.75\% = .0575$

$.0575 \times n = \$649$

$n = \$649 \div .0575 = 575\overline{)\$6490000.0}$ (quotient: $\$\ 11286.9$) $\approx \$11{,}287$ 19

unit

29

Problems Involving Per Cent—Discount

Objectives:

By the end of this unit you should:

1. know the meaning of *list price*, *net price*, *discount*, and *rate of discount*.
2. be able to solve problems involving discounts.

①

Discount

Example 1: A desk originally sold for \$100 (this is called the gross or list price.). It now is on sale for \$80 (this is called the net price).

1. How much was the discount?
2. What is the rate of discount?

Solution, *Part 1*

$$\$100 - \$80 = \$20$$

List Price − *Net Price* = *Discount*

The Discount Fraction

The rate of discount is established by the fraction whose numerator is the discount and whose denominator is the list or gross price.

$$\frac{\textbf{DISCOUNT}}{\textbf{LIST PRICE (Gross Price)}}$$

Solution: *Part 2*

$$\frac{\text{Discount}}{\text{List Price}} = \frac{\$20}{\$100} = 20\%$$

Study Exercise One

Find the missing information to the nearest whole per cent:

	List Price	Net Price	Discount	Rate of Discount
1.	$32.99	$28.99		
2.	$120	$82		
3.	$3,450	$3,000		
4.	$500	$475		
5.	$89	$79		

(5)

Finding the Discount

Example 1: A store advertises 20% off. A customer wishes to buy a chair listed at $125. What will the discount be?

Line (a) 20% of $125 = discount

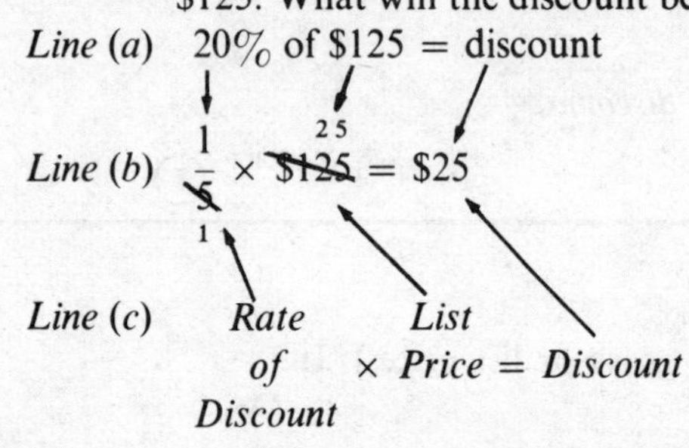

Line (b) $\frac{1}{5} \times \$125 = \$25$
Line (c) Rate of Discount × List Price = Discount

(6)

Study Exercise Two

Find the missing information to the nearest cent:

	List Price	Rate of Discount	Discount
1.	$827	20%	
2.	$46.88	10%	
3.	$32	12%	
4.	$2,781	$33\frac{1}{3}\%$	
5.	$78.62	35%	

(7)

Finding the Net Price

Example 1: A merchant wishes to give his customers a 10% discount on a $290 television set. What should the new price be?

Line (a) 10% of $290 = discount

Line (b) $\frac{1}{10} \times \$290 = \$29$ discount

Line (c) $290 − $29 = $261

List Price − *Discount* = *Net Price*

(8)

A Second Way of Finding the Net Price

If the merchant had a lot of items to discount, he might use this approach.

Gold Bar Representing List Price

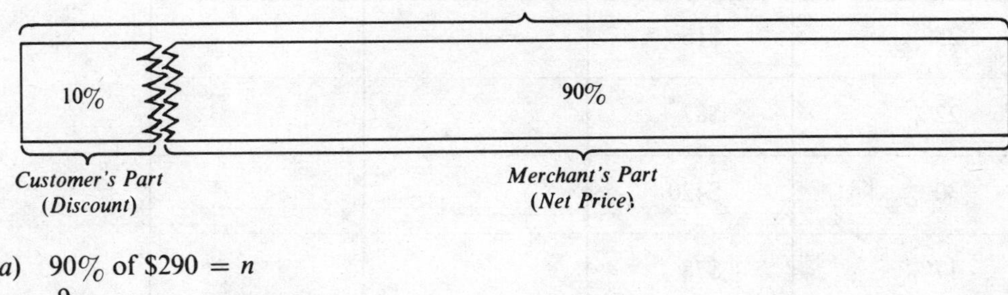

Line (a) 90% of $\$290 = n$

Line (b) $\frac{9}{10} \times \$290 = \$261$ Net Price

(9)

Study Exercise Three

Find the net price without first finding the discount (see Frame 9)

	List Price	Rate of Discount	Net Price
1.	$88	10%	
2.	$724	25%	
3.	$122	18%	
4.	$1,234	32%	

(10)

Finding the List Price When the Discount and the Rate of Discount Are Known

Example 1: A man says, "I bought this article at a discount of 10% and saved $22."

1. What was the list price of the article?
2. How much did he pay?

Line (a) 10% of List Price $= \$22$

Line (b) $\frac{1}{10} \times$ List Price $= \$22$

List Price $= \$22 \div \frac{1}{10} = \$22 \times \frac{10}{1} = \$220$ List Price

Line (c) $\$220 - \$22 = \$198$

List Price − *Discount* = *Net Price*

(11)

Study Exercise Four

Find the missing information to the nearest whole dollar:

	Rate of Discount	Discount	List Price	Net Price
1.	25%	$16		
2.	22%	$87		
3.	30%	$420		
4.	17%	$78		

Finding the List Price When the Net Price and Rate of Discount Are Known

Example 1: A merchant claims that the refrigerator he is selling has had a discount of 20% and that the net price is $480.

1. What was the original price?
2. How much is the discount?

Gold Bar Representing List Price of Refrigerator

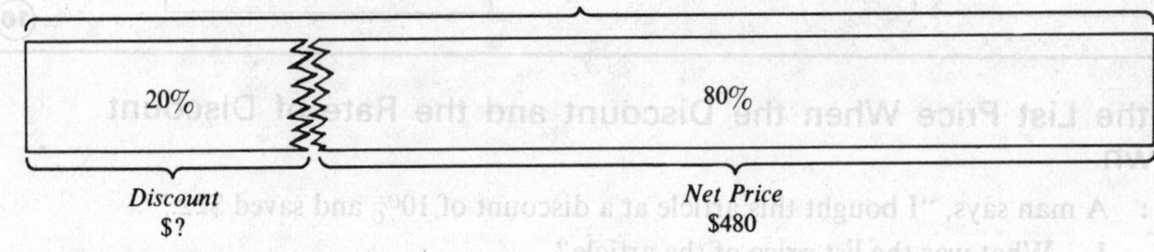

Line (a) 80% of the list price = net price

Line (b) $\frac{8}{10} \times \text{List Price} = \$480$

Line (c) $\text{List Price} = \$480 \div \frac{8}{10} = \overset{60}{\cancel{480}} \times \frac{10}{\underset{1}{\cancel{8}}} = \$600 \text{ List Price}$

Line (d) $600 − $480 = $120 discount

Study Exercise Five

Find the list price and discount to the nearest whole cent given the following information:

	Rate of Discount	Net Price	List Price	Discount
1.	21%	$589		
2.	15%	$28.90		
3.	$33\frac{1}{3}\%$	$14		

(15)

REVIEW EXERCISES

1. The original price of an article is called the ___________ price or ___________ price.
2. If we multiply the rate of discount times the list price, we obtain the ___________.
3. If we subtract the rate of discount from 100% and then multiply the list price by the result, we obtain the ___________.
4. The discount + the ___________ = ___________.
5. The rate of discount is established by the fraction whose numerator is ___________ and whose denominator is ___________.
6. Find the discount and net price if the rate of discount is 13% and the original price is $19.
7. A merchant wishes to give a 4% discount on every item in his store. To get the net price without finding the discount he should multiply each list price by ___________.
8. An article which originally sold for $22 now sells for $18. To the nearest whole per cent, what is the rate of discount?
9. Find the list price if the rate of discount is 30% and the net price is $420.
10. Find the list price if the discount is $8.40 and the rate of discount is 12%.

Find the missing information to the nearest whole per cent or cent:

	List Price	Net Price	Discount	Rate of Discount
11.	$42		$14	
12.	$82	$78		
13.	$120			22%
14.	$18		$2.00	
15.	$120	$110		
16.		$30	$10	
17.		$30		15%
18.			$30	15%
19.	$30			15%

SOLUTIONS TO REVIEW EXERCISES

1. list price or gross price

2. discount

3. net price

4. discount + net price = list price

5. See Frame 3

6. $.13 \times \$19 = \$2.47$ discount
$\$19.00 - \$2.47 = \$16.53$ net price

7. 96%

8. $\$22 - \$18 = \$4$ discount

$$\frac{\$4 \div 2}{\$22 \div 2} = \frac{2}{11} = 11\overset{.181}{\overline{)2.000}} \approx 18\% \text{ rate of discount}$$

9. $100\% - 30\% = 70\%$
70% of list price = \$420

$$\frac{7}{10} \times \text{list price} = \$420$$

$$\text{list price} = \$420 \div \frac{7}{10} = \overset{60}{\cancel{\$420}} \times \frac{10}{\underset{1}{\cancel{7}}} = \$600 \text{ list price}$$

10. 12% of list price = discount
$.12 \times$ list price = \$8.40

$$\text{list price} = \$8.40 \div .12 = .12\overset{\$\ 70}{\overline{)\$8.40}} = \$70 \text{ list price}$$

11. $$\frac{\$14}{\$42} = \frac{14 \div 7}{42 \div 7} = \frac{2}{6} = \frac{1}{3} = 33\frac{1}{3}\% \approx 33\% \text{ rate of discount}$$

$\$42 - \$14 = \$28$ net price

12. $\$82 - \$78 = \$4$ discount

$$\frac{\$4 \div 2}{\$82 \div 2} = \frac{2}{41} = 41\overset{.048}{\overline{)2.000}} \approx 5\% \text{ rate of discount}$$

13. 22% of \$120 = discount
$.22 \times \$120 = \$26.40$ discount
$\$120 - \$26.40 = \$93.60$ net price

14. $$\frac{\$2 \div 2}{\$18 \div 2} = \frac{1}{9} = 9\overset{.111}{\overline{)1.000}} \approx 11\% \text{ rate of discount}$$

$\$18 - \$2 = \$16$ net price

15. $\$120 - \$110 = \$10$ discount

$$\frac{\$10 \div 10}{\$120 \div 10} = \frac{1}{12} = 12\overset{.083}{\overline{)1.000}} \approx 8\% \text{ rate of discount}$$

16. $\$30 + \$10 = \$40$ list price

$$\frac{\$10 \div 10}{\$40 \div 10} = \frac{1}{4} = 25\% \text{ rate of discount}$$

SOLUTIONS TO REVIEW EXERCISES, CONTD.

17. $100\% - 15\% = 85\%$
85% of list price = net price
$.85 \times$ list price = $30

$$\text{list price} = \$30 \div 85 = 85\overline{)\$3000.000}\text{ (quotient } \$35.294) \approx \$35.29 \text{ list price}$$

$35.29 − $30.00 = $5.29 discount

18. 15% of list price = discount
$.15 \times$ list price = $30

$$\text{list price} = \$30 \div 15 = 15\overline{)\$3000}\text{ (quotient } \$200) = \$200 \text{ list price}$$

$200 − $30 = $170 net price

19. 15% of $30 = discount
$.15 \times \$30 = \$4.50$ discount
$30 − $4.50 = $25.50 net price 17

SUPPLEMENTARY PROBLEMS

A. **1.** The __________ + the net price = list price.

2. The rate of discount times the __________ price yields the discount.

3. A lawn mower selling for $62 was offered at a discount of 12%. What was the net price?

4. A dress formerly marked at $29 was offered at $23. What was the rate of discount?

5. The net price on a chair was $122 and the discount was $28. What was the rate of discount?

6. Bricks which were selling at $.08 each were offered at a discount of 10% if 1000 were purchased. What is the net price per 1000 bricks?

7. A stove listed at $248 was offered at a 20% cash discount. What is the net price if cash is paid?

8. One service station advertises regular gasoline at 33.5¢ a gallon. A second advertises regular gasoline at 36.9¢ a gallon with a 10% discount if 10 gallons or more are purchased. You need at least 10 gallons of gas. Which is the better buy?

9. A sand and gravel company offers a 12% discount if a bill is paid within 10 days. The bill is $628. How much can be saved by taking advantage of the discount?

10. One dealer lists a car at $3,728 with a 15% end-of-the-model-year discount. A second dealer lists the same vehicle at $3,150. Which is the better buy and by how much?

SUPPLEMENTARY PROBLEMS, CONTD.

B. Complete the following chart to the nearest whole per cent or dollar:

	Rate of Discount	List Price	Net Price	Discount
11.	14%	$720		
12.	6%		$425	
13.	28%			$163
14.		$98		$22
15.			$635	$55
16.		$358	$342	
17.	10%	$88		
18.	10%		$88	
19.	10%			$88
20.			$88	$12

SOLUTIONS TO STUDY EXERCISES

Study Exercise One (Frame 5)

1. $32.99 − $28.99 = $4.00 discount

$$\frac{\$4.00}{\$32.99} = 3299\overline{)400.000}^{\;.121} \approx 12\% \text{ rate of discount}$$

2. $120 − $82 = $38 discount

$$\frac{\$38 \div 2}{\$120 \div 2} = \frac{19}{60} = 60\overline{)19.000}^{\;.316} \approx 32\% \text{ rate of discount}$$

3. $3450 − $3,000 = $450 discount

$$\frac{\$450 \div 10}{\$3450 \div 10} = \frac{45 \div 15}{345 \div 15} = \frac{3}{23} = 23\overline{)3.000}^{\;.130} \approx 13\% \text{ rate of discount}$$

SOLUTIONS TO STUDY EXERCISES, CONTD.

Study Exercise One (Frame 5, contd.)

4. $500 − $475 = $25 discount

$$\frac{\$25}{\$500} = \frac{\$25 \div 5}{\$500 \div 5} = \frac{5}{100} = 5\% \text{ rate of discount}$$

5. $89 − $79 = $10 discount

$$\frac{\$10}{\$89} = 89\overline{)10.000} \; (.112) \approx 11\% \text{ rate of discount}$$

5A

Study Exercise Two (Frame 7)

1. 20% of $827 = .2 × $827 = $165.40 discount
2. 10% of $46.88 = .1 × $46.88 = $4.688 ≈ $4.69 discount
3. 12% of $32 = .12 × $32 = $3.84 discount
4. $33\frac{1}{3}\%$ of $2,781 = $\frac{1}{\not{3}_1} \times \not{\$2781}^{927}$ = $927 discount
5. 35% of $78.62 = .35 × $78.62 = $27.5170 ≈ $27.52

7A

Study Exercise Three (Frame 10)

1. 100% − 10% = 90%
90% of $88 = $\frac{9}{\not{10}_1} \times \not{\$88}^{8.8}$ = $79.20 net price

2. 100% − 25% = 75%
75% of $724 = $\frac{3}{\not{4}_1} \times \not{\$724}^{181}$ = $543 net price

3. 100% − 18% = 82%
82% of $122 = .82 × $122 = $100.04 net price

4. 100% − 32% = 68%
68% of $1,234 = .68 × $1,234 = $839.12 net price

10A

Study Exercise Four (Frame 12)

1. 25% of list price = discount

$\frac{1}{4}$ × list price = $16

list price = $16 ÷ $\frac{1}{4}$ = $16 × $\frac{4}{1}$ = $64 list price

$64 − $16 = $48 net price

2. 22% of list price = discount
.22 × list price = $87

list price = $87 ÷ .22 = $22\overline{)\$8700.0}$ ($ 395.4) ≈ $395 list price
$395 − $87 = $308 net price

SOLUTIONS TO STUDY EXERCISES, CONTD.

Study Exercise Four (Frame 12, contd.)

3. 30% of list price = discount
.3 × list price = \$420

list price = \$420 ÷ .3 = $3\overline{)\$4200.0}$ with quotient \$1400.0 = \$1400 list price
\$1400 — \$420 = \$980 net price

4. 17% of list price = discount
.17 × list price = \$78

list price = \$78 ÷ 17 = $17\overline{)\$7800.0}$ with quotient \$ 458.8 ≈ \$459 list price
\$459 − \$78 = \$381 net price

(12A)

Study Exercise Five (Frame 15)

1. 100% − 21% = 79%
79% of list price = net price
.79 × list price = \$589

list price = \$589 ÷ .79 = $79\overline{)\$58900.000}$ with quotient \$ 745.569 ≈ \$745.57 list price
\$745.57 − \$589.00 = \$156.57 discount

2. 100% − 15% = 85%
85% of list price = net price
.85 × list price = \$28.90

list price = \$28.90 ÷ .85 = $85\overline{)\$2890.000}$ with quotient \$ 34.000 = \$34 list price
\$34.00 − \$28.90 = \$5.10 discount

3. $100\% - 33\frac{1}{3}\% = 66\frac{2}{3}\%$

$66\frac{2}{3}\%$ of list price = net price

$\frac{2}{3}$ × list price = \$14

list price = \$14 ÷ $\frac{2}{3}$ = $\overset{7}{\cancel{\$14}} \times \frac{3}{\underset{1}{\cancel{2}}}$ = \$21 list price

\$21 − \$14 = \$7 discount

(15A)

---

unit

30

Problems Involving Per Cent—Commission

Objectives:

By the end of this unit you should:

1. know the meaning of rate of commission, commission, and net proceeds.
2. be able to compute a rate of commission from a basic fraction.
3. be able to solve problems involving commission.

(1)

Commission

Example 1: A salesman sold a used car for \$1,000. (This is called the *sale price*). His rate of commission was 20%.

1. How much did the salesman receive (this is called the *commission*)?
2. How much did the company he works for receive (this is called the *net proceeds*)?

Solution, *Part* 1: 20% of \$1,000 = n

Line (*a*) $\frac{1}{5} \times \$1{,}000 = n$

Line (*b*) $\$200 = n$

Line (*c*) $n = \$200$ *Answer:* \$200 commission

Solution, *Part* 2:

Line (*a*) $\$1{,}000 - \$200 = \$800$

Answer: \$800 net proceeds

(2)

A Second Approach to Part 2

Gold Bar Worth Exactly \$1,000

Same Bar Cut In Two Pieces

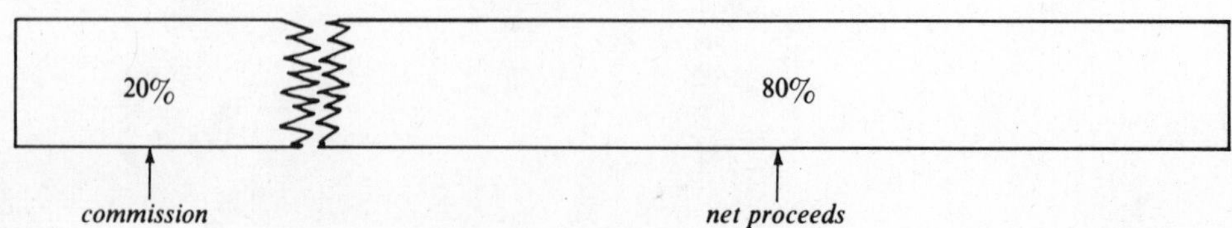

(Frame 3, contd.)

Line (a) 20% of the bar belongs to the salesman
Line (b) 80% of the bar belongs to his company
Line (c) 80% of \$1,000 = net proceeds
Line (d) $\frac{8}{10} \times \$1{,}000 = \$800$

③

Example 1: This time the salesman sells a car for \$2,000. Using the same rate of commission (20%) how much does the company receive (net proceeds)?

Line (a) 80% of \$2,000 = net proceeds
Line (b) $\frac{8}{\cancel{10}_{1}} \times \cancel{\$2{,}000}^{200} = \$1{,}600$
Line (c) *Answer:* \$1,600 net proceeds

④

Study Exercise One

Find the missing information to the nearest cent:

	Rate of Commission	Sale Price	Commission	Net Proceeds
1.	$33\frac{1}{3}\%$	\$2,700		
2.	$6\frac{1}{4}\%$	\$1,728		
3.	28%	\$3,784	✕	
4.	25%	\$24.32	✕	
5.	30%	\$16.87		✕

⑤

Determining the Rate of Commission

The rate of commission is established by the fraction whose numerator is the commission and whose denominator is the sale price. This fraction is then converted to a per cent.

$$\frac{\text{COMMISSION}}{\text{SALE PRICE}}$$

⑥

Example 1: Find the rate of commission on a sale of \$300, if the salesman received \$39.

Solution:

$$\frac{\text{Commission}}{\text{Sale Price}} = \frac{\$39}{\$300} = \frac{39 \div 3}{300 \div 3} = \frac{13}{100} = 13\%$$

Answer: 13%

(Frame 7, contd.)

Example 2: Find the rate of commission to the nearest whole per cent on a sale of \$525 if the salesman received \$75.

Line (a) $\frac{\text{Commission}}{\text{Sale Price}} = \frac{\$75}{\$525}$

Line (b) $\$525 \overline{)75.000}$ with quotient .142

Line (c) 14.2% ≈ 14% *Answer:* 14%

Study Exercise Two

Complete the following chart expressing all rates to the nearest whole per cent:

	Commission	Sale Price	Rate of Commission
1.	\$382	\$1280	
2.	\$56	\$434	
3.	\$.90	\$10.20	
4.	\$87	\$1,000	

(8)

When the Sale Price is Not Known

Example 1: A salesman whose rate of commission is 25% wishes to make \$300 per week. What must his total weekly sales be to achieve this?

Solution:

Line (a) 25% of Total Sales = \$300

Line (b) $.25 \times s = \$300$

Line (c) $s = \$300 \div .25$

Line (d) $s = \$1200$ *Answer:* \$1,200 total sales

(9)

In order to stay in business, a manufacturer determines that his net proceeds must be a minimum of \$230,000 per week. He pays his sales personnel a rate of commission of 32%. What is the smallest weekly sales total needed?

Gold Bar Representing Total Sales

32% Commission	68% Net Proceeds
(belongs to the salesman)	(belongs to the manufacturer)

(10)

Find:

Line (a) 68% of total sales = net proceeds
Line (b) 68% of total sales = \$230,000
Line (c) $.68 \times s =$ \$230,000
Line (d) $s =$ \$230,000 $\div$ .68
Line (e) $s =$ \$338,235 + a remainder less than \$1.00
Line (f) $s =$ \$338,236 *Answer:* \$338,236 Total Sales
Line (g) *Check:* \$338,236 $\times$ 68% = \$230,000.48

(11)

Study Exercise Three

Find the missing information to the nearest whole dollar:

	Rate of Commission	Commission	Net Proceeds	Total Sales
1.	25%	\$72		
2.	$33\frac{1}{3}\%$		\$1,844	
3.	22%	\$84		
4.	16%		\$2,870	
5.	50%	\$420		

(12)

REVIEW EXERCISES

A. A car was sold for \$2,000. The salesman received \$200 and the dealer received \$1,800.

1. The \$200 is the ____________.
2. The \$2,000 is the ____________.
3. The \$1,800 is the ____________.
4. What was the rate of commission on this sale?

B.

5. The rate of commission is established by the fraction whose numerator is ____________ and whose denominator is ____________.
6. If the commission on a sale of \$840 is \$48, find the rate of commission to the nearest tenth of a per cent.
7. A salesman receives \$60 per week plus 4% commission on sales. What are his total weekly earnings if his sales this week are \$1,562. (To nearest cent)
8. A salesman whose rate of commission is 27% wishes to make at least \$200 per week. What must his minimum total sales be to achieve that goal? (To nearest whole dollar)
9. A salesman sold a used car for \$1,700. His rate of commission was 16%. How much did his employer receive from the sale?

SOLUTIONS TO REVIEW EXERCISES

A. 1. commission 2. sale price 3. net proceeds

4. $\frac{\$200}{\$2000} = \frac{\$200 \div 20}{\$2000 \div 20} = \frac{10}{100} = 10\%$

B. 5. See Frame 6

6. $\frac{48}{840} = \frac{48 \div 12}{840 \div 12} = \frac{4 \div 2}{70 \div 2} = \frac{2}{35} = 35\overline{)2.000}$ with quotient $.057\frac{1}{7}$ $\approx 5.7\%$

7. $.04 \times \$1{,}562 = \$62.48 \qquad \$62.48 + \$60.00 = \$122.48$

8. 27% of $n = \$200$

   $.27 \times n = \$200$

   $n = \$200 \div .27 = 27\overline{)\$20000.0}$ with quotient $\$\ 740.7$ $\approx \$741$

9. *Solution 1:*

   $100\% - 16\% = 84\% = .84$

   $.84 \times \$1{,}700 = \$1{,}428$

   Solution 2:

   16% of $\$1{,}700 = \$272$

   $\$1{,}700 - \$272 = \$1{,}428$

SUPPLEMENTARY PROBLEMS

1. How much will the commission be on a \$28,500 house if the realtor charges 6%?
2. How much will the net proceeds be on a car costing \$3,400 if the rate of commission is 18%?
3. A salesman who is paid a rate of commission of 16% wishes to make \$200 per week. What must his sales total be to achieve this goal?

SUPPLEMENTARY PROBLEMS, CONTD.

Find the missing information to the nearest cent:

	Rate of Commission	Sale Price	Commission	Net Proceeds
4.	10%	$1,287		
5.	32%			$64
6.	15%	$287		
7.	19%		$42	
8.	40%	$162		

Find the missing information to the nearest whole per cent:

	Commission	Net Proceeds	Rate of Commission
9.	$48	$62	
10.	$32	$98	
11.	$5	$18	
12.	$72	$360	

13. If you know the commission you can find the net proceeds by ___________ the commission from ___________.

14. If you subtract the rate of commission from 100% and then multiply the result times the total sales you obtain ___________.

SOLUTIONS TO STUDY EXERCISES

Study Exercise One (Frame 5)

1. $33\frac{1}{3}\% = \frac{1}{3}$

$\frac{1}{3} \times \$2,700 = \$900$ commission

$$\begin{array}{r} \$2700 \\ -900 \\ \hline \$1800 \end{array}$$ net proceeds

2. $6\frac{1}{4}\% = \dfrac{6\frac{1}{4}}{100} = 6\frac{1}{4} \times \frac{1}{100} = \frac{\overset{1}{\cancel{25}}}{4} \times \frac{1}{\underset{4}{\cancel{100}}} = \frac{1}{16}$

$\frac{1}{16} \times \$1728 = \$108$ commission

$$\begin{array}{r} \$1728 \\ -108 \\ \hline \$1620 \end{array}$$ net proceeds

SOLUTIONS TO STUDY EXERCISES, CONTD.

Study Exercise One (Frame 5, contd.)

3. $100\% - 28\% = 72\%$
72% of \$3,784 = .72 × \$3,784 = \$2724.48 net proceeds

4. $100\% - 25\% = 75\% = \frac{3}{4}$

$\frac{3}{\cancel{4}_1} \times \cancel{\$24.32}^{6.08} = \$18.24$ net proceeds

5. 30% of \$16.87 = .3 × \$16.87 = \$5.06

5A

Study Exercise Two (Frame 8)

1. $\frac{\$382}{\$1280} = \frac{382 \div 2}{1280 \div 2} = \frac{191}{640} = 640\overline{)191.000}^{.298} \approx 30\%$

2. $\frac{\$56}{\$434} = \frac{56 \div 7}{434 \div 7} = \frac{8 \div 2}{62 \div 2} = \frac{4}{31} = 31\overline{)4.000}^{.129} \approx 13\%$

3. $\frac{\$\ .90}{\$10.20} = \frac{\$\ .90 \times 100}{\$10.20 \times 100} = \frac{90 \div 10}{1020 \div 10} = \frac{9 \div 3}{102 \div 3} = \frac{3}{34} = 34\overline{)3.000}^{.088} \approx 9\%$

4. $\frac{\$87}{\$1000} = \frac{\$87 \div 10}{\$1000 \div 10} = \frac{8.7}{100} = 8.7\%$ or 9%

8A

Study Exercise Three (Frame 12)

1. 25% of s = \$72

$\frac{1}{4} \times s = \$72$

$s = \$72 \div \frac{1}{4} = \$72 \times \frac{4}{1} = \$288$ total sales

\$288 − \$72 = \$216 net proceeds

2. $100\% - 33\frac{1}{3}\% = 66\frac{2}{3}\% = \frac{2}{3}$

$\frac{2}{3} \times s = \$1,844$

$s = \$1,844 \div \frac{2}{3} = \cancel{\$1,844}^{922} \times \frac{3}{\cancel{2}_1} = \$2,766$ total sales

\$2,766 − \$1,844 = \$922 commission

3. 22% of s = \$84
$.22 \times s = \$84$

$s = \$84 \div .22 = 22\overline{)\$8,400.}^{\$381.8} \approx \$382$ total sales
\$382 − \$84 = \$298 net proceeds

SOLUTIONS TO STUDY EXERCISES, CONTD.

Study Exercise Three (Frame 12, contd.)

4. $100\% - 16\% = 84\%$
84% of $s = \$2{,}870$
$.84 \times s = \$2{,}870$

$s = \$2{,}870 \div .84 = 84\overline{)\$287000.}$ with quotient $\$\ 3416.6$ $\approx \$3{,}417$ total sales

$\$3{,}417 - \$2{,}870 = \$547$ commission

5. 50% of $s = \$420$

$\frac{1}{2} \times s = \$420$

$s = \$420 \div \frac{1}{2} = \$420 \times \frac{2}{1} = \$840$ total sales

$\$840 - \$420 = \$420$ net proceeds

unit

31

Word Problems Involving Per Cent—Increase and Decrease

Objectives:

By the end of this unit you should be able to:

1. find the per cent increase.
2. solve problems involving per cent increase.
3. find the per cent decrease.
4. solve problems involving per cent decrease.

Per Cent Increase

Example 1: Yesterday a bicycle sold for \$100. Today it sells for \$110.

1. How large is the increase?
2. What is the per cent increase?

Solution, *Part 1:*

Line (a)	\$110	−	\$100	=	\$10	*Answer:* \$10 increase
	↑		↑		↑	
Line (b)	*new price*	−	*original price*	=	*increase*	

Determining the Per Cent Increase

The per cent increase is established by the fraction whose numerator is the increase and whose denominator is the original or starting amount.

$$\frac{\textbf{INCREASE}}{\textbf{ORIGINAL AMOUNT}}$$

Solution, *Part 2*

Line (a) Increase = \$10

Line (b) Original price = \$100

Line (c) $\frac{\$10}{\$100} = \frac{1}{10}$ or 10%

Check:

Line (a) 10% of \$100 = \$10

↑ ↑ ↑

Line (b) *per cent increase* of *original amount* = *increase*

4

Study Exercise One

Find the missing information to the nearest whole per cent:

	Original Price	New Price	Increase	Per Cent Increase
1.	\$249	\$298		
2.	\$75	\$100		
3.	\$128	\$172		
4.	\$1,400	\$1,587		

Using Per Cent Increase

A factory produces 240 television sets per day. If production is increased 20%, how many will it produce per day?

Line (a) *per cent increase* of *original amount* = *increase*

↓ ↓ ↓

Line (b) 20% of 240 = 48

Line (c) *original amount* + *increase* = *new amount*

Line (d) 240 + 48 = 288

Line (e) *Answer:* 288 television sets

Another Approach to the Problem

Line (a) 100% of original amount = original amount

Line (b) +20% of original amount = increase

Line (c) 120% of original amount = original amount + increase

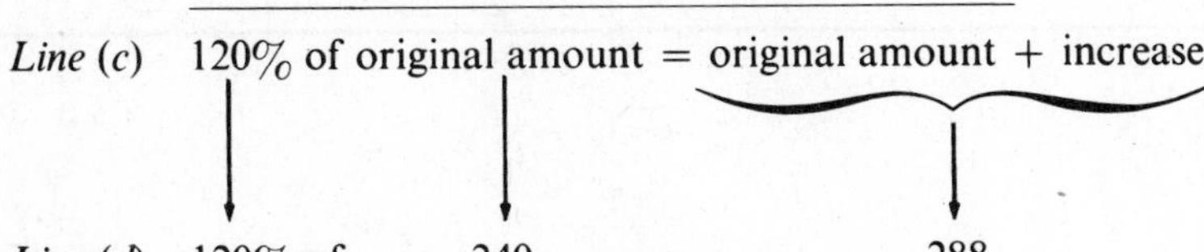

Line (d) 120% of 240 = 288

Line (e) original amount + increase = new amount = 288

Study Exercise Two

1. An article costs \$18.45. If the price is increased $33\frac{1}{3}\%$, find the new price.
2. A man's pay is raised 7%. If his old pay was \$4.88, what will his new pay be? (To nearest whole cent)
3. A factory which formerly produced 728 articles per day installed new machinery designed to boost production 23%. How many articles should it now be producing? (To nearest whole article)

Per Cent Decrease

Example 1: Yesterday a bicycle sold for \$110. Today it sells for \$100.

1. How large is the decrease?
2. What is the per cent decrease?

Solution: *Part 1*

Line (a) \$110 − \$100 = \$10

↑ ↑ ↑

Line (b) *Original Price* − *New Price* = *Decrease*

Determining the Per Cent Decrease

The per cent decrease is established by the fraction whose numerator is the decrease and whose denominator is the original or starting amount.

$$\frac{\text{DECREASE}}{\text{ORIGINAL AMOUNT}}$$

Solution: *Part 2*

Line (a) Decrease = \$10

Line (b) Original Price = \$110

Line (c) $\frac{\$10}{\$110} = \frac{1}{11} = 9\frac{1}{11}\%$

Check:

Line (a) $9\frac{1}{11}\%$ of \$110 = \$10

Line (b) $\frac{1}{11} \times \$110 = \$10$

Line (c) $\frac{\textit{Per Cent}}{\textit{Decrease}} \times \frac{\textit{Original}}{\textit{Amount}} = \textit{Decrease}$

Study Exercise Three

Find the missing information to the nearest whole per cent:

	Original Price	New Price	Decrease	Per Cent Decrease
1.	$75	$50		
2.	$328	$311		
3.	$186	$142		
4.	$1,879	$1,800		

Using Per Cent Decrease

A school with an enrollment of 1,200 students expects a decrease of 20% next year. How many students will it then have?

Line (a) per cent decrease of original amount = decrease

Line (b) 20% of 1,200 = 240

Line (c) original amount − decrease = new amount

Line (d) 1,200 − 240 = 960

Line (e) *Answer:* 960 students

Another Approach to the Problem

Line (a) 100% of original amount = original amount

Line (b) −20% of original amount = − decrease

Line (c) 80% of original amount = original amount − decrease

Line (d) 80% of 1,200 = 960

Line (e) original amount − decrease = new amount = 960

Study Exercise Four

1. An article costs $21.60. If the price is decreased $33\frac{1}{3}\%$, find the new price.
2. A man's pay is cut 8%. If his old pay was $5.40, what will his new pay be (to nearest cent)?
3. A factory which formerly produced 728 articles per day reduced its output by 23%. How many articles will it now produce (to nearest whole article)?

REVIEW EXERCISES

1. Per cent increase is established by a fraction whose numerator is ___________ and whose denominator is ___________.
2. Per cent decrease is established by a fraction whose numerator is ___________ and whose denominator is ___________.
3. A man bought a house for $20,000 and sold it for $23,500. What was his per cent increase to the nearest tenth of a per cent?
4. Another man bought a house for $23,500 and sold it for $20,000. What was his per cent decrease to the nearest tenth of a per cent?
5. The price of butter dropped from $.94 a lb. to $.87. Find the per cent decrease to the nearest whole per cent.
6. A dress marked at $32.00 is to be decreased by 12%. What will the new sale price be?
7. A suit formerly sold for $79. Its price will be increased by 17%. What will the new price be?
8. If we subtract the original price from the new price, we will find the ___________.
9. If we multiply the per cent increase times the ___________, we get the ___________.

Find the missing information to the nearest whole per cent:

	Original Price	New Price	Increase or Decrease	Per Cent Increase or Decrease
10.	$110	$100		
11.	$54	$50		
12.	$108	$128		
13.	$630	$840		
14.	$300	$213		

16

SOLUTIONS TO REVIEW EXERCISES

1. See Frame 3
2. See Frame 10
3. $\dfrac{\$3{,}500 \div 100}{\$20{,}000 \div 100} = \dfrac{35 \div 2}{200 \div 2} = \dfrac{17.5}{100} = 17.5\%$ increase
4. $\dfrac{\$3{,}500 \div 100}{\$23{,}500 \div 100} = \dfrac{35 \div 5}{235 \div 5} = \dfrac{7}{47} = 47\overline{)7.0000}^{\,.1489} \approx 14.9\%$
5. $\dfrac{\$.07 \times 100}{\$.94 \times 100} = \dfrac{7}{94} = 94\overline{)7.000}^{\,.074} \approx 7\%$
6. **Solution 1:**
   $100\% - 12\% = 88\% = .88$
   $.88 \times \$32.00 = \$28.16$ new price
   Solution 2:
   12% of $\$32.00 = \$3.84$
   $\$32.00 - \$3.84 = \$28.16$ new price

SOLUTIONS TO REVIEW EXERCISES, CONTD.

7. **Solution 1:**
   $100\% + 17\% = 117\% = 1.17$
   $1.17 \times \$79 = \$92.43$
   Solution 2:
   17% of $\$79 = \$13.43$
   $\$79 + \$13.43 = \$92.43$
8. See Frame 2
9. See Frame 4
10. $\$110 - \$100 = \$10$ decrease
   $$\frac{\$10 \div 10}{\$110 \div 10} = \frac{1}{11} = 11\overline{)1.000}^{\;.090} \approx 9\% \text{ decrease}$$
11. $\$54 - \$50 = \$4$ decrease
   $$\frac{\$4 \div 2}{\$54 \div 2} = \frac{2}{27} = 27\overline{)2.000}^{\;.074} \approx 7\% \text{ decrease}$$
12. $\$128 - \$108 = \$20$ increase
   $$\frac{\$20 \div 4}{\$108 \div 4} = \frac{5}{27} = 27\overline{)5.000}^{\;.185} \approx 19\% \text{ increase}$$
13. $\$840 - \$630 = \$210$ increase
   $$\frac{\$210 \div 10}{\$630 \div 10} = \frac{21 \div 7}{63 \div 7} = \frac{3 \div 3}{9 \div 3} = \frac{1}{3} = 33\frac{1}{3}\% \approx 33\% \text{ increase}$$
14. $\$300 - \$213 = \$87$ decrease
   $$\frac{\$87}{\$300} = \frac{87 \div 3}{300 \div 3} = \frac{29}{100} = 29\% \text{ decrease}$$

17

SUPPLEMENTARY PROBLEMS

1. What will the total price be on a lawn mower costing \$78 if the sales tax is 5%?
2. The price of eggs increased from \$.82 a dozen to \$.87. What was the per cent increase to the nearest tenth of a percent?
3. Production in a factory decreased 23% this week. If it produced 822 articles the week before, how many did it produce this week?
4. It was reported that the population of a city increased 17% over its last census. If at that time the population was 879,000, what is the new population?
5. At one time bread cost \$.10 a loaf. Today it costs \$.41. What was the per cent increase to the nearest whole per cent?
6. A factory which formerly employed 2,080 men is today only employing 890 men. What was the per cent decrease to the nearest whole per cent?
7. A chair which originally sold for \$120 was decreased by 15%. Later its new price was increased by 10%. What was the final price?
8. Another chair which also originally sold for \$120 was first increased by 10%. Later its new price was decreased by 15%. What was the final price?

SUPPLEMENTARY PROBLEMS, CONTD.

Find the missing information to the nearest whole per cent:

	Original Price	New Price	Increase or Decrease	Per Cent Increase or Decrease
9.	$72	$84		
10.	$84	$72		
11.	$680	$720		
12.	$720	$680		
13.	$20	$17		

SOLUTIONS TO STUDY EXERCISES

Study Exercise One (Frame 5)

1. $298 − $249 = $49 increase

$$\frac{\$49}{\$249} = 249\overline{)49.000}^{\,.196} \approx 20\% \text{ increase}$$

2. $100 − $75 = $25 increase

$$\frac{\$25 \div 25}{\$75 \div 25} = \frac{1}{3} \approx 33\% \text{ increase}$$

3. $172 − $128 = $44 increase

$$\frac{\$44 \div 4}{\$128 \div 4} = \frac{11}{32} = 32\overline{)11.000}^{\,.343} \approx 34\% \text{ increase}$$

4. $1,587 − $1,400 = $187 increase

$$\frac{\$\ 187}{\$1{,}400} = 1{,}400\overline{)187.000}^{\,.133} \approx 13\% \text{ increase}$$

5A

Study Exercise Two (Frame 8)

1. $33\frac{1}{3}\% = \frac{1}{3}$

$\frac{1}{3} \times \$18.45 = \$6.15$

$18.45 + $6.15 = $24.60 new price

or

$100\% + 33\frac{1}{3}\% = 133\frac{1}{3}\% = 1\frac{1}{3}$

$1\frac{1}{3} \times \$18.45 = \frac{4}{\cancel{3}_1} \times \cancel{\$18.45}^{6.15} = \$24.60$

2. 7% = .07

.07 × $4.88 = $.3416

$4.88 + .3416 = $5.2216 ≈ $5.22 new pay

or

100% + 7% = 107% = 1.07

1.07 × $4.88 = $5.2216 ≈ $5.22

Study Exercise Two (Frame 8, contd.)

3. $23\% = .23$
$.23 \times 728 = 167.44 \approx 167$
$728 + 167 = 895$ articles
or
$100\% + 23\% = 123\% = 1.23$
$1.23 \times 728 = 895.44 \approx 895$

Study Exercise Three (Frame 12)

1. $\$75 - \$50 = \$25$ decrease
$\dfrac{\$25}{\$75} = \dfrac{25 \div 25}{75 \div 25} = \dfrac{1}{3} \approx 33\%$ decrease

2. $\$328 - \$311 = \$17$ decrease
$\dfrac{\$17}{\$328} = 328\,\overline{)17.000}$ (quotient $.051$) $\approx 5\%$ decrease

3. $\$186 - \$142 = \$44$ decrease
$\dfrac{\$44}{\$186} = \dfrac{44 \div 2}{186 \div 2} = \dfrac{22}{93} = 93\,\overline{)22.000}$ (quotient $.236$) $\approx 24\%$ decrease

4. $\$1{,}879 - \$1{,}800 = \$79$ decrease
$\dfrac{\$79}{\$1{,}879} = 1{,}879\,\overline{)79.000}$ (quotient $.042$) $\approx 4\%$ decrease

(12A)

Study Exercise Four (Frame 15)

1. $33\frac{1}{3}\% = \dfrac{1}{3}$
$\dfrac{1}{3} \times \$21.60 = \$7.20$
$\$21.60 - \$7.20 = \$14.40$ new price
or
$100\% - 33\frac{1}{3}\% = 66\frac{2}{3}\% = \dfrac{2}{3}$
$\dfrac{2}{\cancel{3}_{1}} \times \cancel{\$21.60}^{7.20} = \$14.40$

2. $8\% = .08$
$.08 \times \$5.40 = \$.4320$
$\$5.40 - \$.4320 = \$4.9680 \approx \$4.97$ new pay
or
$100\% - 8\% = 92\% = .92$
$.92 \times \$5.40 = \$4.9680 \approx \$4.97$

3. $23\% = .23$
$.23 \times 728 = 167.44$
$728 - 167.44 = 560.56 \approx 561$ articles

(15A)

---

unit

32

Problems Involving Per Cent—Profit

Objectives:

By the end of this unit you should be able to:

1. define *profit*, *selling price*, *cost price*, *mark-up*, and *margin*.
2. determine the rate of profit based on the selling price.
3. determine the rate of profit based on the cost. (1)

Profit Based on Selling Price

Example 1: A merchant bought a chair for $75 and sold it for $100. What was the rate of profit based on the selling price?

$100 =	Selling Price
–$ 75 =	–Cost Price
$ 25 =	Profit (margin or mark-up)

Basic Fraction

The rate of profit based on the selling price is established by the fraction whose denominator is the selling price and whose numerator is the profit.

$$\frac{\textbf{PROFIT}}{\textbf{SELLING PRICE}}$$

(3)

Selling Price – *Cost Price* = *Profit*

↓ ↓ ↓

$100 – $75 = $25

$$\frac{\text{Profit}}{\text{Selling Price}} = \frac{\$25}{\$100} = \frac{25 \div 25}{100 \div 25} = \frac{1}{4} = 25\%$$

Study Exercise One

Find the missing information to the nearest whole per cent:

	Selling Price	Cost Price	Profit	Rate of Profit*
1.	$80	$60		
2.	$90		$30	
3.		$40	$20	
4.	$35	$25		
5.	$60		$24	
6.		$270	$130	

*Based on selling price. (5)

Example 1: A dry goods store determines that it must make a 30% profit on the *selling* price in order to cover reasonable expenses and still be competitive with others selling the same type of goods. What should an article sell for which costs $6.30?

Line (a) 30% of the Selling Price = Profit

Line (b) *Gold Bar Representing Selling Price*

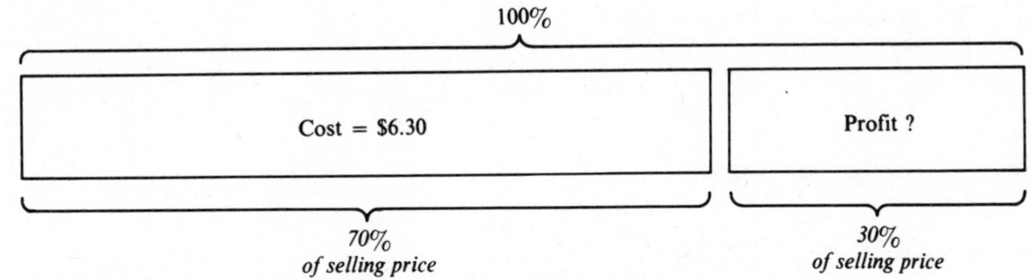

Line (c) 70% of Selling Price = Cost

Line (d) $\frac{7}{10} \times s = \$6.30$

Line (e) $s = \$6.30 \div \frac{7}{10} = \overset{90}{\cancel{\$6.30}} \times \frac{10}{\underset{1}{\cancel{7}}} = \$9.00$

Line (f) $9.00 = Selling Price (6)

Let's Check the Problem

Line (a) *Selling Price* − *Cost* = *Profit*

Line (b) \$9.00 − \$6.30 = \$2.70

Line (c) $\dfrac{\text{Profit}}{\text{Selling Price}}$ = Rate of Profit Based on Selling Price

Line (d) $\dfrac{\$2.70}{\$9.00} = \dfrac{270}{900} = \dfrac{270 \div 9}{900 \div 9} = \dfrac{30}{100} = 30\%$

(7)

Another Example

A merchant wishes to make a 25% profit on the selling price. An article costs him \$32.88.

1. What should he sell it for?
2. What will his margin be?

Line (a) 25% of the Selling Price = Profit

Line (b) *Gold Bar Representing Selling Price*

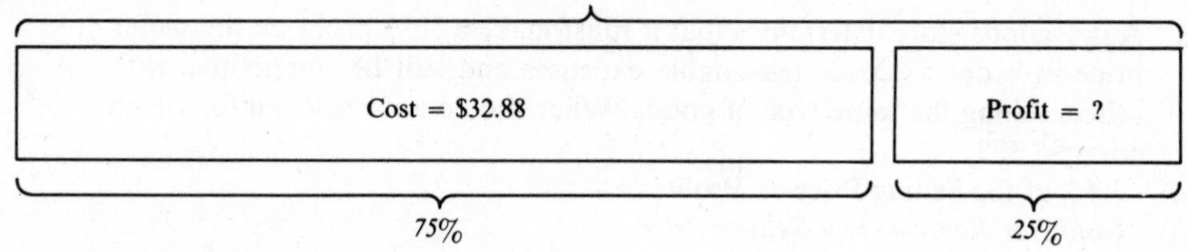

Line (c) 75% of the Selling Price = Cost

Line (d) $\dfrac{3}{4} \times s = \$32.88$

Line (e) $s = \$32.88 \div \dfrac{3}{4} = \overset{10.96}{\cancel{\$32.88}} \times \dfrac{4}{\underset{1}{\cancel{3}}} = \$43.84$

Line (f) \$43.84 = Selling Price

Line (g) Selling Price − Cost = Profit (Margin)

Line (h) \$43.84 − \$32.88 = \$10.96

(8)

Study Exercise Two

Find the missing information to the nearest cent or whole per cent:

	Selling Price	Cost Price	Profit	Rate of Profit*
1.		$24		20%
2.		$32		$33\frac{1}{3}\%$
3.		$150		40%
4.		$3.12		32%

*Based on selling price.

When the Selling Price and Rate Are Known

Example 1: A man sells an article for $48. His rate of profit based on the selling price is 22%.

Part 1: What was the profit?
Part 2: What was the cost?

Solution to Part 1

Line (a) Rate Based On Selling Price of Selling Price = Profit

Line (b) 22% of $48 = Profit

.22 × $48 = $10.56

Solution to Part 2

Line (a) *Selling Price* − *Profit* = *Cost*

Line (b) $48.00 − $10.56 = $37.44

Another Method for Part 2

Line (a) Selling Price = $48

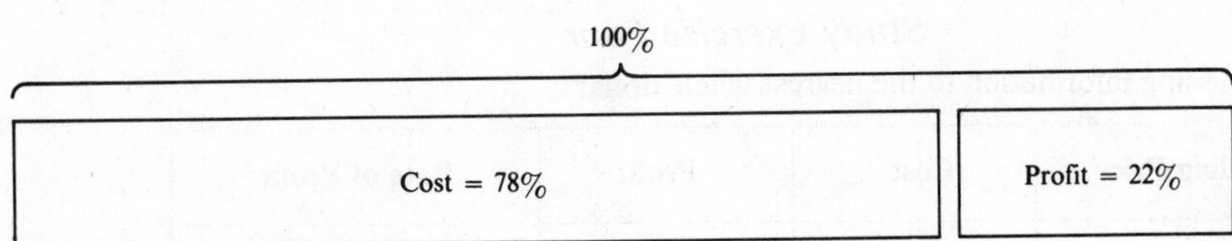

Line (b) 78% of Selling Price = Cost

Line (c) .78 × $48 = $37.44

13

Study Exercise Three

Find the missing information to the nearest cent:

	Selling Price	Cost	Profit	Rate of Profit*
1.	$56			30%
2.	$14			28%
3.	$36			40%
4.	$28			15%

*Based on selling price.

When the Profit and the Rate Are Known

Example 1: The rate of profit based on the selling price is 20%. The profit is $30.

Part 1: What is the selling price?
Part 2: What is the cost?

Solution to Part 1

Line (a) 20% of Selling Price = Profit

Line (b) $\frac{1}{5} \times s = \$30$

Line (c) $s = \$30 \div \frac{1}{5} = \$30 \times \frac{5}{1} = \$150$ Selling Price

Solution to Part 2

Line (a) Selling Price − Profit = Cost

Line (b) $150 − $30 = $120

Study Exercise Four

Find the missing information to the nearest whole dollar:

	Selling Price	Cost	Profit	Rate of Profit*
1.			$22	10%
2.			$432	$33\frac{1}{3}\%$
3.			$62	17%

*Based on selling price.

Profit Based on Cost

Example 1: A merchant bought a chair for $75 and sold it for $100.

1. What is the profit?
2. What is the rate of profit based on the cost?

$$\begin{array}{rl} \$100 = & \text{Selling Price} \\ -\$\ 75 = & -\ \text{Cost Price} \\ \hline \$\ 25 = & \text{Profit} \end{array}$$

Basic Fraction

The rate of profit based on the cost is established by the fraction whose denominator is the cost price and whose numerator is the profit.

$$\frac{\text{PROFIT}}{\text{COST PRICE}}$$

$$\frac{\text{PROFIT}}{\text{COST PRICE}} = \frac{\$25}{\$75} = \frac{25 \div 25}{75 \div 25} = \frac{1}{3} = 33\tfrac{1}{3}\%$$

Study Exercise Five

Go back to Frame 5, Study Exercise One and find the missing information using the rate of profit based on the *Cost Price* instead of the *Selling Price*.

Example:

A dry goods store determines that it must make a 30% profit on the *cost* price in order to cover reasonable expenses and still be competitive with others selling the same type of goods. What should an article sell for which costs $6.30?

Solution:

Line (a)	*Rate Based On Cost*	*of*	*Cost*	= *Profit*
	↓		↓	↓
Line (b)	30%	of	$6.30	= Profit
	↓		↓	↓
Line (c)	.3	×	$6.30	= $1.89
Line (d)	Cost	+	Profit	= Selling Price
	↓		↓	↓
Line (e)	$6.30	+	$1.89	= $8.19

Another Example

A merchant wishes to make a 25% profit on the *cost* price. An article costs him $32.88.

1. What will the margin be?
2. What should he sell it for?

Line (a) 25% of Cost = Profit (Margin)

Line (b) $\frac{1}{4} \times \$32.88 = \$8.22$

Line (c) Cost + Profit = Selling Price

Line (d) $32.88 + $8.22 = $41.10

Study Exercise Six

Go back to Frame 9, Study Exercise Two, and find the missing information using the rate of profit based on the *cost* price instead of the *selling* price.

When the Selling Price and Rate Are Known

Example 1: A man sells an article for \$48. His rate of profit based on the *cost* price is 22%.
Part 1: What was the cost?
Part 2: What was the profit?

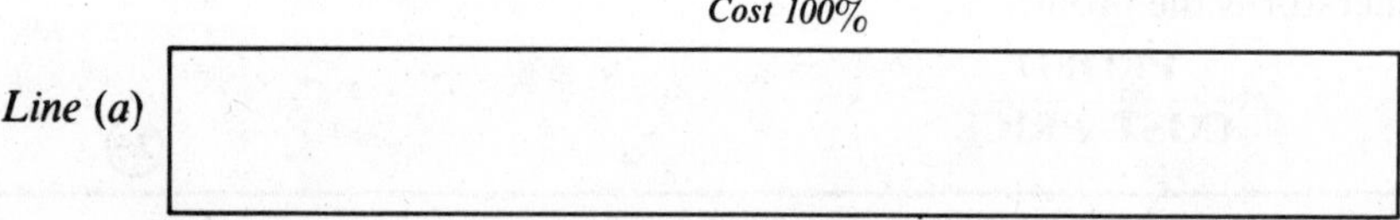

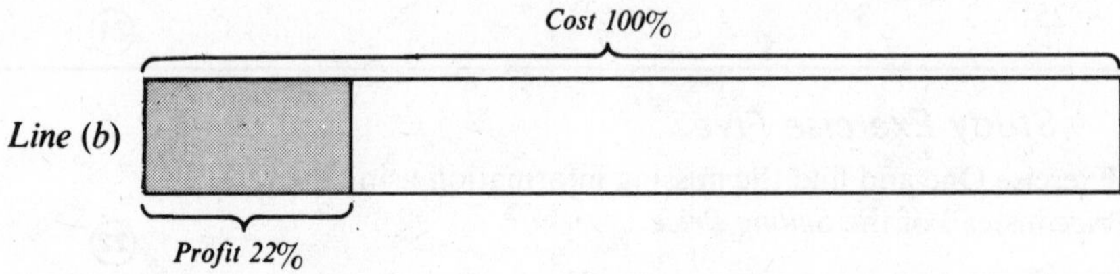

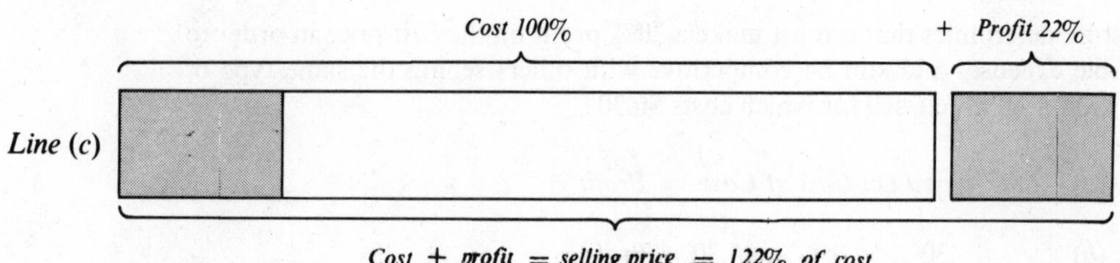

Line (d) 122% of Cost Price = Selling Price

Line (e) $1.22 \times C = \$48$

Line (f) $C = \$48 \div 1.22 = 122\overline{)\$4800.00}$
$= \$39.34$

Line (g) Selling Price − Cost Price = Profit

Line (h) \$48.00 − \$39.34 = \$8.66

Study Exercise Seven

Go back to Frame 14, Study Exercise Three, and find the missing information using the rate of profit based on the *cost* price instead of the *selling* price.

When the Profit and the Rate Are Known

Example 1: The rate of profit based on the *cost* price is 20%. The profit is $30.

1. What is the cost price?
2. What is the selling price?

Solution

Line (a) 20% of Cost Price = Profit

Line (b) $\frac{1}{5} \times C = \$30$

Line (c) $C = \$30 \div \frac{1}{5} = \$30 \times \frac{5}{1} =$

$= \$150$ Cost

Line (d) Cost + Profit = Selling Price

Line (e) $150 + $30 = $180

Study Exercise Eight

Go back to Frame 18, Study Exercise Four, and find the missing information using the rate of profit based on the *cost* price instead of the *selling* price.

REVIEW EXERCISES

A. Complete the following:

1. The rate of profit based on the ___________ price is determined by a fraction whose numerator is the ___________ and whose denominator is the selling price.
2. The rate of profit based on the cost price is determined by a fraction whose denominator is the ___________ price and whose numerator is the ___________.
3. If the cost price is multiplied by the rate of profit based on the cost the result is the ___________.
4. If the selling price is multiplied by the rate of profit based on the ___________ price, the result is the profit.
5. ___________ + ___________ = Selling Price.
6. If the rate of profit based on the selling price is subtracted from 100% and the result then multiplied by the selling price, the answer will be the ___________.
7. Two other terms for profit are ___________ and ___________.

B. In the following problems the rate of profit is based on the selling price: (all answers to nearest cent or whole per cent).

8. An article costs $20 and is sold for $32. What is the profit? What is the rate of profit.
9. A television set costs a dealer $88.40. At what price should he sell it in order to make a profit of 45%? What will the profit be?
10. The rate of profit on a dress was 30%. The profit was $20. What was the selling price? What was the cost price?
11. A profit of $624.60 was made on a car selling for $3,470. What was the rate of profit? What was the cost price?
12. An article that cost $40 was sold at a $10 profit. What was the selling price? What was the rate of profit?
13. Using a rate of profit of 40%, what was the cost of a watch selling for $120? What was the profit?

C. **14-19.** Do problems 8 thru 13 in section B above using a rate of profit based on the *cost* instead of on the *selling* price.

SOLUTIONS TO REVIEW EXERCISES

A.
1. Selling Price, Profit
2. Cost Price, Profit
3. Profit
4. Selling Price
5. Cost + Profit
6. Cost Price
7. Margin, Mark-up

B. (Rate of Profit based on Selling Price)

8. $32 − $20 = $12 Profit

$$\frac{\$12}{\$32} = \frac{12 \div 4}{32 \div 4} = \frac{3}{8} \approx 38\% \text{ Rate of Profit based on Selling Price}$$

9. 100% − 45% = 55%

55% of Selling Price = Cost

$.55 \times s = \$88.40$

$s = \$88.40 \div .55 \approx \$160.73$ Selling Price

$160.73 − $88.40 = $72.33 Profit

SOLUTIONS TO REVIEW EXERCISES, CONTD.

(Frame 33, contd.)

10. 30% of Selling Price = Profit

$.3 \times s = \$20$

$s = \$20 \div .3 \approx \$66.67$ Selling Price

$\$66.67 - \$20.00 = \$46.67$ Cost

11. $\frac{\text{Profit}}{\text{Selling Price}} = \frac{\$624.60}{\$3{,}470} = 18\%$ Rate of Profit based on Selling Price

$\$3{,}470 - \$624.60 = \$2{,}845.40$ Cost

12. $\$40 + \$10 = \$50$ Selling Price

$\frac{\text{Profit}}{\text{Selling Price}} = \frac{\$10}{\$50} = \frac{10 \times 2}{50 \times 2} = \frac{20}{100} = 20\%$ Rate of Profit based on Selling Price

13. 40% of \$120 = Profit

$\frac{2}{5} \times \$120 = \$48$

$\$120 - \$48 = \$72$ Cost

C. (Rate of Profit based on Cost Price)

14. $\$32 - \$20 = \$12$ Profit

$\frac{\text{Profit}}{\text{Cost}} = \frac{\$12}{\$20} = \frac{12 \div 2}{20 \div 2} = \frac{6}{10} = 60\%$ Rate of Profit based on Cost

15. Rate of Profit based on Cost × Cost = Profit

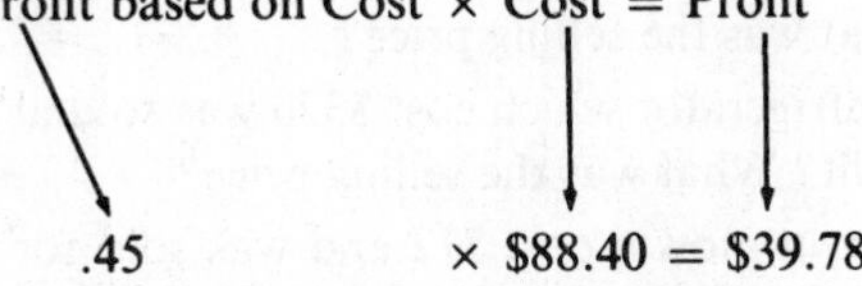

$.45 \times \$88.40 = \$39.78$

$\$88.40 + \$39.78 = \$128.18$ Selling Price

16. Rate of Profit based on Cost × Cost = Profit

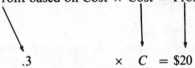

$.3 \times C = \$20$

$C = \$20 \div .3 \approx \$66.67$ Cost

$\$66.67 + \$20.00 = \$86.67$ Selling Price

17. $\$3{,}470 - \$624.60 = \$2845.40$ Cost

$\frac{\$624.60}{\$2{,}845.40} \approx 22\%$ Rate of Profit based on Cost

18. $\$40 + \$10 = \$50$ Selling Price

$\frac{\$10}{\$40} = \frac{1}{4} = 25\%$ Rate of Profit based on Cost

SOLUTIONS TO REVIEW EXERCISES, CONTD.

(Frame 33, contd.)

19.

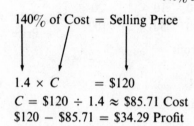

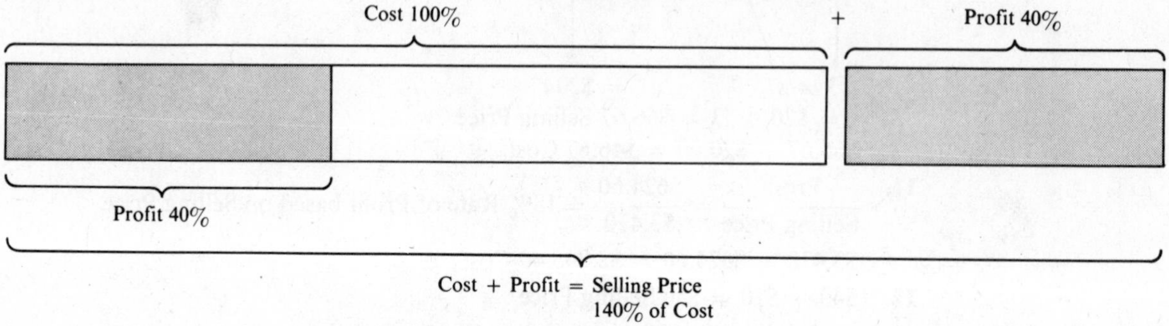

Cost 100% + Profit 40%

Profit 40%

Cost + Profit = Selling Price
140% of Cost

140% of Cost = Selling Price

$1.4 \times C \quad = \$120$

$C = \$120 \div 1.4 \approx \$85.71$ Cost

$\$120 - \$85.71 = \$34.29$ Profit

33

SUPPLEMENTARY PROBLEMS

(All answers to the nearest cent or whole per cent)

A. Solve the following problems using rate of profit based on selling price:

1. A sweater was sold at a 25% rate of profit. The profit was $6. What was the cost? What was the selling price?
2. A refrigerator which cost $320 was sold at a rate of profit of 22%. What was the profit? What was the selling price?
3. A lawn mower cost $72 and was sold for a profit of $27. What was the selling price? What was the rate of profit?
4. The rate of profit on a suit selling for $84 was $33\frac{1}{3}\%$. What was the profit? What was the cost?
5. A pen whose cost was 39¢ was sold for 49¢. What was the profit? What was the rate of profit?
6. A profit of 4¢ was made on a candy bar selling for 10¢. What was the cost? What was the rate of profit?

B. Do problems 1 thru 6 in section A above using a rate of profit based on the *cost* instead of on the *selling* price.

SOLUTIONS TO STUDY EXERCISES

Study Exercise One (Frame 5)

Rate of Profit Based on Selling Price

1. $80 − $60 = $20 Profit

$$\frac{\$20}{\$80} = \frac{20 \div 20}{80 \div 20} = \frac{1}{4} = 25\% \text{ Rate of profit}$$

2. $90 − $30 = $60 Cost price

$$\frac{\$30}{\$90} = \frac{30 \div 30}{90 \div 30} = \frac{1}{3} \approx 33\% \text{ Rate of profit}$$

3. $40 + $20 = $60 Selling price

$$\frac{\$20}{\$60} = \frac{20 \div 20}{60 \div 20} = \frac{1}{3} \approx 33\% \text{ Rate of profit}$$

4. $35 − $25 = $10 Profit

$$\frac{\$10}{\$35} = \frac{10 \div 5}{35 \div 5} = \frac{2}{7} = 7\overline{)2.00}^{\,.28\frac{4}{7}} \approx 29\% \text{ Rate of profit}$$

5. $60 − $24 = $36 Cost price

$$\frac{\$24}{\$60} = \frac{24 \div 6}{60 \div 6} = \frac{4}{10} = 40\% \text{ Rate of profit}$$

6. $270 + $130 = $400 Selling price

$$\frac{\$130}{\$400} = \frac{130 \div 4}{400 \div 4} = \frac{32\frac{1}{2}}{100} \approx 33\% \text{ Rate of profit}$$

Study Exercise Two (Frame 9)

1. 80% of Selling Price = $24

$.8 \times s = \$24$

$s = \$24 \div .8 = 8\overline{)\$240} = \$30$ Selling Price

$30 − $24 = $6 Profit

2. $66\frac{2}{3}\%$ of Selling Price = $32

$\frac{2}{3} \times s = \$32$

$s = \$32 \div \frac{2}{3} = \overset{16}{\cancel{\$32}} \times \frac{3}{\underset{1}{\cancel{2}}} = \$48$ Selling Price

$48 − $32 = $16 Profit

3. 60% of Selling Price = $150

$\frac{3}{5} \times s = \$150$

$s = \$150 \div \frac{3}{5} = \overset{50}{\cancel{\$150}} \times \frac{5}{\underset{1}{\cancel{3}}} = \$250$ Selling Price

$250 − $150 = $100 Profit

SOLUTIONS TO STUDY EXERCISES, CONTD.

Study Exercise Two (Frame 9, contd.)

4. 68% of Selling Price = \$3.12
   $.68 \times s = \$3.12$
   $s = \$3.12 \div .68 = 68\overline{)312} = \$4.59$ Selling Price
   \$4.59 − \$3.12 = \$1.47 Profit

9A

Study Exercise Three (Frame 14)

1. 30% of \$56 = Profit
   $.3 \times \$56 = \$16.80$ Profit
   \$56.00 − \$16.80 = \$39.20 Cost
2. $.28 \times \$14 = \$3.92$ Profit
   \$14.00 − \$3.92 = \$10.08 Cost
3. $.4 \times \$36 = \$14.40$ Profit
   \$36.00 − \$14.40 = \$21.60 Cost
4. $.15 \times \$28 = \$4.20$ Profit
   \$28.00 − \$4.20 = \$23.80 Cost

14A

Study Exercise Four (Frame 18)

1. 10% of Selling Price = Profit
   $.1 \times s = \$22$
   $s = \$22 \div .1 = \$220$ Selling Price
   \$220 − \$22 = \$198 Cost
2. $33\frac{1}{3}\%$ of Selling Price = Profit
   $\frac{1}{3} \times s = \$432$
   $s = \$432 \div \frac{1}{3} = \$432 \times \frac{3}{1} = \$1{,}296$ Selling Price
   \$1,296 − \$432 = \$864 Cost
3. 17% of Selling Price = Profit
   $.17 \times s = \$62$
   $s = \$62 \div .17 = 17\overline{)6200} \approx \$365$ Selling Price
   \$365 − \$62 = \$303 Cost

18A

Study Exercise Five (Frame 22)

Rate of Profit Based on Cost Price

1. \$80 − \$60 = \$20 Profit
   $\frac{\$20}{\$60} = \frac{20 \div 20}{60 \div 20} = \frac{1}{3} \approx 33\%$ Rate of Profit
2. \$90 − \$30 = \$60 Cost
   $\frac{\$30}{\$60} = \frac{30 \div 30}{60 \div 30} = \frac{1}{2} = 50\%$ Rate of Profit
3. \$40 + \$20 = \$60 Selling Price
   $\frac{\$20}{\$40} = \frac{20 \div 20}{40 \div 20} = \frac{1}{2} = 50\%$ Rate of Profit

SOLUTIONS TO STUDY EXERCISES, CONTD.

Study Exercise Five (Frame 22, contd.)

Rate of Profit Based on Cost Price, contd.

4. $35 − $25 = $10 Profit

$$\frac{\$10}{\$25} = \frac{10 \times 4}{25 \times 4} = \frac{40}{100} = 40\% \text{ Rate of Profit}$$

5. $60 − $24 = $36 Cost

$$\frac{\$24}{\$36} = \frac{24 \div 12}{36 \div 12} = \frac{2}{3} \approx 67\% \text{ Rate of Profit}$$

6. $270 + $130 = $400 Selling Price

$$\frac{\$130}{\$270} = \frac{130 \div 10}{270 \div 10} = \frac{13}{27} = 27\overline{)13.00}\ \ .48\frac{4}{27} \approx 48\% \text{ Rate of Profit}$$

22A

Study Exercise Six (Frame 25)

1. .2 × $24 = $4.80 Profit
$24 + $4.80 = $28.80 Selling Price

2. $\frac{1}{3}$ × $32 = $10.67 Profit
$32 + $10.67 = $42.67 Selling Price

3. $\frac{2}{5}$ × $150 = $60 Profit
$150 + $60 = $210 Selling Price

4. .32 × $3.12 ≈ $1.00 Profit
$3.12 + $1.00 = $4.12 Selling Price

25A

Study Exercise Seven (Frame 28)

1. 130% of Cost = $56
1.3 × C = $56
C = $56 ÷ 1.3 = $43.08 Cost
$56 − $43.08 = $12.92 Profit

2. 128% of Cost = $14
1.28 × C = $14
C = $14 ÷ 1.28 = $10.94 Cost
$14 − $10.94 = $3.06 Profit

3. 140% of Cost = $36
1.4 × C = $36
C = $36 ÷ 1.4 = $25.71 Cost
$36.00 − $25.71 = $10.29 Profit

4. 115% of Cost = $28
1.15 × C = $28
C = $28 ÷ 1.15 = $24.35 Cost
$28 − $24.35 = $3.65 Profit

28A

SOLUTIONS TO STUDY EXERCISES, CONTD.

Study Exercise Eight (Frame 31)

1. 10% of Cost = \$22

$\frac{1}{10} \times C = \$22$

$C = \$22 \div \frac{1}{10} = \$22 \times \frac{10}{1} = \$220$ Cost

\$220 + \$22 = \$242 Selling Price

2. $33\frac{1}{3}\%$ of Cost = \$432

$\frac{1}{3} \times C = \$432$

$C = \$432 \div \frac{1}{3} = \$432 \times \frac{3}{1} = \$1{,}296$ Cost

\$1,296 + \$432 = \$1,728 Selling Price

3. 17% of Cost = \$62

$.17 \times C = \$62$

$C = \$62 \div .17 = 17\overline{)\$6200.0}$ with quotient \$ 364.7 $\approx \$365$

\$365 + \$62 = \$427

31A

unit

33

Problems Involving Per Cent—Interest

Objectives:

By the end of this unit you should:

1. know the meaning of the following terms:
   (a) *interest* (b) *principal*
   (c) *rate of interest* (d) *amount*
   (e) *simple interest* (f) *compound interest*
   (g) *Truth-in-Lending Act* (h) *revolving charge account*
2. be able to compute simple interest from a formula.
3. be able to compute compound interest from a table.
4. be able to determine actual annual interest on the balance of two types of revolving charge accounts:
   (a) those which do not subtract payments and credits prior to computing finance charges.
   (b) those which do perform this subtraction first. ①

Example 1: A man borrows $1,000 at a rate of 8% per year and agrees to pay back the $1,000 plus an additional $160 two years later. At the end of the second year, he returns $1,160.

Line (a) $1,000 = principal
Line (b) $160 = interest
Line (c) $1,160 = principal + interest = amount
Line (d) 8% = annual interest rate ②

Computing Simple Interest

Line (a) 8% of $1,000 = .08 × $1,000 = $80 per year
Line (b) $80 × 2 years = $160
or
Line (c) 8% of $1,000 × 2 =
Line (d) .08 × $1,000 × 2 =
Line (e) .08 × $2,000 = $160 interest ③

A Simple Interest Formula

Annual Rate × Principal × Time (in years) = Interest

④

Applying the Formula

Example 1: $100 is borrowed for 8 months at an annual rate of 6%. What is the interest?

Solution:

Line (a) 8 months $= \frac{8}{12} = \frac{2}{3}$ year

Line (b) 6% = .06

Line (c) *Rate* × *Principal* × *Time* = *Interest*

↓ ↓ ↓ ↓

Line (d) $.06 \times \$100 \times \frac{2}{3} = \$4$

Example 2: $300 is borrowed for 1 year and 6 months at an annual rate of 12%. What is the interest?

Solution:

Line (a) 1 year 6 months $= 1\frac{1}{2}$ yr. $= \frac{3}{2}$ yr.

Line (b) *Rate* × *Principal* × *Time* = *Interest*

↓ ↓ ↓ ↓

Line (c) $.12 \times \$300 \times \frac{3}{2} = \$54$

Example 3: $500 is borrowed for 3 years at an annual interest rate of 6%. What is the amount?

Solution:

Line (a) Principal + Interest = Amount

Line (b) *Rate* × *Principal* × *Time* = *Interest*

↓ ↓ ↓ ↓

Line (c) .06 × $500 × 3 = $90

Line (d) *Principal* + *Interest* = *Amount*

↓ ↓ ↓

Line (e) $500 + $90 = $590

Study Exercise One

Find the interest and amount given the following information:

	Annual Rate	Principal	Time	Interest	Amount
1.	4%	$1,000	1 year		
2.	12%	$420	7 months		
3.	$6\frac{3}{4}\%$	$1,600	3 years		
4.	7%	$720	2 months		
5.	$5\frac{1}{4}\%$	$3,200	1 year, 9 months		
6.	$8\frac{1}{2}\%$	$484	4 months		

7. What is the interest on $100 for one year at a monthly interest rate of $\frac{3}{4}\%$? (Hint: multiply $\frac{3}{4}\% \times 12$ to get annual rate.)

Compound Interest

Example 1: Mr. Doe has $1,000 to lend for a period of two years and finds that he has two choices.

Choice 1: He can lend the money to Jones at an annual rate of 6% for the two years.
Choice 2: He can lend the money to Smith for one year at 6% and then lend the amount he receives from Smith to Brown for the second year at 6%.

What is his best course of action?

Analysis of Choice One

Line (a) *Rate* × *Principal* × *Time* = *Interest*

Line (b) .06 × $1,000 × 2 = $120

At the end of two years Mr. Doe has $120 more than his original $1,000.

Analysis of Choice Two

Line (a) *Rate* × *Principal* × *Time* = *Interest*

Line (b) .06 × $1,000 × 1 = $60 received from Mr. Smith at the end of one year

Line (c) *Principal* + *Interest* = *Amount*

Line (d) $1,000 + $60 = $1,060 at the end of one year

(Frame 9, contd.)

Line (e) *Rate* × *Principal* × *Time* = *Interest*

↓ ↓ ↓ ↓

Line (f) .06 × \$1,060 × 1 = \$63.60 received from Mr. Brown at the end of the second year

Line (g) Mr. Smith's Interest + Mr. Brown's Interest = Total Interest on the Original \$1,000

↓ ↓ ↓

Line (h) \$60 + \$63.60 = \$123.60

At the end of two years Mr. Doe has \$123.60 more than his original \$1,000 or \$123.60 − \$120.00 = \$3.60 more by selecting choice two over choice one.

Choice Two is an example of compound interest

Compound Interest is interest paid on both the principal and the interest earned previously. (9)

Mr. Doe, however, suddenly finds that he can lend his \$1,000 to four people in succession at the same annual rate for 2 years. Each person will have the money for six months or $\frac{1}{2}$ year. The interest gained is computed as follows:

Line (a) Each person has the money 6 months or $\frac{1}{2}$ year.

Line (b) First Man: $.06 \times \$1{,}000 \times \frac{1}{2} = \$30$

Line (c) Amount at the end of first six months: \$1,000 + \$30 = \$1,030

Line (d) Second Man: $.06 \times \$1{,}030 \times \frac{1}{2} = \$30.90$

Line (e) Amount at the end of second six months: \$1,030 + \$30.90 = \$1,060.90

Line (f) Third Man: $.06 \times \$1{,}060.90 \times \frac{1}{2} = \$31.827$

Line (g) Amount at the end of third six months: \$1,060.90 + \$31.827 = \$1,092.727

Line (h) Fourth Man: $.06 \times \$1{,}092.727 \times \frac{1}{2} = \$32.78181$

Line (i) Amount at the end of fourth six months: \$1,092.727 + \$32.78181 = \$1,125.50881

Line (j) $\$1{,}125.50881 \approx \$1{,}125.51$

Line (k) \$1,125.51 − \$1,000.00 = \$125.51 interest (10)

Using a Compound Interest Table

Line (a) $6\% \text{ of } \$1{,}000 \times \frac{1}{2}\text{ year} = 6\% \times \frac{1}{2}\text{ year} \times \$1{,}000$

$= 3\% \times \$1{,}000 = \$30$

Line (b) The interest rate per period is 3%.

To find the interest rate per period divide the annual interest rate by the number of periods per year.

Example 1: What is the interest rate per period on a loan which is to be compounded quarterly at 8%?

Line (a) $8\% \div 4 = 2\%$ per period

(Frame 11, contd.)

Example 2: What is the interest rate per period on a loan which is to be compounded monthly at 6% for 3 years?
What is the total number of periods?

Line (a) $6\% \div 12 = \frac{1}{2}\%$ per period

Line (b) 12 periods per year × 3 years = 36 periods

Solution of Frame Ten by Table

Line (a) Interest is to be compounded every 6 months or twice a year.
Line (b) $6\% \div 2 = 3\%$ per period
Line (c) 2 periods per year × 2 years = 4 periods
Line (d) Using Table II at the back of the book, find the number in the 3% column and the 4 periods' row. $1.1255—this is what $1 will amount to in the two years.
Line (e) Multiply $1,000 × 1.1255 = $1,125.50
This is what $1,000 will amount to in the two years.
Line (f) Subtract $1,125.50 − $1,000.00 = $125.50
This is the interest on $1,000 at 6% compounded semi-annually over 2 years.

Summary

Step 1: Find the interest rate per period by dividing the annual rate by the number of periods per year.
Step 2: Find the number of periods by multiplying the periods per year times the number of years.
Step 3: Find how much $1 will amount to by using the interest rate per period column and the number of periods' row.
Step 4: Multiply the number obtained in Step 3 by the principal and you will obtain the amount.
Step 5: Subtract the principal from the amount found in Step 4 and you will have the interest.

Another Example

Example 1: What is the interest on $400 compounded quarterly at 6% per year for 2 years?

Step 1: $6\% \div 4 = 1\frac{1}{2}\%$

Step 2: 4 periods per year × 2 = 8 periods

Step 3: $1\frac{1}{2}\%$ column, 8 period row ⟶ 1.1265

Step 4: $400 × 1.1265 = $450.60
Step 5: $450.60 − $400.00 = $50.60 Interest

Study Exercise Two

Find the interest on the following:

1. $200 compounded annually for 4 years at 5%
2. $500 compounded semi-annually for 5 years at 4%
3. $700 compounded monthly for $\frac{1}{2}$ year at 6%
4. $800 compounded 3 times a year for 1 year at 6%
5. $1,000 compounded annually for 25 years at 3%

15

True Annual Interest (An Approximation Method)

Mr. Taken borrows $1,000 for four months at 6% per year. He agrees to pay the bank in four equal monthly payments including interest. The bank computes the size of the payments as follows:

Line (a) 4 months $= \frac{4}{12} = \frac{1}{3}$ year.

Line (b) *Rate* × *Principal* × *Time* = *Interest*

Line (c) $.06 \times \$1{,}000 \times \frac{1}{3} = \$20$

Line (d) *Principal* + *Interest* = *Amount*

Line (e) $1,000 + $20 = $1,020

Line (f) $1,020 ÷ 4 = $255 per month

Line (g) Each payment of $255 contains $250 towards the principal and $5 towards the interest.
$250 × 4 = $1,000 $5 × 4 = $20

Did Mr. Taken Really Borrow $1,000 for Four Months?

Line (a) $1,000 (the original amount of the loan; owed during the first month only.)

Line (b) $750 ($1,000 − $250; owed during the second month only.)

Line (c) $500 ($750 − $250; owed during the third month only.)

Line (d) $250 ($500 − $250; owed during the fourth month only.)

Line (e)

$$\begin{array}{r} \$1{,}000 \\ 750 \\ 500 \\ 250 \end{array}$$

$$4\overline{)2{,}500}$$

$625 average amount owed during this period, not $1,000.

A Short Cut

To find the average amount owed when it is to be payed off in equal amounts, add the original amount owed to the last amount owed and divide by 2.

$$\frac{\$1{,}000 + \$250}{2} = \frac{\$1{,}250}{2} = \$625$$

Line (a) *Rate* × *Principal* × *Time* = *Interest*

$$\downarrow \quad \downarrow \quad \downarrow \quad \downarrow$$

Line (b) $r \times \$625 \times \frac{1}{3} = \$20$

Line (c) $r \times \frac{\$625}{3} = \$20$

Line (d) $r = 20 \div \frac{625}{3} = 20 \times \frac{3}{625} = \frac{60}{625}$

Line (e) $r = 625\overline{)60.000}$ (quotient .096) $= 9.6\%$

Line (f) Stated annual interest 6%

Line (g) Approximate true annual interest 9.6%

Determining An Approximation of True Annual Interest

Step 1: Use the simple interest formula and the annual rate given to find the interest to be paid.

Step 2: Find the average amount owed using the short cut in Frame 18.

Step 3: Use the number found in Step 2 for the principal and the number found in Step 1 for the interest and apply the simple interest formula.

Example 1: Approximate the true annual interest on a loan of $600 at 8% which is to be paid in 10 equal monthly installments of $60 each plus interest.

Step 1: $.08 \times \$600 \times \frac{10}{12} = \$40$ Interest

Step 2: $\frac{\$600 + \$60}{2} = \frac{\$660}{2} = \$330$ average amount loaned

Step 3: *Rate* × *Principal* × *Time* = *Interest*

$$\downarrow \quad \downarrow \quad \downarrow \quad \downarrow$$

$$r \times \$330 \times \frac{10}{12} = \$40$$

$$r \times \frac{3300}{12} = \$40$$

$$r = 40 \div \frac{3300}{12} = 40 \times \frac{12}{3300} = \frac{480}{3300} = \frac{48}{330}$$

$r = 330\overline{)48.000}$ (quotient .145) $\approx 14.5\%$ Approximate True Annual Interest

Study Exercise Three

Find the true annual interest on the following loans: (nearest tenth of a per cent)

1. $1,000 at 5% paid in 10 monthly installments of $100 each plus interest.
2. $120 at 9% paid in 5 monthly installments of $24 each plus interest.
3. $48 at 8% paid in 3 monthly installments of $16 each plus interest.

The Truth-In-Lending Act

The Federal Truth-In-Lending Act went into effect July 1, 1969 and requires conspicuous disclosures of comparable true annual interest rates on the vast majority of credit contracts, private house first mortgages and time-sale agreements. You should use this figure when shopping for the lowest price money available to you.

The Revolving-Credit Charge Account

Stores A and B each charge $1\frac{1}{2}\%$ per month on the unpaid balance and state a true Annual Percentage Rate of 18% as required by the Truth-In-Lending legislation. However, Store A calculates interest on the previous balance *before* deducting payments and credits. Store B calculates interest on the previous balance *after* deducting payments and credits.

Example 1: Mr. Lucky and Mr. Unlucky each bought $350 worth of garden plants in the same month and each returned $50 worth of unused plants to the store for credit. The next month each decides to pay $150 toward his bill. Mr. Unlucky trades at Store A and Mr. Lucky trades at Store B.

A Comparison

Opening Balance	Payments plus Credits	Monthly Rate	Finance Charge
Store A—$350	$200	$1\frac{1}{2}\%$	$.015 \times \$350 = \$5.25$
Store B—$350	$200	$1\frac{1}{2}\%$	$.015 \times \$150 = \$2.25$

Store B's True Rate

Line (a) *Rate* × *Principal* × *Time* = *Interest*

↓ ↓ ↓ ↓

Line (b) $r \times \$150 \times \frac{1}{12} = \$2.25$

Line (c) $r \times \frac{150}{12} = 2.25$

Line (d) $r = 2.25 \div \frac{150}{12} = \cancelto{.09}{2.25} \times \frac{\cancelto{2}{12}}{\cancelto{\cancelto{1}{25}}{150}} = .18 = 18\%$

Store A's True Rate

Line (a) *Rate* × *Principal* × *Time* = *Interest*

↓ ↓ ↓ ↓

Line (b) $r \times \$150 \times \frac{1}{12} = \$5.25$

Line (c) $r \times \frac{150}{12} = 5.25$

(Frame 26, contd.)

Line (d) $r = 5.25 \div \dfrac{150}{12} = \dfrac{5.25 \times 12}{150} = \dfrac{63}{150}$

Line (e) $r = 150\overline{\smash{)}63.00}^{\;.42} = 42\%$ actual annual rate of interest

Study Exercise Four

Compute the actual annual rate of interest on the following account in two ways. First by deducting the payments and credits before computing the interest on the balance. Second, compute the interest on the balance before deducting payments and credits.

	Opening Balance	Payments and Credits	Monthly Rate	Finance Charge
Mr. Jones	\$200	\$100	$1\frac{1}{2}\%$	

27

REVIEW EXERCISES

A. Find the interest and the amount given the following information:

	Annual Rate	Principal	Time	Interest	Amount
1.	5%	$200	2 years		
2.	$10\frac{1}{2}\%$	$1,800	8 months		
3.	$5\frac{3}{4}\%$	$900	1 year, 4 months		
4.	9%	$720	3 months		

B. **5.** An interest rate of 18% per annum may be thought of as __________ % per month.

6. Interest + __________ = __________.

7. What is meant by compound interest?

8. State the simple interest formula.

9. The amount borrowed is called the __________ while the amount returned which is in excess of what was borrowed is called the __________.

10. When shopping for a loan, your best guide is the statement of __________ which must be indicated according to the Federal __________ Act.

11. Explain why two revolving-charge plans each stating an actual annual interest rate of 18% may still not be the same.

C. Using a table, determine the interest on the following:

12. $600 compounded semi-annually for 3 years at 4% annual rate.

13. $800 compounded quarterly for 2 years at 6% annual rate.

14. When his daughter was born, a man deposited $1,000 in a bank which paid 6% compounded annually. How much was on deposit when she entered college 20 years later?

15. Approximately how long would it take for money deposited at 6% compounded annually to increase by 50%?

16. Which produces more interest: a dollar invested at 4% compounded quarterly for 2 years or a dollar invested at 6% compounded three times a year for 1 year and 4 months? (4% and 6% are annual rates.)

D. **17.** What is the average amount borrowed on a $4,800 loan which was to be paid back in 30 equal installments?

18. Find the true annual interest on a loan of $96 at 8% paid in 3 monthly installments of $32 each plus interest.

19. What was the true annual rate of interest on a revolving-charge account if the monthly rate was $1\frac{1}{2}\%$, the balance $400 and a payment of $300 was not subtracted from the balance prior to the interest computation?

SOLUTIONS TO REVIEW EXERCISES

1. $\frac{5}{\underset{1}{\cancel{100}}} \times \overset{2}{\cancel{\$200}} \times 2 = \$20$ interest, $\$200 + \$20 = \$220$ amount

2. $\frac{21}{\underset{1}{\cancel{200}}} \times \overset{3}{\overset{\cancel{9}}{\cancel{\$1,800}}} \times \frac{\overset{2}{\cancel{8}}}{\underset{1}{\underset{\cancel{4}}{\cancel{12}}}} = \$126$ interest, $\$1,800 + \$126 = \$1,926$ amount

3. $\frac{23}{\underset{1}{\cancel{400}}} \times \overset{3}{\cancel{\$900}} \times \frac{\overset{1}{\cancel{4}}}{\underset{1}{\cancel{3}}} = \$69$ interest, $\$900 + \$69 = \$969$ amount

4. $\frac{9}{100} \times \overset{60}{\cancel{\$720}} \times \frac{\overset{1}{3}}{\underset{1}{\cancel{12}}} = \$16.20$ interest, $\$720 + \$16.20 = \$736.20$ amount

5. $18\% \div 12 = 1\frac{1}{2}\%$
6. See Frame 2
7. See Frame 9
8. See Frame 4
9. See Frame 2
10. Actual Annual Interest; Truth-in-Lending Act
11. See Frames 24–26
12. *Step 1:* $4\% \div 2 = 2\%$
    Step 2: 2 periods per year $\times$ 3 years $=$ 6 periods
    Step 3: 1.1262
    Step 4: $\$600 \times 1.1262 = \$675.72$ amount
    Step 5: $\$675.72 - \$600 = \$75.72$ interest
13. *Step 1:* $6\% \div 4 = 1\frac{1}{2}\%$
    Step 2: 4 periods per year $\times$ 2 years $=$ 8 periods
    Step 3: 1.1265
    Step 4: $\$800 \times 1.1265 = \$901.20$ amount
    Step 5: $\$901.20 - \$800 = \$101.20$ interest
14. *Step 1:* $6\% \div 1 = 6\%$
    Step 2: 1 period per year $\times$ 20 years $=$ 20 periods
    Step 3: 3.2071
    Step 4: $\$1,000 \times 3.2071 = \$3,207.10$ amount
15. 50% of \$1 = \$.50 Therefore, a 50% increase means a dollar deposited must amount to \$1.50 some time in the future. Looking in the 6% column we find 1.5036 in 7th period or year. Approximately 7 years.
16. A dollar at 4% compounded quarterly for 2 years amounts to \$1.0829.

    A dollar at 6% compounded three times a year for $1\frac{1}{3}$ years amounts to \$1.0824.

    The 4% is slightly better but most of us would probably choose the 6% because of the shorter time involved.
17. $30\overline{)\$4800}$ = \$ 160 (the final zeros of 30 and \$4800 are cancelled)

    $\frac{\$4800 + \$160}{2} = \frac{\$4960}{2} = \$2480$ average amount borrowed

SOLUTIONS TO REVIEW EXERCISES, CONTD.

18. *Step 1:* $\frac{8}{100} \times \overset{8}{\cancel{\$96}} \times \frac{3}{\underset{1}{\cancel{12}}} = \$1.92$ interest

Step 2: $\frac{\$96 + \$32}{2} = \frac{\$128}{2} = \$64$ average amount loaned

Step 3: $r \times \overset{16}{\cancel{\$64}} \times \frac{\overset{1}{\cancel{3}}}{\underset{\underset{1}{\cancel{4}}}{\cancel{12}}} = \$1.92$

$r \times 16 = 1.92$

$r = 1.92 \div 16 = 16\overline{)1.92}^{\,.12} = 12\%$ approximate annual interest

19. Interest was charged on $400.

$1\frac{1}{2}\%$ of $\$400 = \frac{3}{\underset{1}{\cancel{200}}} \times \overset{2}{\cancel{400}} = \$6$ interest

Interest should have been charged on $400 − $300 = $100

Rate × Principal × Time = Interest

↓ ↓ ↓ ↓

$r \times \$100 \times \frac{1}{12} = \$6$

$r \times \frac{100}{12} = 6$

$r = 6 \div \frac{100}{12} = 6 \times \frac{12}{100} = \frac{72}{100} = 72\%$ actual rate

(29)

SUPPLEMENTARY PROBLEMS

A. Using the simple interest formula find the interest on the following:

1. $1,000 for 2 years at $6\frac{1}{4}\%$
2. $480 for 3 months at 9%
3. $600 for 4 months at 8%
4. $2,400 for 4 years at 3%

B. Using the compound interest table find the interest on the following:

5. $1,000 compounded quarterly at 4% for 3 years.
6. $800 compounded semi-annually at 6% for 5 years.
7. $500 compounded annually at 5% for 20 years.
8. $200 compounded quarterly at 6% for 2 years.

C. **9.** If $1,000 were invested the day you were born at 6% interest compounded annually, how much would it amount to on your following birthdays:
10th birthday?
15th birthday?
20th birthday?
25th birthday?

D. **10.** What was the average amount borrowed on a loan of $1,710 paid back in 36 equal monthly installments?

SOLUTIONS TO STUDY EXERCISES

Study Exercise One (Frame 6)

1. .04 × \$1,000 × 1 = \$40 Interest, \$1,000 + \$40 = \$1,040 Amount

2. $\frac{\overset{1}{\cancel{12}}}{10\cancel{0}} \times \$42\cancel{0} \times \frac{7}{\underset{1}{\cancel{12}}} = \frac{294}{10} = \$29.40$ Interest

\$420 + \$29.40 = \$449.40 amount

3. $\frac{27}{\underset{1}{\cancel{4}}} \times \frac{1}{\cancel{100}} \times \overset{4}{\cancel{\$1,600}} \times 3 = \$324$ Interest

\$1,600 + \$324 = \$1,924 amount

4. $\frac{7}{10\cancel{0}} \times \overset{12}{\cancel{\$72}\cancel{0}} \times \frac{\overset{1}{\cancel{2}}}{\underset{\underset{1}{\cancel{6}}}{\cancel{12}}} = \frac{84}{10} = \$8.40$ Interest

\$720 + \$8.40 = \$728.40 amount

5. $\frac{21}{\underset{1}{\cancel{4}}} \times \frac{1}{\cancel{100}} \times \overset{\overset{2}{\cancel{8}}}{\cancel{\$3,200}} \times \frac{7}{\underset{1}{\cancel{4}}} = \$294$ Interest

\$3,200 + \$294 = \$3,494 amount

6. $\frac{17}{\underset{1}{\cancel{2}}} \times \frac{1}{100} \times \overset{\overset{121}{\cancel{242}}}{\cancel{\$484}} \times \frac{\overset{2}{\cancel{4}}}{\underset{\underset{3}{\cancel{6}}}{\cancel{12}}} = \frac{4114}{300} \approx \$13.71$ Interest

\$484 + \$13.71 = \$497.71 amount

7. $\frac{3}{4}\% \times 12 = 9\%$

.09 × \$100 × 1 = \$9 Interest

6A

Study Exercise Two (Frame 15)

1. *Step 1:* 5% ÷ 1 = 5%
Step 2: 1 period per year × 4 years = 4 periods
Step 3: 1.2155
Step 4: \$200 × 1.2155 = \$243.10 amount
Step 5: \$243.10 − \$200.00 = \$43.10 Interest

2. *Step 1:* 4% ÷ 2 = 2%
Step 2: 2 periods per year × 5 years = 10 periods
Step 3: 1.219
Step 4: \$500 × 1.219 = \$609.50 amount
Step 5: \$609.50 − \$500.00 = \$109.50 Interest

3. *Step 1:* $6\% \div 12 = \frac{1}{2}\%$
Step 2: 12 periods per year × $\frac{1}{2}$ year = 6 periods
Step 3: 1.0304
Step 4: \$700 × 1.0304 = \$721.28 amount
Step 5: \$721.28 − \$700 = \$21.28 Interest

4. *Step 1:* 6% ÷ 3 = 2%
Step 2: 3 periods per year × 1 year = 3 periods
Step 3: 1.0612
Step 4: \$800 × 1.0612 = \$848.96 amount
Step 5: \$848.96 − \$800 = \$48.96 Interest

SOLUTIONS TO STUDY EXERCISES

Study Exercise Two (Frame 15, contd.)

5. *Step 1:* $3\% \div 1 = 3\%$
Step 2: 1 period per year × 25 years = 25
Step 3: 2.0938
Step 4: $1,000 × 2.0938 = $2,093.80 amount
Step 5: $2,093.80 − $1,000 = $1,093.80 Interest

15A

Study Exercise Three (Frame 22)

1. *Step 1:* $$\frac{5}{100} \times \$1{,}000 \times \frac{10}{12} = \frac{\$500}{12} = \$41.67$$

Step 2: $$\frac{\$1{,}000 + \$100}{2} = \frac{\$1{,}100}{2} = \$550$$

Step 3: Rate × Principal × Time = Interest
$$r \times \$550 \times \frac{10}{12} = \$41.67$$
$$r \times \frac{5500}{12} = \$41.67$$
$$r = \$41.67 \times \frac{12}{5500} = 5500\overline{)500.04}$$

.0909 ≈ 9.1%

```
       .0909 ≈ 9.1%
5500 |500.04
      495 00
      ------
        5 0400
        4 9500
        ------
           900
```

2. *Step 1:* $$\frac{9}{100} \times \overset{1}{\cancel{\$120}}\,\underset{10}{} \times \frac{5}{\underset{1}{\cancel{12}}} = \frac{45}{10} = \$4.50$$

Step 2: $$\frac{\$120 + \$24}{2} = \frac{\$144}{2} = \$72$$

Step 3: Rate × Principal × Time = Interest
$$r \times \overset{6}{\cancel{\$72}} \times \frac{5}{\underset{1}{\cancel{12}}} = \$4.50$$
$$r \times 30 = 4.50$$
$$r = 4.50 \div 30 = 30\overline{)4.50} \quad .15 = 15\%$$

3. *Step 1:* $$\frac{8}{100} \times \overset{4}{\cancel{\$48}} \times \frac{3}{\underset{1}{\cancel{12}}} = \$.96$$

Step 2: $$\frac{\$48 + \$16}{2} = \frac{\$64}{2} = \$32$$

Step 3: Rate × Principal × Time = Interest
$$r \times \overset{8}{\cancel{\$32}} \times \frac{\overset{1}{\cancel{3}}}{\underset{1}{\cancel{\underset{4}{\cancel{12}}}}} = \$.96$$
$$r \times 8 = .96$$
$$r = .96 \div 8 = .12 = 12\%$$

22A

SOLUTIONS TO STUDY EXERCISES, CONTD.

Study Exercise Four (Frame 27)

Mr. Jones: *With Deduction*

$$\$200 - \$100 = \$100 \qquad \frac{3}{2} \times \frac{1}{\cancel{100}_1} \times \cancel{\$100}^1 = \$1.50$$

Rate × Principal × Time = Interest

$$r \times \$100 \times \frac{1}{12} = \$1.50$$

$$r \times \frac{100}{12} = 1.50$$

$$r = 1.50 \div \frac{100}{12} = 1.5 \times \frac{12}{100} = \frac{18.0}{100} = 18\%$$

Without Deductions

$$\frac{3}{\cancel{2}_1} \times \frac{1}{\cancel{100}_1} \times \cancel{\$200}^{\cancel{2}\,1} = \$3.00$$

Rate × Principal × Time = Interest

$$r \times \$100 \times \frac{1}{12} = \$3.00$$

$$r \times \frac{100}{12} = 3$$

$$r = 3 \div \frac{100}{12} = 3 \times \frac{12}{100} = \frac{36}{100} = 36\%$$

27A

Practice Test—Per Cent—Units 26–33

1. Change to per cents:

a. .32 b. $2\frac{1}{6}$ c. $\frac{2}{5}$

2. Change to decimal fractions:

a. 67% b. 105% c. $\frac{1}{4}\%$

3. Change $14\frac{2}{7}\%$ to a common fraction in lowest terms.

4. Express $233\frac{1}{3}\%$ as a mixed numeral.

5. Find 6.2% of 70.

6. Find 8% of $22.43 to the nearest cent.

7. What per cent of 65 is 13?

8. To the nearest thousandth of a per cent, what per cent of 7 is 1?

9. 125% of what number is 30?

10. The rate of discount is established by the fraction whose numerator is the ____________ and whose denominator is ____________.

11. If the list price is $500 and the net price is $430, what is the rate of discount?

12. The list price of an auto is $2,860 and the rate of discount is 12%.
   a. What is the discount?
   b. What is the net price?

13. A man says, "I bought this suit at a discount of 10% and saved $12."
   a. What was the list price of the suit?
   b. How much did he pay for it?

14. A merchant claims that the television set he is selling has had a discount of 15% and that the net price is $400.
   a. To the nearest dollar, what was the original price?
   b. To the nearest dollar, how much is the discount?

15. A salesman whose rate of commission is 3% sold a house for $24,000.
   a. What was his commission?
   b. What are the net proceeds?

16. The net proceeds on a sale were $2,000 and the salesman's rate of commission was 5%.
   a. To the nearest dollar, what was the total sale?
   b. To the nearest dollar, what was his commission?

17. The original price of an article was $80 and it later sold for $100.
   a. What was the per cent increase?
   b. What would be the per cent decrease if it first sold for $100 and later was reduced to $80?

18. The rate of profit based on the selling price is established by the fraction whose denominator is the ____________ and whose numerator is the ____________.

Practice Test—Per Cent, contd.

19. A dress which cost $40 was sold for $50.
   a. What was the profit?
   b. What was the rate of profit based on the cost?
   c. What was the rate of profit based on the selling price?

20. A suit sold for $100 and the merchant made a $30 profit.
   a. What was the cost?
   b. What was the rate of profit based on the cost? (To nearest whole per cent)
   c. What was the rate of profit based on the selling price?

21. A chair costs $20. Find the selling price to the nearest whole dollar if the rate of profit, 30%, is based:
   a. on the cost.
   b. on the selling price.

22. If the selling price is $30 and the rate of profit is 10%, find the cost and profit if the rate is based:
   a. on the selling price.
   b. on the cost.

23. Find the interest on a loan of $1,200, if the rate is $5\frac{1}{4}\%$ per year and the money is borrowed for 8 months.

24. Find the interest on $900 compounded semi-annually for 5 years at 6%. Use Table II.

25. $300 was to be returned in six equal monthly payments at 8% annual rate plus interest.
   a. How much will the interest be?
   b. What was the average amount borrowed?
   c. Approximate the true annual interest rate to the nearest whole per cent.

Answers—Practice Test—Per Cent

1. a. 32%	b. $216\frac{2}{3}$%	c. 40%
2. a. .67	b. 1.05	c. .0025
3. $\frac{1}{7}$	**4.** $2\frac{1}{3}$	**5.** 4.34
6. $1.79	**7.** 20%	**8.** 14.286%
9. 24	**10.** discount, list or gross price	**11.** 14%
12. a. $343.20	b. $2,516.80	
13. a. $120	b. $108	
14. a. $471	b. $71	
15. a. $720	b. $23,280	
16. a. $2,105	b. $105	
17. a. 25%	b. 20%	
18. selling price, profit		
19. a. $10	b. 25%	c. 20%
20. a. $70	b. 43%	c. 30%
21. a. $26	b. $29	
22. a. cost = $27 profit = $3	b. cost = $27.27 (to nearest cent) profit = $2.73 (to nearest cent)	
23. $42	**24.** $309.51	
25. a. $12	b. $175	c. 14%

unit

34

Introduction to Denominate Numerals

Objectives:

By the end of this unit you should:

1. know the meaning of a denominate numeral.
2. be able to use a table of measure.
3. be able to change denominate numerals to smaller units of measure.
4. be able to change denominate numerals to larger units of measure.
5. be able to convert metric units to English units.
6. be able to convert English units to metric units.
7. be able to make conversions within the metric system. (1)

A *denominate numeral* is a numeral which has a specified name. Abstract numerals have no such designation.

	Denominate Numerals	Abstract Numerals
Line (*a*)	1 inch	1
Line (*b*)	2 acres	2
Line (*c*)	3.7 cubic yards	3.7
Line (*d*)	$\frac{1}{2}$ pint	$\frac{1}{2}$
Line (*e*)	5 months	5
Line (*f*)	6.9 centimeters	6.9
Line (*g*)	$13\frac{1}{4}$ knots	$13\frac{1}{4}$

(2)

Using a Table of Measure

Turn to Table I on page 433. The tape will tell you how to use this table. (3)

Study Exercise One

This exercise will be given on the tape. (4)

Changing to Smaller Units of Measure

To change a given number of units of one denomination to units of a smaller denomination:

1. find the number of units of the smaller denomination that is equivalent to one unit of the larger denomination.
2. multiply the given number of units of the larger denomination by this number which is sometimes called the conversion factor.

Example 1: Change 7 feet to inches.

Solution:

Line (a) 12 inches = 1 foot (from Table I)
Line (b) 12 is our conversion factor
Line (c) 12 × 7 = 84 inches

Example 2: Change 5 pounds to ounces

Solution:

Line (a) 16 ounces = 1 pound (from Table I)
Line (b) 16 is our conversion factor
Line (c) 16 × 5 = 80 ounces

Example 3: Change 4 hours to minutes

Solution:

Line (a) 60 minutes = 1 hour (from Table I)
Line (b) 60 is our conversion factor
Line (c) 60 × 4 = 240 minutes

(5)

More Examples

Example 1: Change 8 square yards to square feet.

Solution:

Line (a) 9 square feet = 1 square yard (from Table II)
Line (b) 9 is our conversion factor
Line (c) 9 × 8 = 72 square feet

Example 2: Change 3 cubic feet to cubic inches

Solution:

Line (a) 1,728 cubic inches = 1 cubic foot (from Table II)
Line (b) 1,728 is our conversion factor
Line (c) 1,728 × 3 = 5,184 cubic inches

(6)

Study Exercise Two

1.	Change	4 days	to	hours
2.	Change	30 yards	to	feet
3.	Change	4 bushels	to	pecks
4.	Change	13 statute miles	to	feet
5.	Change	5 quarts	to	pints
6.	Change	3 long tons	to	pounds
7.	Change	5 fathoms	to	feet
8.	Change	6 gallons	to	quarts
9.	Change	4 square miles	to	acres
10.	Change	15 square yards	to	square feet

(7)

Sometimes we must take more than one step to complete a change.

Example 1: Change 3 quarts to ounces.

Solution:

Line (a) 1 quart = 2 pints
Line (b) $2 \times 3 = 6$ pints
Line (c) 1 pint = 16 ounces
Line (d) $16 \times 6 = 96$ ounces
Line (e) 3 quarts = 96 ounces

(8)

Another Approach

Example 1: Change 3 quarts to ounces

Solution:

Line (a) 1 quart = 2 pints
Line (b) 1 pint = 16 ounces
Line (c) 2 pints = 32 ounces
Line (d) 1 quart = 2 pints = 32 ounces
Line (e) $3 \times 32 = 96$ ounces

(9)

Example: Change 2 weeks to hours

Solution:

Line (a) 1 week = 7 days
Line (b) $2 \times 7 = 14$ days
Line (c) 1 day = 24 hours
Line (d) $14 \times 24 = 336$ hours
or
Line (e) 1 week = 7 days = $7 \times 24 = 168$ hours
Line (f) 2 weeks = $2 \times 168 = 336$ hours

(10)

Study Exercise Three

1.	Change	5 weeks	to	hours
2.	Change	3 gallons	to	ounces
3.	Change	6 yards	to	inches
4.	Change	$\frac{1}{2}$ cup	to	teaspoons
5.	Change	2 bushels	to	pints

(11)

Changing to Larger Units of Measure

To change a given number of units of one denomination to units of a larger denomination:

1. find the number of units of the smaller denomination that is equivalent to one unit of the larger denomination.
2. divide the given number of units of the smaller denomination by this conversion factor.

Example 1: Change 45 feet to yards

Solution:

Line (*a*) 3 feet = 1 yard
Line (*b*) 3 is our conversion factor
Line (*c*) 45 ÷ 3 = 15 yards

Example 2: Change 43 ounces to the nearest tenth of a pound.

Solution:

Line (*a*) 16 ounces = 1 pound
Line (*b*) 16 is our conversion factor
Line (*c*)

$$\begin{array}{r} 2.68 \\ 16\overline{)43.0} \\ 32 \\ \hline 11\,0 \\ 9\,6 \\ \hline 1\,40 \\ 1\,28 \\ \hline 12 \end{array}$$

Line (*d*) 2.7 pounds

(12)

Example 3: Change 40 months to years and months

Solution:

Line (*a*) 12 months = 1 year
Line (*b*) 12 is our conversion factor
Line (*c*)

$$\begin{array}{r} 3 \\ 12\overline{)40} \\ 36 \\ \hline 4 \end{array}$$

Line (*d*) 3 years, 4 months

Example 4: Change 33,420 minutes to weeks, days and hours.

Solution:

Line (a) 60 minutes = 1 hour

Line (b) $60\!\!\!\backslash \overline{)3342 0\!\!\!\backslash}$ = 557 hours (quotient 557)

Line (c) 24 hours = 1 day

Line (d) $24\,\overline{)557}$ = 23 days, 5 hours (quotient 23)

```
      23
 24 |557
     48
     --
     77
     72
     --
      5
```

Line (e) 7 days = 1 week

Line (f) $7\,\overline{)23}$ = 3 weeks, 2 days

Line (g) 33,420 minutes = 3 weeks, 2 days, 5 hours

Study Exercise Four

A. All answers should be rounded to the nearest tenth.

1. Change 29 feet to yards.
2. Change 22 pints to quarts.
3. Change 14,000 pounds to short tons.
4. Change 368 square inches to square feet.
5. Change 208 cubic feet to cubic yards.
6. Change 420 minutes to degrees.
7. Change 22 tablespoons to cups and tablespoons.
8. Change 795 hours to weeks, days and hours.
9. Change 178 ounces to quarts, pints and ounces.

Metric to English Units

Example 1: Change 3 kilometers to miles

Solution:

Line (a) 1 kilometer = .62 miles

Line (b) .62 × 3 = 1.86 miles

Example 2: Change 4 kilograms to pounds.

Solution:

Line (a) 1 kilogram = 2.2 pounds

Line (b) 2.2 × 4 = 8.8 pounds

Example 3: Change 7 centimeters to inches

Solution:

Line (a) 1 centimeter = .39 inches

Line (b) .39 × 7 = 2.73 inches

Study Exercise Five

1. Change 6 meters to yards.
2. Change 5 liters to liquid quarts.
3. Change 14 kilograms to pounds.
4. Change 32 kilometers to miles.
5. Change 48 grams to ounces.

English to Metric Units

Example 1: Change 5 feet to meters.

Solution:

Line (a) 1 foot = .3 meter
Line (b) .3 × 5 = 1.5 meter

Example 2: Change 60 miles per hour to kilometers per hour.

Solution:

Line (a) 1 mile = 1.61 kilometers
Line (b) 1.61 × 60 = 96.60 kilometers per hour

Example 3: Change 2 liquid quarts to liters.

Solution:

Line (a) 1 liquid quart = .95 liter
Line (b) .95 × 2 = 1.90 liters

Study Exercise Six

1. Change 6 pounds to kilograms.
2. Change 7 miles to kilometers.
3. Change 3 ounces to grams.
4. Change 5 inches to millimeters.
5. Change 1 fathom to meters.

Conversions within the Metric System

Example 1: Change 3.4 meters to decimeters.

Solution:

Line (a) 1 meter = 10 decimeters
Line (b) 10 × 3.4 = 34 decimeters

Example 2: Change 2.2 kilometers to meters.

Solution:

Line (a) 1 kilometer = 10 hectometers
Line (b) 10 × 2.2 = 22 hectometers
Line (c) 1 hectometer = 10 dekameters
Line (d) 10 × 22 = 220 dekameters
Line (e) 1 dekameter = 10 meters
Line (f) 10 × 220 = 2,200 meters

A Shorter Method

Example 1: Change 2.2 kilometers to meters.

Solution:

Line (*a*) kilo means thousand
Line (*b*) 1 kilometer means 1,000 meters
Line (*c*) $1000 \times 2.2 = 2{,}200$ meters

(21)

Useful Prefixes

Greek	English Equivalent	Metric	Meaning
deca	ten	dekameter	ten meters
hecto	hundred	hectometer	one hundred meters
kilo	thousand	kilometer	one thousand meters

Latin	English Equivalent	Metric	Meaning
milli	one-thousandth	millimeter	one-thousandth of a meter
centi	one-hundredth	centimeter	one-hundredth of a meter
deci	one-tenth	decimeter	one-tenth of a meter

(22)

Example 1: Change 5 meters into hectometers.

Solution:

Line (*a*) 1 hectometer = 100 meters
Line (*b*) $5 \div 100 = .05$ hectometers

Example 2: Change 5 meters into centimeters.

Solution:

Line (*a*) 1 centimeter = $\frac{1}{100}$ of a meter
Line (*b*) 100 centimeters = 1 meter
Line (*c*) $5 \times 100 = 500$ centimeters

(23)

Study Exercise Seven

Change 4 meters to:

1. millimeters **2.** kilometers **3.** dekameters

REVIEW EXERCISES

You may use Table I for reference.

1. 1 foot = ___________ inches.
2. 1 square mile = ___________ acres.
3. ___________ teaspoons = one tablespoon.
4. 5 quarts = ___________ pints.
5. 5 kilograms = ___________ pounds.
6. 14 grams = ___________ ounces.
7. 30 kilometers per hour = ___________ miles per hour.
8. Which is moving faster: a car at 5 miles per hour or a boat at 5 knots? How much faster? (Answer in feet per hour.)
9. A carpet layer has 29 square yards of carpet. He needs to cover an area of 270 square feet. Does he have enough material?
10. A gallon of paint is designed to cover 600 square feet. The area to be painted contains 46 square yards. How large an area may be painted with the remaining paint? (Answer in square feet and square yards.)
11. Which contains more liquid: a quart bottle or a liter bottle?
12. Find the number of yards in:
    (a) 2 statute miles (b) 6 rods (c) 954 feet
13. Find the number of pints in:
    (a) 7 quarts (b) 80 ounces (c) 3 cups
14. Find the number of hours in:
    (a) 1 week (b) 5,400 seconds (c) 270 minutes
15. Find the number of meters in:
    (a) 6 kilometers (b) 2 dekameters (c) 1.7 centimeters

(25)

SOLUTIONS TO REVIEW EXERCISES

1. 12 inches
2. 640 acres
3. three
4. $5 \times 2 = 10$ pints
5. $5 \times 2.2 = 11$ pounds
6. $14 \times .04 = .56$ ounces
7. $30 \times .62 = 18.6$ miles per hour
8. 1 knot = 1 nautical mile per hour = 6,080 feet per hour
   1 mile per hour = 5,280 feet per hour
   $6{,}080 - 5{,}280 = 800$ feet
   The boat is moving $800 \times 5 = 4{,}000$ feet per hour faster.
9. $29 \times 9 = 261$ square feet
   $270 - 261 = 9$ square feet or 1 square yard short of the amount needed.
10. $600 \div 9 = 66\frac{2}{3}$ square yards $\qquad 66\frac{2}{3} - 46 = 20\frac{2}{3}$ square yards
    $\frac{62}{\not{3}_1} \times \not{9}^3 = 186$ square feet
11. The liter has about 2 ounces more than the quart.
12. (a) $2 \times 1{,}760 = 3{,}520$ yards
    (b) $6 \times 5\frac{1}{2} = 33$ yards
    (c) $954 \div 3 = 318$ yards

SOLUTIONS TO REVIEW EXERCISES, CONTD.

13. (a) $7 \times 2 = 14$ pints
(b) $80 \div 16 = 5$ pints
(c) $3 \times 8 = 24$ ounces; $24 \div 16 = 1\frac{1}{2}$ pint

14. (a) 1 week = 7 days; $7 \times 24 = 168$ hours
(b) $5{,}400 \div 60 = 90$ minutes; $90 \div 60 = 1\frac{1}{2}$ hours
(c) $270 \div 60 = 4\frac{1}{2}$ hours

15. (a) $6 \times 1{,}000 = 6{,}000$ meters
(b) 1 dekameter = 10 meters
2 dekameters = 20 meters
(c) 100 centimeters = 1 meter

$1 \text{ centimeter} = \frac{1}{100} \text{ meter}$

$1.7 \times \frac{1}{100} = .017 \text{ meter}$

SUPPLEMENTARY PROBLEMS

Convert each of the following to the units indicated: (round answers to the nearest tenth.)

1. 72 inches to feet.
2. 9,800 lb. to long tons.
3. 2,582 hours to weeks, days, hours.
4. 16,000 feet to miles, yards, feet.
5. 300 hours to days.
6. 125 months to years, months.
7. 13,320 feet to fathoms.
8. 10,561 seconds to hours, minutes, seconds.
9. 425 pints to gallons, quarts, pints.
10. 3 nautical miles to feet.
11. 10 liters to gallons. (liquid measure)
12. 12 inches to millimeters.
13. A city is 100 feet above sea level. How many meters is this?
14. Which is larger—100 meters or 110 yards? By how much?
15. Convert the speed of sound for a given temperature from 335 meters per second to feet per second.
16. Convert the speed of light from 186,000 miles per second to meters per second.
17. If the speedometer on a foreign car indicates 25 kilometers per hour and the driver is going through a speed zone marked 15 m.p.h. is he violating the law?
18. Who is taller: a man whose height is 5 feet 8 inches, or one whose height is 1.8 meters?
19. The Leaning Tower of Pisa is about 179 feet high. How high is it in meters?
20. The Eiffel Tower in Paris is about 300 meters high. How high is this in feet?

SOLUTIONS TO STUDY EXERCISES

Study Exercise One (Frame 4)

These are given on the tape.

SOLUTIONS TO STUDY EXERCISES, CONTD.

Study Exercise Two (Frame 7)

1. $24 \times 4 = 96$ hours
2. $3 \times 30 = 90$ feet
3. $4 \times 4 = 16$ pecks
4. $5{,}280 \times 13 = 68{,}640$ feet
5. $2 \times 5 = 10$ pints
6. $2{,}240 \times 3 = 6{,}720$ pounds
7. $6 \times 5 = 30$ feet
8. $4 \times 6 = 24$ quarts
9. $640 \times 4 = 2{,}560$ acres
10. $9 \times 15 = 135$ square feet

7A

Study Exercise Three (Frame 11)

1. $5 \times 7 = 35$ days; $35 \times 24 = 840$ hours
2. $3 \times 4 = 12$ quarts; $12 \times 2 = 24$ pints; $24 \times 16 = 384$ ounces
3. $6 \times 3 = 18$ feet; $18 \times 12 = 216$ inches
4. $\frac{1}{2} \times 16 = 8$ tablespoons; $8 \times 3 = 24$ teaspoons
5. $2 \times 4 = 8$ pecks; $8 \times 8 = 64$ quarts; $64 \times 2 = 128$ pints

11A

Study Exercise Four (Frame 15)

A. 1. $29 \div 3 \approx 9.7$ yards
2. $22 \div 2 = 11$ quarts
3. $14{,}000 \div 2{,}000 = 7$ short tons
4. $368 \div 144 \approx 2.6$ square feet
5. $208 \div 27 \approx 7.7$ cubic yards
6. $420 \div 60 = 7$ degrees
7. $22 \div 16 = 1$ cup, 6 tablespoons
8. $795 \div 24 = 33$ days, 3 hours; $33 \div 7 = 4$ weeks, 5 days
   795 hours = 4 weeks, 5 days, 3 hours
9. $178 \div 16 = 11$ pints, 2 ounces; $11 \div 2 = 5$ quarts, 1 pint
   178 ounces = 5 quarts, 1 pint, 2 ounces

15A

Study Exercise Five (Frame 17)

1. $6 \times 1.09 = 6.54$ yards
2. $5 \times 1.06 = 5.30$ liquid quarts
3. $14 \times 2.2 = 30.8$ pounds
4. $32 \times .62 = 19.84$ miles
5. $48 \times .04 = 1.92$ ounces

17A

Study Exercise Six (Frame 19)

1. $6 \times .45 = 2.70$ kilograms
2. $7 \times 1.61 = 11.27$ kilometers
3. $3 \times 28.35 = 85.05$ grams
4. $5 \times 25.4 = 127$ millimeters
5. 1 fathom = 6 feet
   $6 \times .3 = 1.8$ meters

SOLUTIONS TO STUDY EXERCISES, CONTD.

Study Exercise Seven (Frame 24)

1. 1,000 millimeters = 1 meter
$4 \times 1{,}000 = 4{,}000$ millimeters

2. 1 kilometer = 1,000 meters

$\frac{1}{1{,}000}$ of a kilometer = 1 meter

$$4 \times \frac{1}{1{,}000} = \frac{4}{1{,}000} = .004 \text{ kilometer}$$

3. 10 meters = 1 dekameter

1 meter = $\frac{1}{10}$ dekameter

$$4 \times \frac{1}{10} = \frac{4}{10} = .4 \text{ dekameter}$$

24A

unit

35

Operations With Denominate Numbers

Objectives:

By the end of this unit you should be able to:

1. do the following four fundamental operations with denominate numbers:
   a) Addition
   b) Subtraction
   c) Multiplication
   d) Division

2. express your answers in simplified form.

(1)

Simplified Form

Whenever we are given a number of units which may in turn be converted to larger units we shall make this conversion and say that our answer is in simplified form.

Example 1: 35 ounces (weight) may be converted to:

Line (a) 35 oz. ÷ 16 = 2 lb. 3 oz.

Example 2: 15 pints (liquid) may be converted to:

Line (a) 15 pt. ÷ 2 = 7 qt. 1 pt.
but 7 qt. ÷ 4 = 1 gal. 3 qt.

Line (b) 15 pt. = 1 gal. 3 qt. 1 pt.

(2)

Study Exercise One

Simplify the following:

1. 39 oz. (liquid)

2. 420 in.

3. 320 sec. (time)

4. 280 sq. ft.

5. 420″ (angle)

(3)

Addition of Denominate Numbers

Example 1:

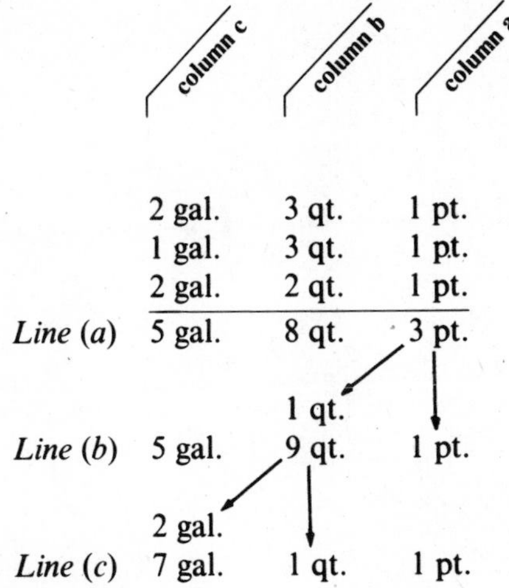

column c
column b
column a
2 gal. 3 qt. 1 pt.
1 gal. 3 qt. 1 pt.
2 gal. 2 qt. 1 pt.
Line (a) 5 gal. 8 qt. 3 pt.
1 qt.
Line (b) 5 gal. 9 qt. 1 pt.
2 gal.
Line (c) 7 gal. 1 qt. 1 pt.

④

Example 2:

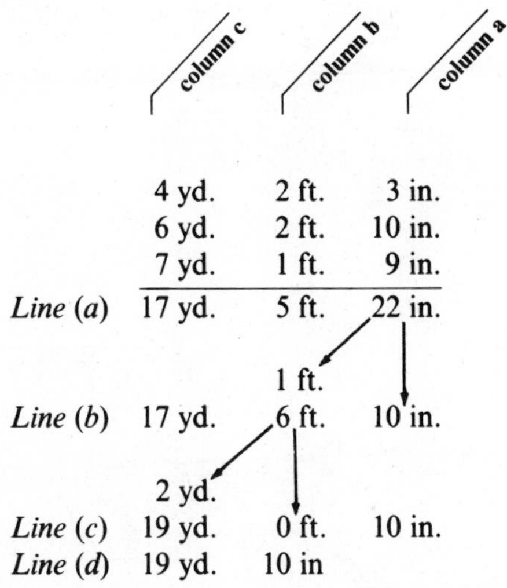

column c
column b
column a
4 yd. 2 ft. 3 in.
6 yd. 2 ft. 10 in.
7 yd. 1 ft. 9 in.
Line (a) 17 yd. 5 ft. 22 in.
1 ft.
Line (b) 17 yd. 6 ft. 10 in.
2 yd.
Line (c) 19 yd. 0 ft. 10 in.
Line (d) 19 yd. 10 in

⑤

Study Exercise Two

Add the following and simplify:

1. 2 ft. 5 in.
4 ft. 10 in.

2. 3 yd. 18 in.
4 yd. 26 in.

3. 3 mi. 600 yd.
4 mi. 440 yd.
7 mi. 900 yd.

4. 1 c. 2 tbsp. 1 tsp.
3 c. 5 tbsp. 2 tsp.
4 c. 10 tbsp. 1 tsp.

5. 5 hr. 20 min. 51 sec.
3 hr. 22 min. 46 sec.
8 hr. 17 min. 14 sec.

6. 8° 34′ 40″
41° 52′ 48″

Subtraction of Denominate Numbers

Example 1: Subtract

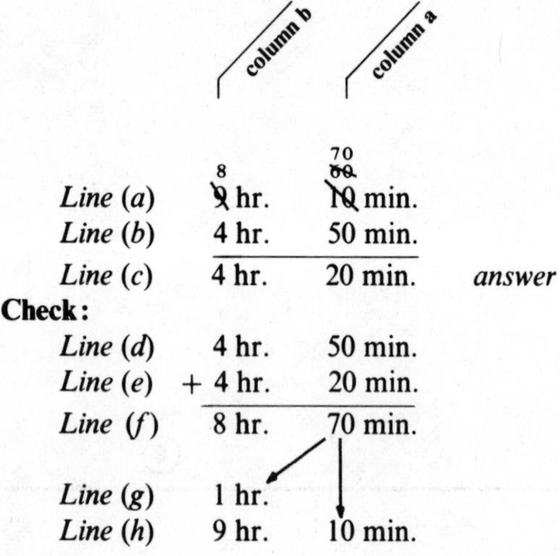

	column b	column a	
	8	70 ~~60~~	
Line (*a*)	~~9~~ hr.	~~10~~ min.	
Line (*b*)	4 hr.	50 min.	
Line (*c*)	4 hr.	20 min.	*answer*

Check:

Line (*d*)	4 hr.	50 min.
Line (*e*)	+ 4 hr.	20 min.
Line (*f*)	8 hr.	70 min.
Line (*g*)	1 hr.	
Line (*h*)	9 hr.	10 min.

(7)

Example 2:

	column c	column b	column a	
	3	30 ~~24~~ ~~6~~	80 ~~60~~	
Line (*a*)	~~4~~ da.	~~7~~ hr.	~~20~~ min.	
Line (*b*)	2 da.	9 hr.	40 min.	
Line (*c*)	1 da.	21 hr.	40 min.	*answer*

Check:

Line (*d*)	2 da.	9 hr.	40 min.
Line (*e*)	1 da.	21 hr.	40 min.
Line (*f*)	3 da.	30 hr.	80 min.
Line (*g*)		1 hr.	
Line (*h*)	3 da.	31 hr.	20 min.
Line (*i*)	1 da.		
Line (*j*)	4 da.	7 hr.	20 min.

(8)

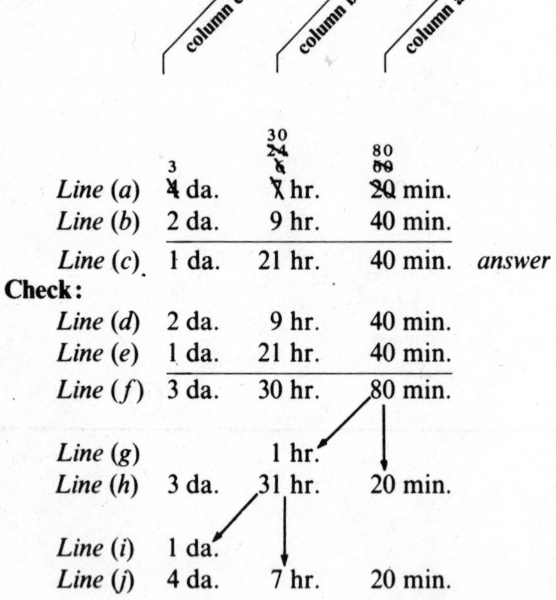

Study Exercise Three

Subtract the following and check your work:

1. 3 gal. 1 qt.
1 gal. 3 qt.

2. 3 qt. 1 pt. 4 oz.
1 qt. 1 pt. 6 oz.

3. 4 lb. 2 oz.
1 lb. 7 oz.

4. 5 bu. 2 pk. 1 qt.
2 bu. 2 pk. 7 qt.

5. 180°
92° 30′ 45″

(9)

Multiplication of Denominate Numbers

Example 1: Multiply

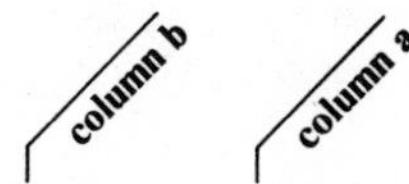

Line (a)	4 yd.	2 ft.
Line (b)		5
Line (c)	20 yd.	10 ft.
Line (d)	3 yd.	
Line (e)	23 yd.	1 ft. *answer*

10

Example 2: Multiply

column c column b column a

Line (a)	4 gal.	2 qt.	1 pt.
Line (b)			7
Line (c)	28 gal.	14 qt.	7 pt.
Line (d)		3 qt.	
Line (e)	28 gal.	17 qt.	1 pt.
Line (f)	4 gal.		
Line (g)	32 gal.	1 qt.	1 pt. *answer*

11

Study Exercise Four

Multiply the following and simplify:

1. 3 yd. 2 ft. 7 in. × 5

2. 7 gal. 3 qt. × 4

3. 1 lb. 11 oz. × 5

4. 8 yr. 3 mo. × 6

5. 22° 15′ × 8

12

Division of Denominate Numbers

Example 1:

Line (a)		3 lb. 7 oz.	
Line (b)	5	17 lb. 3 oz.	
Line (c)		15 lb.	
Line (d)		2 lb. 3 oz. =	35 oz.
Line (e)			35 oz.
Line (f)			0 oz.

13

Example 2:

Line (a)		5 lb. 10 oz.	
Line (b)	5	28 lb. 4 oz.	
Line (c)		25 lb.	
Line (d)		3 lb. 4 oz. =	52 oz.
Line (e)			50 oz.
Line (f)			2 oz. remainder

14

Example 3:

Line (a)		2 yd. 1 ft.		4 in.	
Line (b)	4	9 yd. 2 ft.		5 in.	
Line (c)		8 yd.			
Line (d)		1 yd. 2 ft. =	5 ft.		
Line (e)			4 ft.		
Line (f)			1 ft. 5 in. =		17 in.
Line (g)					16 in.
Line (h)					1 in. remainder

15

A Check of the Problem

Line (a)	2 yd.	1 ft.	4 in.
Line (b)			4
Line (c)	8 yd.	4 ft.	16 in.
Line (d)			1 in. remainder
Line (e)	8 yd.	4 ft.	17 in.
		1 ft.	
Line (f)	8 yd.	5 ft.	5 in.
	1 yd.		
Line (g)	9 yd.	2 ft.	5 in.

16

Study Exercise Five

Divide and Check the following:

1. 4 |26 yd. 2 ft.

2. 10 |31 gal. 2 qt. 1 pt.

3. 5 |26 1.ton 1600 lb.

4. 2 |3 c. 5 tbsp.

5. 3 |18° 31′ 40″

17

REVIEW EXERCISES

A. Add and Simplify:

1.
6 yd. 2 ft. 7 in.
4 yd. 1 ft. 9 in.
2 yd. 2 ft. 10 in.

2.
6 gal. 3 qt.
5 gal. 1 qt.
4 gal. 2 qt.

3.
7 lb. 8 oz.
9 lb. 10 oz.
4 lb. 11 oz.

4.
4 hr. 20 min. 30 sec.
3 hr. 10 min. 50 sec.
8 hr. 30 min. 40 sec.

B. Subtract and Check:

5.
170° 25′
40° 35′

6.
7 hr. 20 min. 5 sec.
3 hr. 30 min. 40 sec.

7.
9 lb. 4 oz.
4 lb. 10 oz.

8.
4 yd. 1 ft. 6 in.
2 yd. 2 ft. 8 in.

C. Multiply and Simplify:

9.
10 yd. 2 ft.
3

10.
4 gal. 2 qt. 1 pt.
5

11.
9 bu. 2 pk.
6

12.
2 hr. 32 min. 3 sec.
4

D. Divide and Check:

13. 3)27 yr. 9 mo.

14. 4)30° 22′ 5″

15. 6)22 lb. 14 oz.

16. 5)22 gal. 3 qt. 1 pt.

E.

17. Twenty years ago the height of a tree was 4 ft. 8 in. Today its height is 63 ft. 4 in. How much did it grow?

18. If a *gallon* of milk weighs 8.3 lb., how many *quarts* may be obtained from 480 lb. of milk? (Answer to the nearest whole quart.)

19. 48 small boxes of cereal each weighing 1 lb. 3 oz. are to be packed in a carton weighing 2 lb. 4 oz. How much will ten fully packed cartons weigh?

20. The first floor of a ten-story building is 21 ft. 5 in. high. The other 8 floors are each 12 ft. 4 in. high and the last floor is 14 ft. 7 in. high. How high is the building? (18)

SOLUTIONS TO REVIEW EXERCISES

A. 1.
6 yd. 2 ft. 7 in.
4 yd. 1 ft. 9 in.
2 yd. 2 ft. 10 in.
12 yd. 5 ft. 26 in.
2 ft. ↓
12 yd. 7 ft. 2 in.
2 yd. ↓
14 yd. 1 ft. 2 in.

2.
6 gal. 3 qt.
5 gal. 1 qt.
4 gal. 2 qt.
15 gal. 6 qt.
1 gal. ↓
16 gal. 2 qt.

SOLUTIONS TO REVIEW EXERCISES, CONTD.

(Frame 19, contd.)

3.

```
 7 lb.  8 oz.
 9 lb. 10 oz.
 4 lb. 11 oz.
-------------
20 lb. 29 oz.
   1 lb.
-------------
21 lb. 13 oz.
```

4.

```
 4 hr. 20 min.  30 sec.
 3 hr. 10 min.  50 sec.
 8 hr. 30 min.  40 sec.
-----------------------
15 hr. 60 min. 120 sec.
        2 min.
15 hr. 62 min.   0 sec.
 1 hr.
-----------------------
16 hr.  2 min.   0 sec.
```

B. 5.

```
 169°  85′  (60′ crossed out)
 170°  25′  (crossed out)
  40°  35′
-----------
 129°  50′
```

Check:

```
  40° 35′
 129° 50′
---------
 169° 85′
  1°
---------
 170° 25′
```

6.

```
          79  (80 crossed out)       65  (60 crossed out)
 6        19  (crossed out)
 7 hr. 20 min.  5 sec.  (7, 20, 5 crossed out)
 3 hr. 30 min. 40 sec.
----------------------
 3 hr. 49 min. 25 sec.
```

Check:

```
 3 hr. 30 min. 40 sec.
 3 hr. 49 min. 25 sec.
----------------------
 6 hr. 79 min. 65 sec.
        1 min.
 6 hr. 80 min.  5 sec.
 1 hr.
----------------------
 7 hr. 20 min.  5 sec.
```

7.

```
 8       20  (16 crossed out)
 9 lb.  4 oz.  (9, 4 crossed out)
 4 lb. 10 oz.
-------------
 4 lb. 10 oz.
```

Check:

```
 4 lb. 10 oz.
 4 lb. 10 oz.
-------------
 8 lb. 20 oz.
 1 lb.
-------------
 9 lb.  4 oz.
```

8.

```
 3      3  (0 crossed out)  18  (12 crossed out)
 4 yd. 1 ft.  6 in.  (4, 1, 6 crossed out)
 2 yd. 2 ft.  8 in.
-------------------
 1 yd. 1 ft. 10 in.
```

Check:

```
 2 yd. 2 ft.  8 in.
 1 yd. 1 ft. 10 in.
-------------------
 3 yd. 3 ft. 18 in.
       1 ft.
 3 yd. 4 ft.  6 in.
 1 yd.
-------------------
 4 yd. 1 ft.  6 in.
```

SOLUTIONS TO REVIEW EXERCISES, CONTD.

(Frame 19, contd.)

C. **9.**

10 yd. 2 ft.
3
30 yd. 6 ft.
2 yd.
32 yd. 0 ft.

10.

4 gal. 2 qt. 1 pt.
5
20 gal. 10 qt. 5 pt.
2 qt.
20 gal. 12 qt.
3 gal.
23 gal. 0 qt. 1 pt.

11.

9 bu. 2 pk.
6
54 bu. 12 pk.
3 bu.
57 bu. 0 pk.

12.

2 hr. 32 min. 3 sec.
4
8 hr. 128 min. 12 sec.
2 hr.
10 hr. 8 min. 12 sec.

D. **13.**

9 yr. 3 mo.
3 | 27 yr. 9 mo.
27 yr. 9 mo.

Check:

9 yr. 3 mo.
× 3
27 yr. 9 mo.

14.

7° 35′ 31″
4 | 30° 22′ 5″
28°
2° 22′ = 142′
140′
2′ 5″ = 125″
124″
1″ remainder

Check:

7° 35′ 31″
4
28° 140′ 124″
1″ remainder
28° 140′ 125″
2′
28° 142′ 5″
2°
30° 22′ 5″

15.

3 lb. 13 oz.
6 | 22 lb. 14 oz.
18 lb.
4 lb. 14 oz. = 78 oz.
78 oz.
0 oz. remainder

Check:

3 lb. 12 oz.
6
18 lb. 72 oz.
6 oz. remainder
18 lb. 78 oz.
4 lb.
22 lb. 14 oz.

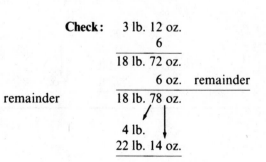

SOLUTIONS TO REVIEW EXERCISES, CONTD.

(Frame 19, contd.)

16.

```
        4 gal. 2 qt. 0 pt.
    5 |22 gal. 3 qt. 1 pt.
       20 gal.
        2 gal. 3 qt. = 11 qt.
                       10 qt.
                        1 qt. 1 pt. = 3 pt.
                                      0 pt.
                                      3 pt.   remainder
```

Check:

```
 4 gal.  2 qt.
              5
20 gal. 10 qt.
         3 pt.   remainder
20 gal. 10 qt. 3 pt.
         1 qt.
20 gal. 11 qt. 1 pt.
 2 gal.
22 gal.  3 qt. 1 pt.
```

E. 17.

```
  62      16
  63 ft.  4 in.    (63 crossed out to 62; 4 crossed out to 16)
-  4 ft.  8 in.
  58 ft.  8 in.
```

18. 1 gal. or 4 qt. weigh 8.3 lb.

```
                   2.075 lbs.
1 qt. weighs   4 |8.300
                  231.3
        2.075 |480000.0  ≈ 231 qt.
```

19.

```
   1 lb.     3 oz.
          × 48
  48 lb.   144 oz.
   9 lb.
  57 lb.     0 oz.
+  2 lb.     4 oz. weight of carton
  59 lb.     4 oz. per carton
          × 10
 590 lb.    40 oz.
   2 lb.
 592 lb.     8 oz.
```

20.

```
 12 ft.   4 in.
        × 8
 96 ft.  32 in.
 21 ft.   5 in. first floor
 14 ft.   7 in. last floor
131 ft.  44 in.
  3 ft.
134 ft.   8 in.
```

(19)

SUPPLEMENTARY PROBLEMS

A. Add:

1. 7 yd. 2 ft. 3 in.
4 yd. 1 ft. 6 in.
6 yd. 3 ft. 2 in.

2. 1 gal. 2 qt. 1 pt.
4 gal. 3 qt. 1 pt.
6 gal. 2 qt. 0 pt.

3. $16^\circ\ 5'$
$42^\circ\ 3'$
$17^\circ\ 22'$

B. Subtract and check:

4. 4 hr. 20 min. 16 sec.
2 hr. 30 min. 17 sec.

5. 3 yd. 2 ft. 6 in.
1 yd. 2 ft. 8 in.

6. 8 lb. 6 oz.
4 lb. 10 oz.

C. Multiply and simplify:

7. $30^\circ\ 22'$
× 3

8. 6 da. 15 hr. 36 min.
× 4

9. 2 c. 6 tbsp.
× 6

D. Divide and check:

10. 3)51 yr. 6 mo.

11. 8)23 lb. 15 oz.

12. 7)$40^\circ\ 22'\ 13''$

E. Find:

13. A baby weighed 8 lb. 12 oz. when born. Today he weighs 17 lb. 8 oz. How much did he gain?

14. Athletic award ribbons each 4 in. long were made from 9 yd. of ribbon. If 22 awards were given, how much of the ribbon remained?

15. A 6-oz. can of frozen orange concentrate requires 3 cans of water to make orange juice. How many such cans would be needed to make 3 gallons?

SOLUTIONS TO STUDY EXERCISES

Study Exercise One (Frame 3)

1. 39 oz. ÷ 16 = 2 pt. 7 oz. = 1 qt. 7 oz.
2. 420 in. ÷ 12 = 35 ft. = 11 yd. 2 ft.
3. 320 sec. ÷ 60 = 5 min. 20 sec.
4. 280 sq. ft. ÷ 9 = 31 sq. yd. 1 sq. ft.
5. $420'' \div 60 = 7'$

3A

SOLUTIONS TO STUDY EXERCISES, CONTD.

Study Exercise Two (Frame 6)

1.

2 ft. 5 in.
4 ft. 10 in.
6 ft. 15 in.
1 ft.
7 ft. 3 in.
2 yd. 1 ft.
2 yd. 1 ft. 3 in.

2.

3 yd. 18 in.
4 yd. 26 in.
7 yd. 44 in.
1 yd.
8 yd. 8 in.

3.

3 mi. 600 yd.
4 mi. 440 yd.
7 mi. 900 yd.
14 mi. 1940 yd.
1 mi.
15 mi. 180 yd.

4.

1 c. 2 tbsp. 1 tsp.
3 c. 5 tbsp. 2 tsp.
4 c. 10 tbsp. 1 tsp.
8 c. 17 tbsp. 4 tsp.
1 tbsp.
8 c. 18 tbsp. 1 tsp.
1 c.
9 c. 2 tbsp. 1 tsp.

5.

5 hr. 20 min. 51 sec.
3 hr. 22 min. 46 sec.
8 hr. 17 min. 14 sec.
16 hr. 59 min. 111 sec.
1 min.
16 hr. 60 min. 51 sec.
1 hr.
17 hr. 0 min. 51 sec.

6.

8° 34′ 40″
41° 52′ 48″
49° 86′ 88″
1′
49° 87′ 28″
1°
50° 27′ 28″

6A

Study Exercise Three (Frame 9)

1.

$\overset{2}{\not 3}$ gal. $\overset{5}{\not 1}$ qt.
1 gal. 3 qt.
1 gal. 2 qt.

Check:

1 gal. 3 qt.
1 gal. 2 qt.
2 gal. 5 qt.
1 gal.
3 gal. 1 qt.

2.

$\overset{2}{\not 3}$ qt. $\overset{2}{\not 1}$ pt. $\overset{20}{\not 4}$ oz.
1 qt. 1 pt. 6 oz.
1 qt. 1 pt. 14 oz.

Check:

1 qt. 1 pt. 6 oz.
1 qt. 1 pt. 14 oz.
2 qt. 2 pt. 20 oz.
1 pt.
2 qt. 3 pt. 4 oz.
1 qt.
3 qt. 1 pt. 4 oz.

SOLUTIONS TO STUDY EXERCISES, CONTD.

Study Exercise Three (Frame 9, contd.)

3.

```
 3    18
 4 lb. 2 oz.
 1 lb. 7 oz.
 ___________
 2 lb. 11 oz.
```

Check:

```
 1 lb.  7 oz.
 2 lb. 11 oz.
 ____________
 3 lb. 18 oz.
 1 lb.
 4 lb.  2 oz.
```

4.

```
 4     5     9
 5 bu. 2 pk. 1 qt.
 2 bu. 2 pk. 7 qt.
 _________________
 2 bu. 3 pk. 2 qt.
```

Check:

```
 2 bu. 2 pk. 7 qt.
 2 bu. 3 pk. 2 qt.
 _________________
 4 bu. 5 pk. 9 qt.
       1 pk.
 4 bu. 6 pk. 1 qt.
 1 bu.
 5 bu. 2 pk. 1 qt.
```

5.

```
 179°  59'  60"
 180°   0'   0"
  92°  30'  45"
 ______________
  87°  29'  15"
```

Check:

```
  92°  30'  45"
  87°  29'  15"
 ______________
 179°  59'  60"
        1'
 179°  60'   0"
   1°
 180°   0'   0"
 ______________
 180°
```

9A

Study Exercise Four (Frame 12)

1.

```
  3 yd.  2 ft.   7 in.
                × 5
 _____________________
 15 yd. 10 ft.  35 in.
         2 ft.
 15 yd. 12 ft.  11 in.
  4 yd.
 19 yd.  0 ft.  11 in.
```

2.

```
  7 gal.   3 qt.
          × 4
 _______________
 28 gal.  12 qt.
  3 gal.
 31 gal.   0 qt.
```

3.

```
 1 lb.  11 oz.
       × 5
 _____________
 5 lb.  55 oz.
 4 lb.
 9 lb.   7 oz.
```

4.

```
  8 yr.   3 mo.
         × 6
 ______________
 48 yr.  18 mo.
  1 yr.
 49 yr.   6 mo.
```

SOLUTIONS TO STUDY EXERCISES, CONTD.

Study Exercise Four (Frame 12, contd.)

5.

```
 22°    15′
      ×  8
----------
176°   120′
  2°
178°     0′
```

12A

Study Exercise Five (Frame 17)

1.

```
    6 yd. 2 ft.
4 | 26 yd. 2 ft.
    24 yd.
     2 yd. 2 ft. = 8 ft.
                   8 ft.
```

Check:

```
 6 yd. 2 ft.
        × 4
-----------
24 yd. 8 ft.
 2 yd.
26 yd. 2 ft.
```

2.

```
     3 gal. 0 qt.        1 pt.
10 | 31 gal. 2 qt.       1 pt.
     30 gal.
      1 gal. 2 qt. = 6 qt.
                     0 qt.
                     6 qt. 1 pt. = 13 pt.
                                   10 pt.
                                    3 pt.  remainder
```

Check:

```
   3 gal.  1 pt.
          × 10
---------------
  30 gal. 10 pt.
           3 pt.  remainder
---------------
  30 gal. 13 pt.
           6 qt.
30 gal. 6 qt. 1 pt.
 1 gal.
31 gal. 2 qt. 1 pt.
```

3.

```
    5 l. ton  768 lb.
5 | 26 l. ton 1600 lb.
    25 l. ton
     1 l. ton 1600 lb. = 3840 lb.
                         3840 lb.
```

```
 5 l. ton  768 lb.
              × 5
------------------
25 l. ton 3840 lb.
 1 l. ton
26 l. ton 1600 lb.
```

SOLUTIONS TO STUDY EXERCISES, CONTD.

Study Exercise Five (Frame 17, contd.)

4.
```
     1 c. 10 tbsp.
  2 | 3 c.  5 tbsp.
      2 c.
      1 c.  5 tbsp. = 21 tbsp.
                      20 tbsp.
                       1 tbsp.  remainder
```

Check:
```
   1 c. 10 tbsp.
        × 2
   2 c. 20 tbsp.
         1 tbsp.  remainder
   2 c. 21 tbsp.
   1 c.
   3 c.  5 tbsp.
```

5.
```
     6° 10′ 33″
  3 | 18° 31′ 40″
      18°
       0° 31′
          30′
           1′ 40″ = 100″
                     99″
                      1″  remainder
```

Check:
```
   6° 10′ 33″
          × 3
   18° 30′ 99″
            1″  remainder
   18° 30′ 100″
        1′
   18° 31′ 40″
```

17A

unit

36

Perimeter and Circumference

Objectives:

By the end of this unit you should:

1. know the meaning of the following terms:

a) *polygon*
b) *regular polygon*
c) *perimeter*
d) *circumference*
e) *diameter*
f) *radius*
g) *rectangle*
h) *square*
i) *triangle*
j) *isosceles triangle*
k) *equilateral triangle*
l) *parallelogram*
m) *trapezoid*
n) *Pi* (π)
o) *quadrilateral*
p) *hexagon*
q) *octagon*
r) *pentagon*
s) *decagon*

2. be able to find the perimeters of given polygons.

3. be able to find the circumference of given circles.

4. be able to find the diameter of a circle when the radius is given. ①

Polygons

A *polygon* is a plane figure made of line segments which begin at a point and end at that point without crossing.

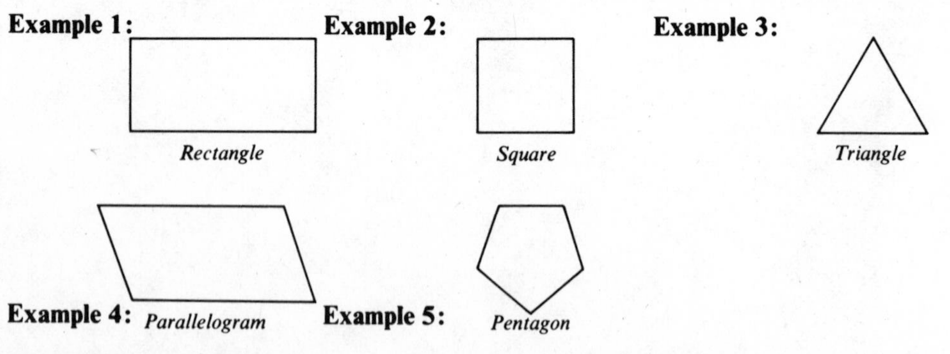

Example 1: *Rectangle*

Example 2: *Square*

Example 3: *Triangle*

Example 4: *Parallelogram*

Example 5: *Pentagon*

②

These are not polygons:

Example 1:

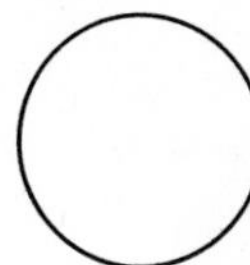

Example 2:

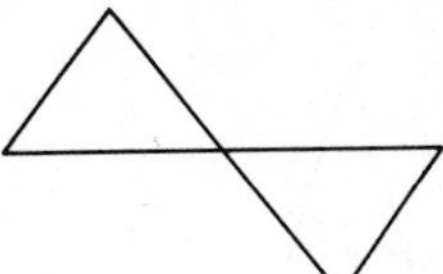

Example 3:

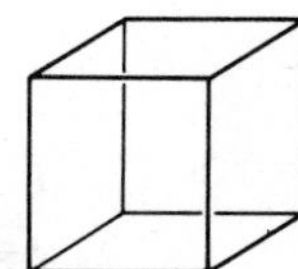

Example 4:

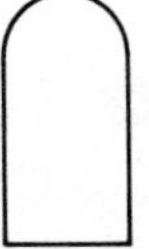

1. A circle is not made of line segments.
2. Segments crossing.
3. Not a plane figure; this is three dimensional.
4. Not made of line segments.

③

Some Common Polygons

Number of Sides	Name
3	triangle
4	quadrilateral
5	pentagon
6	hexagon
8	octagon
10	decagon

④

Special Triangles

Scalene Triangle

No sides are equal.

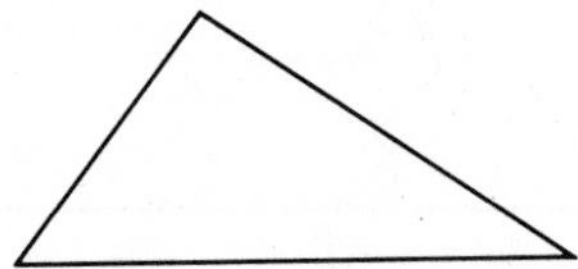

(Frame 5, contd.)

Isosceles Triangle

Two sides are equal.

Equilateral Triangle

All three sides are equal.

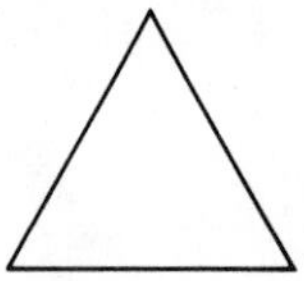

(5)

Special Quadrilaterals

Rectangle

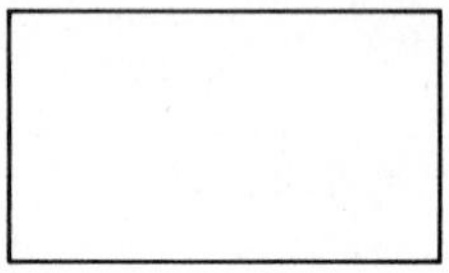

has two pairs of opposite sides which are equal and parallel and four right angles.

Square

has four equal sides with the opposite sides parallel and four right angles.

Parallelogram

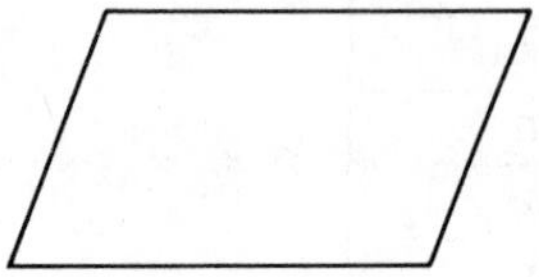

has two pairs of opposite sides which are parallel and equal.

Trapezoid

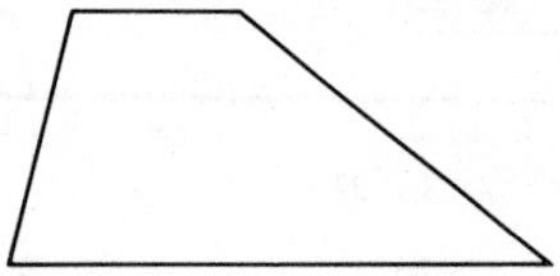

has only one pair of opposite sides that are parallel.

(6)

Regular Polygons

When the sides and angles are equal we use the term *regular*.

Example 1: A six-sided polygon of equal sides and angles is called a *regular hexagon*.

Example 2: A ten-sided polygon of equal sides and angles is called a *regular decagon*.

(7)

Study Exercise One

A. Complete the following chart:

	Number of Sides	Name
1.		quadrilateral
2.		hexagon
3.	10	
4.	3	
5.		octagon
6.	5	

B. Which of the following are not polygons:

7.

8.

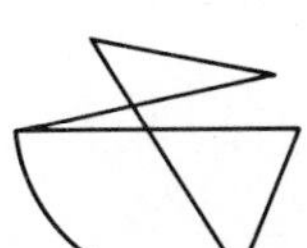

9.

10.

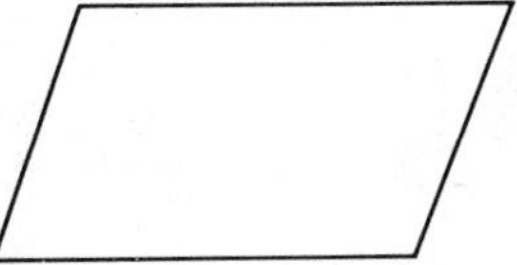

11.

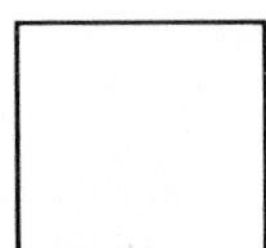

12.

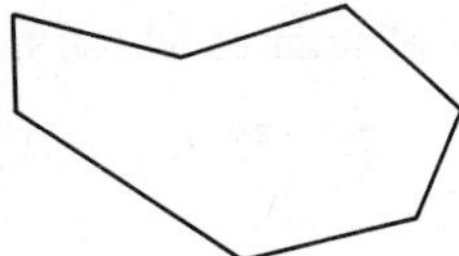

C. Match column B with column A.

Column A

13. has only one pair of opposite parallel sides.
14. has two pairs of opposite equal parallel sides.
15. has 5 equal sides
16. has two pairs of opposite sides which are equal and parallel and four right angles.
17. has 6 sides
18. has 3 sides all of which are equal
19. has 3 sides 2 of which are equal

Column B

(a) regular hexagon
(b) rectangle
(c) parallelogram
(d) regular pentagon
(e) hexagon
(f) trapezoid
(g) isosceles triangle
(h) pentagon
(i) octagon
(j) equilateral triangle

Perimeter

The *perimeter* of a polygon is the sum of the lengths of its sides. It is the distance around the polygon.

Perimeter of a Square

Formula: $p = 4s$

Example 1: Find the perimeter of a square with a side of 6″.

Solution:

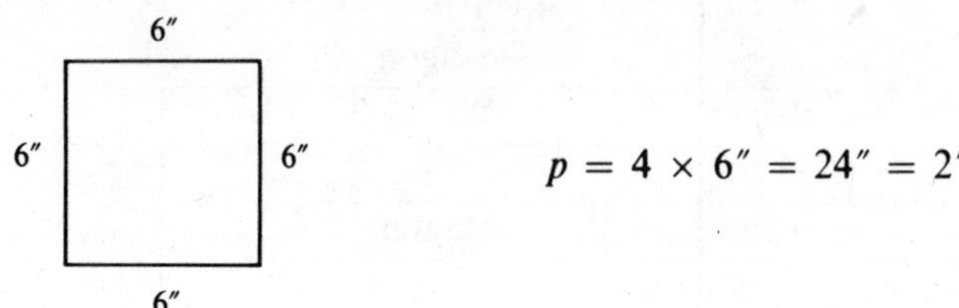

$p = 4 \times 6'' = 24'' = 2'$

Example 2: How many yards of vinyl baseboard are needed for a kitchen floor 12 ft. by 12 ft.? How much will this cost at $.62 per yard?

Solution:

Line (a) $p = 4s$
Line (b) $p = 4 \times 12' = 48'$
Line (c) $48' \div 3' = 16$ yd.
Line (d) 16 yd. × \$.62 = \$9.92

Study Exercise Two

A. Find the perimeter of squares whose sides measure:

1. 8 in.
2. .75 in.
3. $5\frac{3}{4}$ in.
4. 3 ft. 9 in.
5. 4 yd. 2 ft. 7 in.

B.
6. If the distance between bases on a softball diamond is 60 ft. how many ft. does a batter run when he hits a home run? When he hits a triple?
7. How much would it cost at $2.80 per foot to fence a square garden 20 yd. by 20 yd.?

Perimeter of a Rectangle

Formula: $p = 2l + 2w$
$p = 2(l + w)$

Example 1: Find the perimeter of a rectangle with a length of 5 ft. 2 in. and a width of 3 ft. 5 in.:

Solution:

5′ 2″

Line (a) 3′ 5″ $5'\,2'' + 3'\,5'' = 8'\,7''$

(Frame 12, contd.)

Line (b) $p = 2(l + w)$

Line (c) $p = 2 \times 8'\,7'' = 16'\,14''$ or $17'\,2''$

or

Line (d) $p = 2l + 2w$

Line (e) $p = 2 \times 5'\,2'' + 2 \times 3'\,5''$

Line (f) $p = 10'\,4'' + 6'\,10'' = 16'\,14''$ or $17'\,2''$

Study Exercise Three

A. Find the perimeters of rectangles having the following dimensions and simplify:

	1	2	3	4	5
Length	6.2 yd.	$4\frac{3}{4}$ ft.	17 in.	3 ft. 8 in.	16 ft.
Width	4.3 yd.	$3\frac{1}{8}$ ft.	9 in.	1 ft. 3 in.	9 ft. 10 in.

B.

6. If 8 ft. are to be deducted for a doorway, how many feet of baseboard are needed for a room 22 ft. long and 18 ft. wide?
7. A yard measures 60 ft. by 30 ft. How many feet of fencing are needed to enclose it if one 30 ft.-side is bounded by the house? At $3.25 a foot, how much will the fencing cost?
8. A table top 5 ft. 2 in. long and 3 ft. 4 in. wide requires an aluminum molding on the edge. How much molding will be needed? At 18¢ a foot how much will this cost?
9. The rubber stripping on a refrigerator door $2\frac{1}{2}'$ wide and 5′ high is to be replaced. How much stripping is needed? At $1.87 a yard, how much will this cost?

Perimeter of a Triangle

Formula: $p = a + b + c$

Example 1: Find the perimeter of a triangle with sides measuring:

13 in., 12 in., 15 in.

Solution:

Line (a) $p = a + b + c$

Line (b) $p =$ 13 in. + 12 in. + 15 in. = 40 in.

= 3 ft. 4 in.

= 1 yd. 4 in.

(Frame 14, contd.)

Example 2: Find the perimeter of an equilateral triangle whose side measures 4 yd. 2 ft. 7 in.

Solution:

Line (a) $p = 3s$

Line (b)

```
  4 yd. 2 ft.  7 in.
                ×3
 ------------------
 12 yd. 6 ft. 21 in.
              ↙   ↓
          1 ft.
 12 yd. 7 ft.  9 in.
          ↙ ↓
     2 yd.
 14 yd. 1 ft.  9 in.
```

Perimeters of Other Polygons

Example 1: Find the perimeter of a regular pentagon with a side of 20 yd. 2 ft.

Solution:

```
  20 yd.    2 ft.
          × 5
 ----------------
 100 yd.   10 ft.
          ↙   ↓
     3 yd.
 103 yd.    1 ft.
```

Example 2: Find the perimeter of a quadrilateral with sides:

4 ft., 2 ft., 7 ft., 9 ft.

Solution:

Line (a) 4 ft. + 2 ft. + 7 ft. + 9 ft. = 22 ft. = 7 yd. 1 ft.

15

Study Exercise Four

A. Find the perimeters of triangles with sides measuring:

1. 17 in., 31 in., 28 in.

2. 1 ft. 8 in., 2 ft. 3 in., 2 ft. 1 in.

B. Find the perimeters of isosceles triangles with the following dimensions:

	3.	**4.**	**5.**
Base	14 in.	4 ft. 2 in.	1 yd. 2 ft. 6 in.
Each Equal Side	22 in.	7 ft. 8 in.	3 yd. 11 in.

C. **6.** Find the perimeter of a regular decagon with side 3 yd. 2 ft. 10 in.

7. Find the perimeter of an equilateral triangle with sides measuring 3 ft. 5 in.

Circles

A *circle* is the set of points in a plane which are equidistant from a fixed point in the plane called the center.

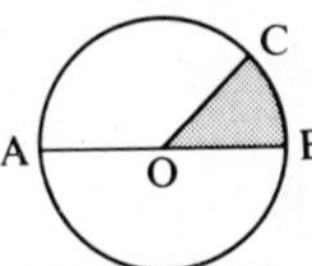

Line (*a*) 0 is the center
Line (*b*) The distance from 0 to C is the *radius*.
Line (*c*) The distance from A to B is the *diameter*.
Line (*d*) $\frac{1}{2}$ diameter (d) = radius (r)
Line (*e*) 2 radii (r) = diameter (d)

Circumference

The *circumference* of a circle is the distance around the circle.

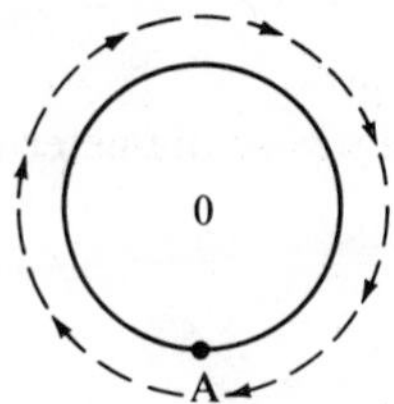

18

A Formula for Finding the Circumference

Line (*a*)

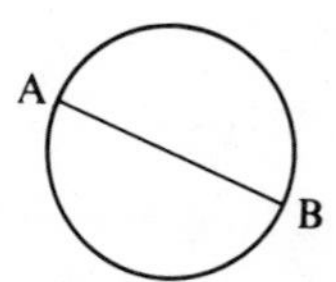

Line (*b*)

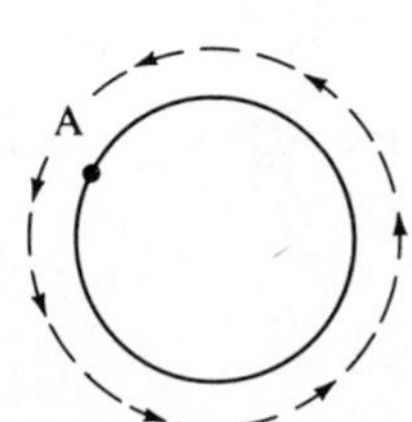

Line (*c*)

|——————————————| ← the circumference

(Frame 19, contd.)

Line (d)

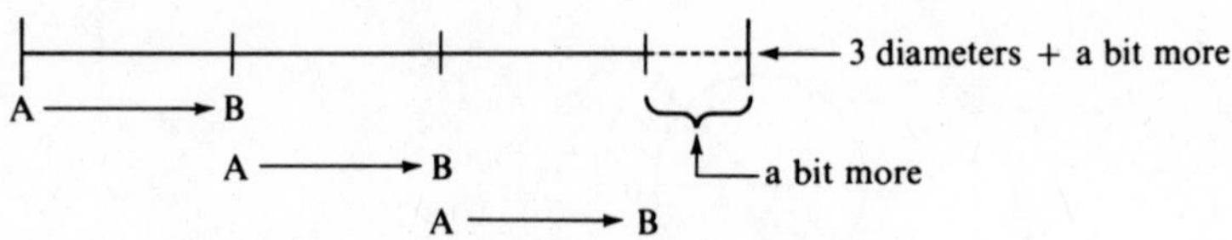

Line (e) 3 × diameter almost equals the circumference.
Line (f) 3.14 × diameter almost equals the circumference.
Line (g) 3.1416 × diameter almost equals the circumference.
Line (h) $\pi \times$ diameter = circumference.
Line (i) $\pi d = c$
Line (j) $\pi = 3.141592653589793238\ldots.$
Line (k) $\frac{22}{7} = 3.142857142857\ldots$
Line (l) $\pi \neq \frac{22}{7}$
Line (m) We will use 3.14 as our approximation unless instructed otherwise.

19

Using the Formula $\pi d = c$

Example 1: Find the circumference of a circle whose diameter is 5 ft.

Solution:

Line (a) $\pi d = c$

Line (b) $3.14 \times 5' = 15.7'$

Example 2: Find the circumference of a circle whose radius is 12 ft.

Solution:

Line (a) $2r = d$

Line (b) $2 \times 12' = 24'$

Line (c) $\pi d = c$

Line (d) $3.14 \times 24' = 75.36'$

Example 3: Find the diameter of a flag pole whose circumference is 12.56 in.

Solution:

Line (a) $\pi d = c$

Line (b) $3.14 \times d = 12.56''$

Line (c) $d = 12.56'' \div 3.14 = 4''$

20

Study Exercise Five

A. Find the circumference of circles with the following diameters:

1. 9 in. **2.** 31 ft. **3.** $4\frac{1}{4}$ ft.

B. Find the circumference of circles with the following radii:

4. 8 ft. **5.** 4 ft. 1 in.

C. **6.** What distance do you travel on one turn of a ferris wheel if you sit 20 ft. from the center?

7. Over what distance will the tip of the minute hand of the clock "Big Ben" in London travel in one hour? The minute hand is about 11 ft. long.

REVIEW EXERCISES

A. Match the number of sides in Column B with the figure in Column A.

	Column A		Column B
1.	quadrilateral	(a)	2
2.	decagon	(b)	3
3.	trapezoid	(c)	4
4.	triangle	(d)	5
5.	rectangle	(e)	6
6.	pentagon	(f)	7
7.	hexagon	(g)	8
8.	parallelogram	(h)	9
9.	octagon	(i)	10

B. **10.** The quadrilateral which has only one pair of sides parallel is called a_________.

11. The formula for the perimeter of an equilateral triangle is ____________.

12. The formula for the circumference of a circle is ___________.

13. Find the diameter of a tree whose circumference is 15.7 ft.

14. The diameter of the earth is 7,918 miles. Find its circumference.

C. Complete the following chart

	Planet	Diameter	Circumference
15.	*Mercury*	3,100 miles	
16.	*Venus*	7,700 miles	
17.	*Mars*		13,188 miles

D. **18.** Find the perimeter of this figure. The top is a semicircle $\left(\frac{1}{2}\text{ circle}\right)$.

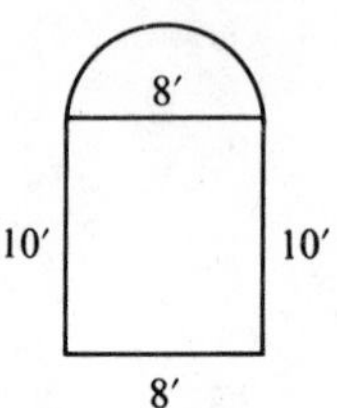

19. Find the perimeter of this figure:

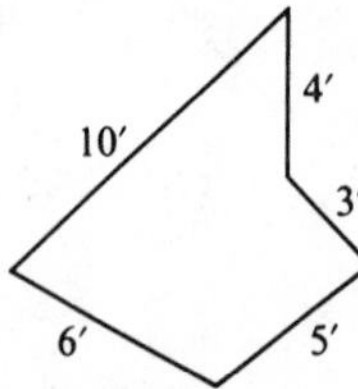

20. Find the perimeter of an equilateral triangle whose side is 3 yd. 1 ft. 9 in.

21. Two times the radius of a circle equals its ___________.

REVIEW EXERCISES, CONTD.

(Frame 22, contd.)

22. The figure below is to be made of ribbon costing $.39 a yard. It is composed of 2 equilateral triangles of the same size measuring 2 ft. 9 in. on a side. How much will this cost?

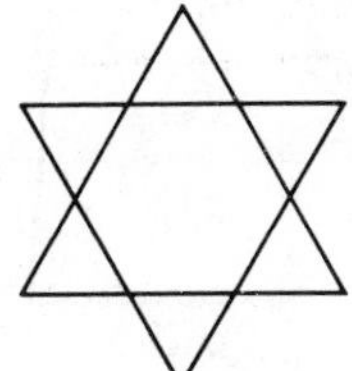

SOLUTIONS TO REVIEW EXERCISES

A. **1.** c **2.** i **3.** c **4.** b **5.** c
6. d **7.** e **8.** c **9.** g

B. **10.** trapezoid **11.** $p = 3s$ **12.** $\pi d = c$

13. $3.14 \times d = 15.7$; $d = 3.14\overline{)15.70}$ (quotient 5) $= 5$ ft.

14. 3.14×7918 miles $\approx 24{,}863$ miles

C. **15.** $3.14 \times 3{,}100$ miles $= 9{,}734$ miles

16. $3.14 \times 7{,}700$ miles $= 24{,}178$ miles

17. $3.14\overline{)1318800}$ (quotient 4200) $= 4{,}200$ miles

D. **18.** Perimeter of Circle $= \pi \times d = 3.14 \times 8' = 25.12'$
Perimeter of Semicircle $= 25.12' \div 2 = 12.56'$
Perimeter of Figure $= 12.56' + 10' + 8' + 10' = 40.56'$

19. $10' + 6' + 5' + 3' + 4' = 28'$

20.

```
  3 yd. 1 ft.  9 in.
             ×  3
 9 yd. 3 ft. 27 in.
       2 ft.
 9 yd. 5 ft.  3 in.
 1 yd.
10 yd. 2 ft.  3 in.
```

21. diameter

22.

```
 2 ft.  9 in.
      ×  6
12 ft. 54 in.
 4 ft.
16 ft.  6 in. = 16.5 ft.
```

$3\overline{)\$.39}$ (quotient $.13)

$.13 per foot × 16.5 ft. = $2.15

23

SUPPLEMENTARY PROBLEMS

1. The diameter of the sun is 864,000 miles. What is its circumference?
2. The planet Saturn has a circumference of 252,681 miles. What is its diameter?
3. Find the perimeter of this pentagon:

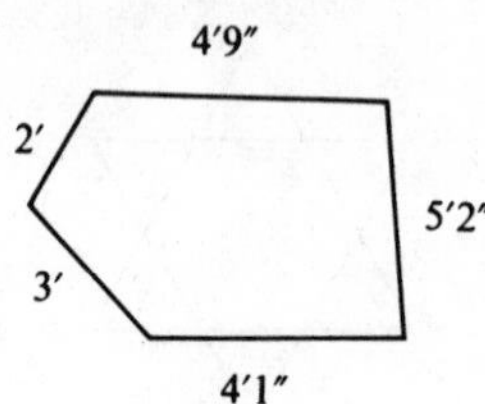

4. A standard sheet of typing paper is $8\frac{1}{2}$ in. by 11 in. What is its perimeter?
5. What is the diameter of a tree whose circumference is 25.12 ft.?
6. The diameter of the face of the clock, "Big Ben" in London is 23 ft. What is its circumference?
7. How much will it cost to fence a garden 30 ft. by 15 ft. at $3.20 a ft.?
8. Which has the greatest perimeter?
   a. A square with side 18.05 ft.
   b. An equilateral triangle with side 3 ft. 1 in.
   c. A regular hexagon with side 12.1 ft.
   d. The clock face in problem 6.

SOLUTIONS TO STUDY EXERCISES

Study Exercise One (Frame 8)

A. **1.** 4 **2.** 6 **3.** decagon **4.** triangle
5. 8 **6.** pentagon

B. 8, 9

C. **13.** f **14.** c **15.** d **16.** b
17. e **18.** j **19.** g

8A

Study Exercise Two (Frame 11)

1. 4×8 in. = 32 in. = 2 ft. 8 in.
2. $4 \times .75$ in. = 3 in.
3. $4 \times 5\frac{3}{4}$ in. $= \overset{1}{\cancel{4}} \times \frac{23}{\underset{1}{\cancel{4}}}$ in. = 23 in. = 1 ft. 11 in.
4. 4×3 ft. 9 in. = 12 ft. 36 in. = 15 ft. or 5 yd.
5. 4×4 yd. 2 ft. 7 in. = 16 yd. 8 ft. 28 in. = 16 yd. 10 ft. 4 in. = 19 yd. 1 ft. 4 in.

B. **6.** 4×60 ft. = 240 ft. for home run; 3×60 ft. = 180 ft. for triple
7. 4×20 yd. = 80 yd.; 80 yd. $\times$ 3 = 240 ft.; 240 ft. $\times$ \$2.80 = \$672

11A

SOLUTIONS TO STUDY EXERCISES, CONTD.

Study Exercise Three (Frame 13)

A. **1.** p = 2(6.2 + 4.3) = 2 × 10.5 = 21 yd.

2. $2\left(4\frac{3}{4} + 3\frac{1}{8}\right) = 2\left(7\frac{7}{8}\right) = 14\frac{14}{8} = 15\frac{6}{8} = 15\frac{3}{4}$ ft.

3. 2(17 + 9) = 2(26) = 52 in. = 4 ft. 4 in.

4. 2(3 ft. 8 in. + 1 ft. 3 in.) = 2(4 ft. 11 in.) = 8 ft. 22 in.
= 9 ft. 10 in.

5. 2(16 ft. + 9 ft. 10 in.) = 2(25 ft. 10 in.) = 50 ft. 20 in.
= 51 ft. 8 in.
= 17 yd. 8 in.

B. **6.** 2(22 + 18) = 2(40) = 80 ft.; 80 ft. − 8 ft. = 72 ft.

7. 2 × 60 + 30 = 120 + 30 = 150 ft.; 150 ft. × \$3.25 = \$487.50

8. 2(5 ft. 2 in. + 3 ft. 4 in.) = 2(8 ft. 6 in.)
= 16 ft. 12 in. = 17 ft.
17 ft. × \$.18 = \$3.06

9. 2(2.5 + 5) = 2(7.5) = 15 ft. = 5 yd.
5 yd. × \$1.87 = \$9.35

13A

Study Exercise Four (Frame 16)

A. **1.** 17 + 31 + 28 = 76 in. = 6 ft. 4 in. or 2 yd. 4 in.

2. 1 ft. 8 in.
2 ft. 3 in.
2 ft. 1 in.
5 ft. 12 in. = 6 ft. or 2 yd.

B. **3.** 2 × 22 + 14 = 44 + 14 = 58 in. = 4 ft. 10 in. or 1 yd. 1 ft. 10 in.

4. 2(7 ft. 8 in.) + 4 ft. 2 in. = 14 ft. 16 in. + 4 ft. 2 in.
= 18 ft. 18 in.
= 6 yd. 1 ft. 6 in.

5. 2(3 yd. 11 in.) + 1 yd. 2 ft. 6 in. = 6 yd. 22 in. + 1 yd. 2 ft. 6 in.
= 7 yd. 2 ft. 28 in. = 7 yd. 4 ft. 4 in.
= 8 yd. 1 ft. 4 in.

C. **6.** 3 yd. 2 ft. 10 in.
× 10
30 yd. 20 ft. 100 in.
8 ft.
30 yd. 28 ft. 4 in.
9 yd.
39 yd. 1 ft. 4 in.

7.

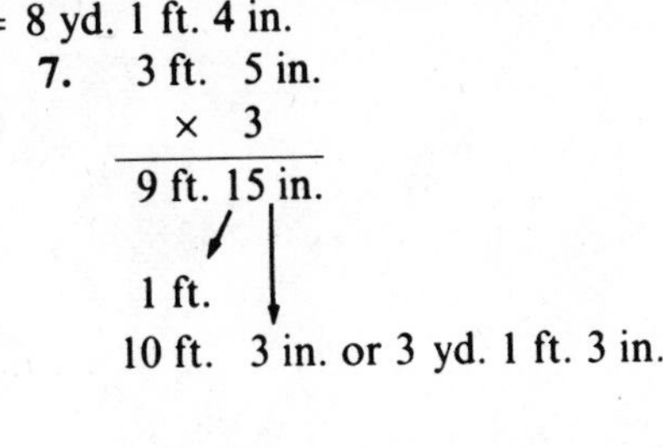

3 ft. 5 in.
× 3
9 ft. 15 in.
1 ft.
10 ft. 3 in. or 3 yd. 1 ft. 3 in.

16A

SOLUTIONS TO STUDY EXERCISES, CONTD.

Study Exercise Five (Frame 21)

A. **1.** 3.14×9 in. = 28.26 in.
2. 3.14×31 ft. = 97.34 ft.
3. 3.14×4.25 ft. = 13.345

B. **4.** 2×8 ft. = 16 ft.; 3.14×16 ft. = 50.24 ft.
5. 2×4 ft. 1 in. = 8 ft. 2 in.;
3.14×8 ft. 2 in. = 25.12 ft. 6.28 in.

C. **6.** 2×20 ft. = 40 ft.; 3.14×40 ft. = 125.6 ft.
7. 2×11 ft. = 22 ft.; 3.14×22 ft. = 69.08 ft.

21A

---

unit

37

Area

Objectives:

By the end of this unit you should:

1. be able to find the areas of:

a. *rectangles* b. *squares*

c. *parallelograms* d. *triangles*

e. *trapezoids* f. *circles*

2. be able to find the total area of:

a. a rectanglular solid b. a cube

3. be able to find the lateral area and the total area of a right circular cylinder.

4. know the area formulas for the above-named figures. ①

Area

The *area* of the interior of any plane figure is given by the number of square units it contains.

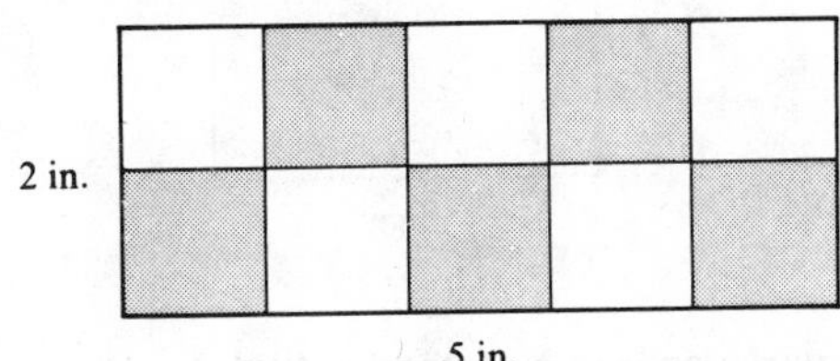

The area of this rectangle (2 in. by 5 in.) is 10 square inches since it contains 10 squares each 1 in. by 1 in. ②

Area of a Rectangle

Formula: $A = lw$

Example 1: Find the area of a rectangle 11 ft. by 17 ft.

Solution:

Line (a) $A = lw$

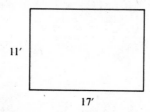

Line (b) A = 17 ft. × 11 ft. = 187 sq. ft.

11′

17′

Example 2: Find the area of a rectangle 3 ft. by 48 in.

Solution:

Line (a) 48 in. ÷ 12 = 4 ft.

Line (b) $A = lw$

Line (c) A = 4 ft. × 3 ft. = 12 sq. ft.

3′

48″

Example 3: Find the number of square yards in a rectangle 4 yd. by 18 in.

Solution:

Line (a) 18 in. ÷ 36 in. = .5 yd.

Line (b) $A = lw$

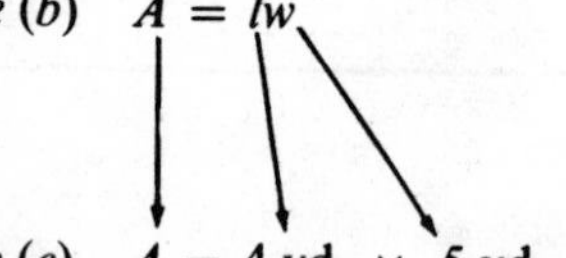

Line (c) A = 4 yd. × .5 yd. = 2 sq. yd.

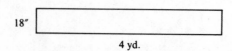

18″

4 yd.

(3)

Example: Find the cost of covering a floor 15 ft. by 13 ft. with carpet that sells for \$8.50 per sq. yd.

Line (a) $A = lw$

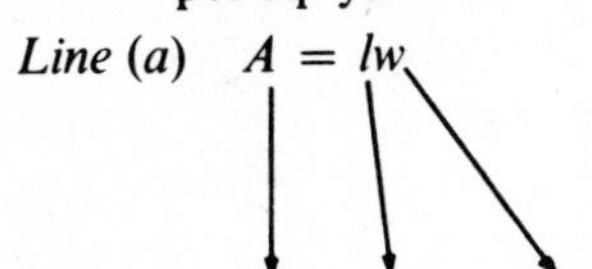

Line (b) $A =$ 15 ft. × 13 ft. = 195 sq. ft.
Line (c) 9 sq. ft. = 1 sq. yd. (From Table I)
Line (d) 195 sq. ft. ÷ 9 sq. ft. = 21.67 ≈ 22 sq. yd.
Line (e) 22 sq. yd. × \$8.50 = \$187 ④

Study Exercise One

Find the area of rectangles having the following dimensions:

	1.	2.	3.	4.	5.
Length	48 ft.	4 ft. 6 in.	9.6 yd.	$7\frac{1}{4}$ in.	4.2 mi.
Width	22 ft.	2 ft. 9 in.	2.3 yd.	$3\frac{3}{8}$ in.	3.6 mi.

6. Find the cost of cementing a sidewalk 16 ft. by 4 ft. at \$.45 per sq. ft.
7. A 9 ft. by 12 ft. rug is placed on a floor 11 ft. by 14 ft. How many square feet of floor space remains uncovered? ⑤

Area of a Square

Formula: $A = s^2$

Line (a) $A = l \times w$ (area of a rectangle)

Line (b) $A = s \times s = s^2$ (area of a square)

Example 1: Find the area of a square whose side is 57 ft.

Solution:

Line (a) $A = s^2$

Line (b) $A = (57 \text{ ft.})^2$
Line (c) $A =$ 57 ft. × 57 ft. = 3,249 sq. ft.

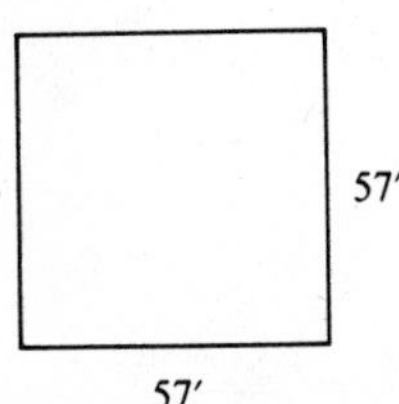

(Frame 6, contd.)

Example 2: Find the area of a square whose side is 3 ft. 5 in.

Solution:

Line (a) $3 \text{ ft. } 5 \text{ in.} = 3\frac{5}{12} \text{ ft.} = \frac{41}{12} \text{ ft.}$

Line (b) $A = s^2$

Line (c) $A = \left(\frac{41}{12} \text{ ft.}\right)^2$

Line (d) $A = \frac{41}{12} \text{ ft.} \times \frac{41}{12} \text{ ft.} = \frac{1681 \text{ sq. ft.}}{144}$

$= 144 \overline{)1681.0}$ (quotient 11.6) ≈ 12 sq. ft.

6

Using a Powers and Roots Table
Table III (pg. 437)

Example 1: $A = (16 \text{ in.})^2 = 256$ sq. in.

Example 2: $A = (47 \text{ ft.})^2 = 2{,}209$ sq. ft.

Example 3: $A = (91 \text{ yd.})^2 = 8{,}281$ sq. yd.

Example 4: $A = (83 \text{ mi.})^2 = 6{,}889$ sq. mi.

Example 5: $A = (8.3 \text{ mi.})^2 = 68.89$ sq. mi.

7

Study Exercise Two

Find the areas of squares whose sides measure:

1. 9 in.
2. 8.7 ft.
3. $5\frac{1}{2}$ ft.
4. 1 ft. 5 in.
5. 440 yd.
6. $\frac{11}{16}$ in. (To nearest hundredth)
7. .65 in.
8. 6 ft. 8 in.

8

Area of a Parallelogram

Formula: $A = bh$

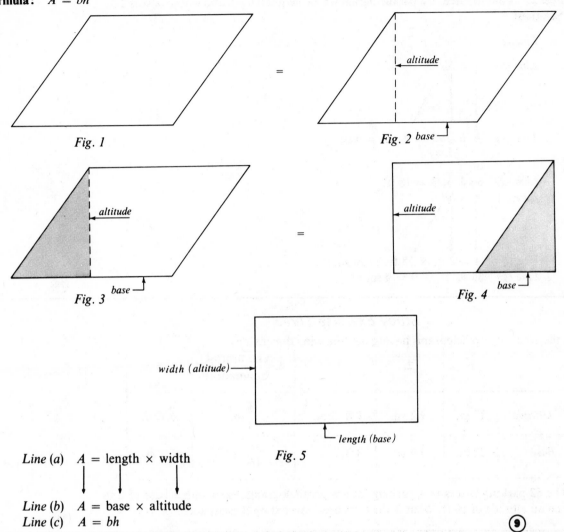

altitude

base

Fig. 1 = *Fig. 2*

altitude

base

altitude

base

Fig. 3 = *Fig. 4* =

width (altitude)

length (base)

Fig. 5

Line (a) $A = \text{length} \times \text{width}$

$\downarrow \quad \downarrow \quad \downarrow$

Line (b) $A = \text{base} \times \text{altitude}$

Line (c) $A = bh$

(9)

Using the Formula $A = bh$

Example 1: Find the area of a parallelogram whose base is 16 in. and whose height is 8 in.

Solution:

Line (a) $A = bh$

Line (b) $A = 16 \text{ in.} \times 8 \text{ in.} = 128 \text{ sq. in.}$

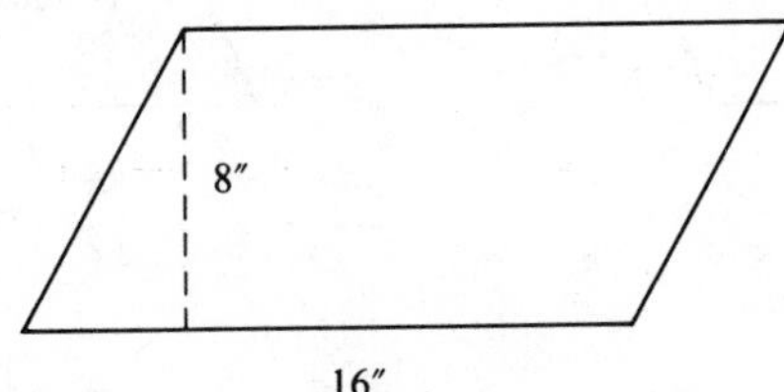

(Frame 10, contd.)

Example 2: Find the area of a parallelogram whose height is 6 yd. and whose base is 2 ft.

Solution:

Line (a) $2 \text{ ft.} \div 3 = \frac{2}{3} \text{ yd.}$

Line (b) $A = bh$

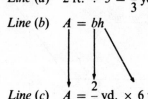

Line (c) $A = \frac{2}{3} \text{ yd.} \times 6 \text{ yd.} = 4 \text{ sq. yd.}$

or

Line (d) $6 \text{ yd.} \times 3 = 18 \text{ ft.}$

Line (e) $A = bh$

Line (f) $A = 2 \text{ ft.} \times 18 \text{ ft.} = 36 \text{ sq. ft.}$

Line (g) $36 \text{ sq. ft.} \div 9 = 4 \text{ sq. yd.}$

Study Exercise Three

Find the areas of parallelograms having the following dimensions:

	1.	**2.**	**3.**	**4.** (To nearest hundredth)	**5.**
Altitude	17 in.	9.2 yd.	1 ft. 3 in.	$\frac{3}{8}$ in.	6.7 mi.
Base	22 in.	3.6 yd.	3 ft.	$\frac{5}{16}$ in.	1.3 mi.

6. The 62 parking spaces in a parking lot are parallelograms, each with a base of 8 ft. and an altitude of 16 ft. What is the total area covered by these spaces?

Area of a Triangle

Formula: $A = \frac{1}{2}bh$

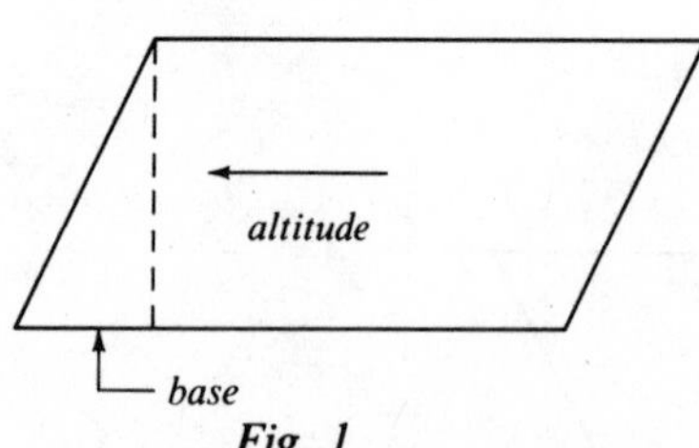

Fig. 1

=

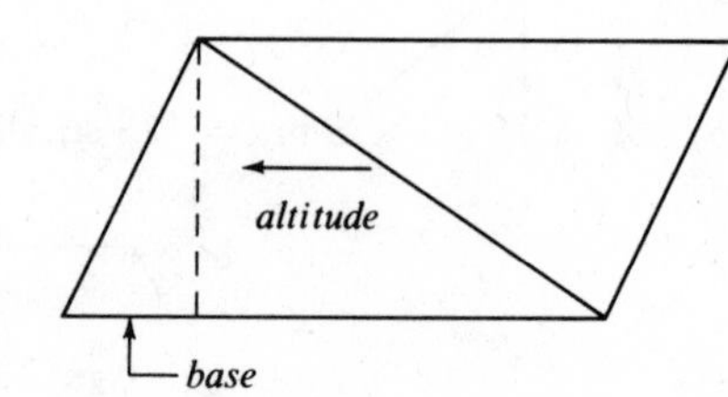

Fig. 2

(Frame 12, contd.)

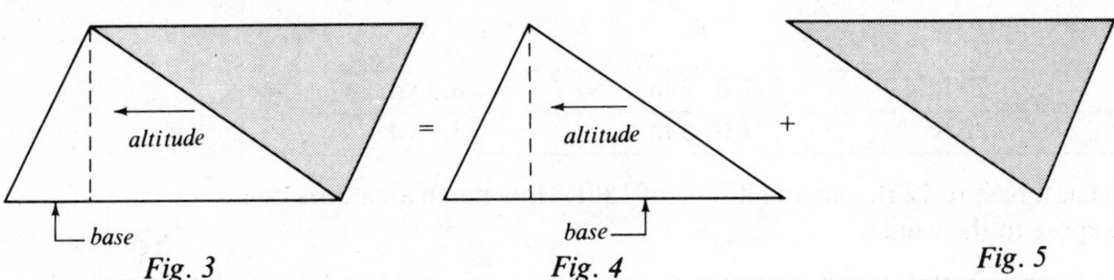

Fig. 3 Fig. 4 Fig. 5

Line (*a*) Area of Fig. 1 $= bh$

Line (*b*) Area of Fig. 4 $= \frac{1}{2}$ of Fig. 1

Line (*c*) Area of Fig. 4 $= \frac{1}{2}bh$

Line (*d*) $A = \frac{1}{2}bh$

⑫

Using the Formula $A = \frac{1}{2}bh$

Example 1: Find the area of a triangle with a base of 24 ft. and a height of 6 ft.

Solution:

Line (*a*) $A = \frac{1}{2}bh$

Line (*b*) $A = \frac{1}{2} \times 24 \times 6 = \frac{1}{\cancel{2}_{1}} \times \cancel{24}^{12} \times 6 = 72$ sq. ft.

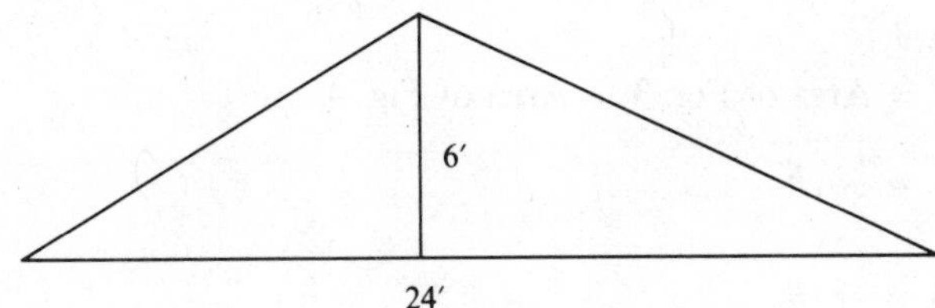

Example 2: Find the area of a triangle with a base of 1 ft. 3 in. and a height of 1 ft. 4 in.

Solution:

Line (*a*) 1 ft. 3 in. $= 1\frac{3}{12}$ ft. $= 1\frac{1}{4}$ ft. $= \frac{5}{4}$ ft.

Line (*b*) 1 ft. 4 in. $= 1\frac{4}{12}$ ft. $= 1\frac{1}{3}$ ft. $= \frac{4}{3}$ ft.

Line (*c*) $A = \frac{1}{2}bh$

Line (*d*) $A = \frac{1}{2} \times \frac{5}{4} \times \frac{4}{3} = \frac{1}{2} \times \frac{5}{\cancel{4}_{1}} \times \frac{\cancel{4}^{1}}{3} = \frac{5}{6}$ sq. ft.

⑬

Study Exercise Four

Find the areas of triangles having the following dimensions:

	1.	2.	3.
Altitude	20 ft.	1 ft. 2 in.	8.2 yd.
Base	22 ft.	1 ft. 8 in.	3.4 yd.

4. A triangular sail has a base of 12 ft. and an altitude of 18 ft. How much area does one side of this sail expose to the wind? (14)

Area of a Trapezoid

Formula: $A = \frac{h}{2}(b_1 + b_2)$

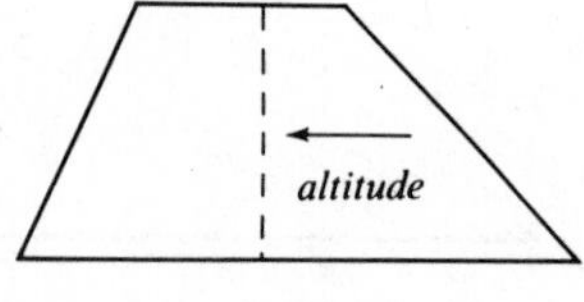

Fig. 1

=

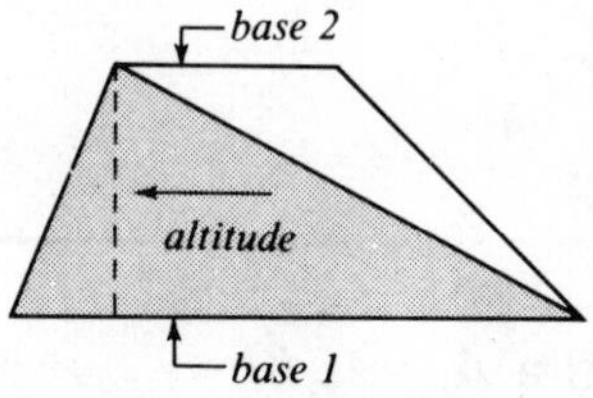

base 2

altitude

base 1

Fig. 2

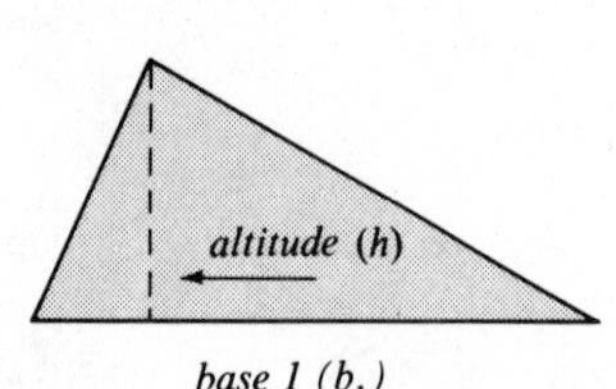

Fig. 3

+

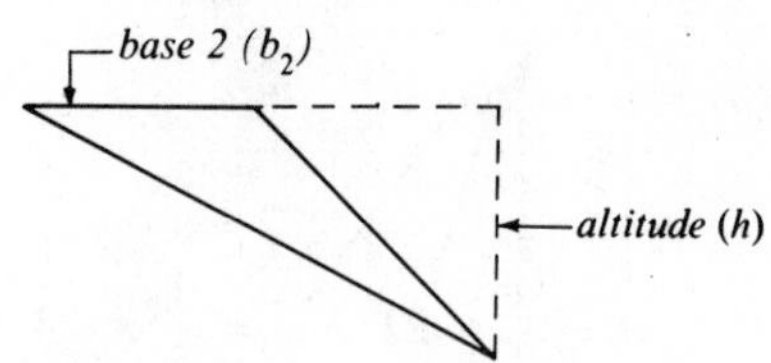

Fig. 4

Line (a) Area of Fig. 1 = Area of Fig. 3 + Area of Fig. 4

Line (b) Area of Fig. 3 $= \frac{1}{2}b_1h$

Line (c) Area of Fig. 4 $= \frac{1}{2}b_2h$

Line (d) Area of Fig. 1 $= \frac{1}{2}b_1h + \frac{1}{2}b_2h$

Line (e) Area of Fig. 1 $= \frac{1}{2} \times h(b_1 + b_2)$

Line (f) $A = \frac{h}{2}(b_1 + b_2)$ (15)

Using the Formula $A = \frac{h}{2}(b_1 + b_2)$

Example 1: Find the area of a trapezoid with bases of 16 ft. and 18 ft. and a height of 9 ft.

Solution:

Line (a) $A = \frac{h}{2}(b_1 + b_2)$

Line (b) $A = \frac{9}{2}(16 + 18)$

Line (c) $A = \frac{9}{2}(34) = \frac{9}{\cancel{2}_1} \times \frac{\cancel{34}^{17}}{1} = 153$ sq. ft.

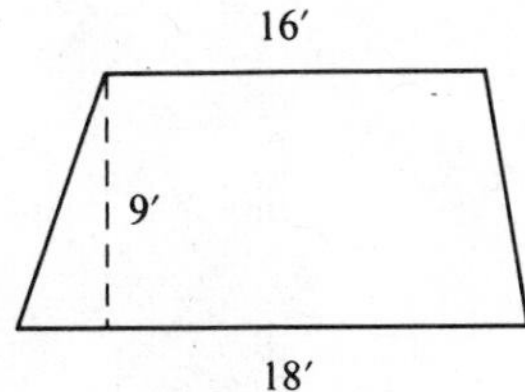

16

Study Exercise Five

Find the areas of trapezoids having the following dimensions:

	1.	2.	3.
Height	8 in.	$6\frac{3}{4}$ ft.	31 yd.
Upper Base	10 in.	$4\frac{1}{2}$ ft.	20 yd.
Lower Base	14 in.	$9\frac{1}{2}$ ft.	14 yd.

4. A roof is shaped in the form of a trapezoid. Its height is 9 ft. and the bases are 14 ft. and 24 ft. If a gallon of paint covers 300 sq. ft. will there be enough to do two coats? 

Area of a Circle

Formula: $A = \pi r^2$

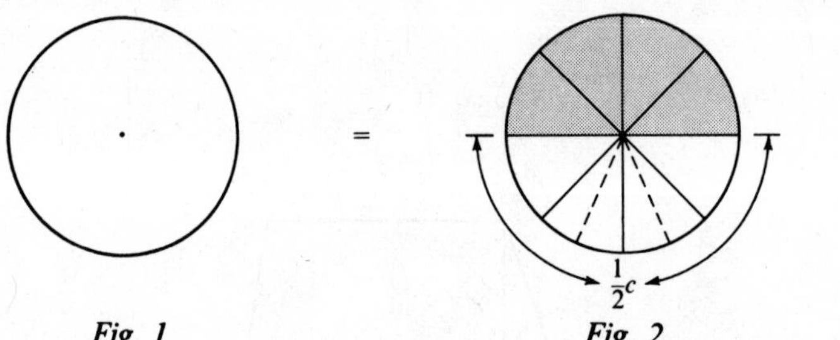

Fig. 1 *Fig. 2*

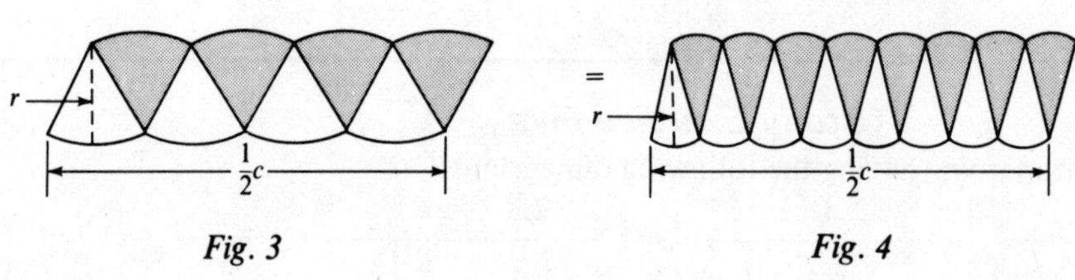

Fig. 3 *Fig. 4*

Line (a) $A = bh$ ← area of a parallelogram

Line (b) $A = \left(\frac{1}{2}c\right)r$

Line (c) $c = \pi d = 2\pi r$ ← circumference of a circle

Line (d) $A = \left(\frac{1}{2} \times 2\pi r\right)r$

Line (e) $A = (\pi r)r = \pi r^2$

(18)

Using the Formula $A = \pi r^2$

Example 1: Find the area of a circle having a radius of 9 inches.

Solution:

Line (a) $A = \pi r^2$

Line (b) $A = 3.14 \times (9 \text{ in.})^2$

Line (c) $A = 3.14 \times 81 \text{ sq. in.} = 254.34 \text{ sq. in.}$

9″

Example 2: Find the area of a circle having a diameter of 24 ft.

Solution:

Line (a) $24 \text{ ft.} \div 2 = 12 \text{ ft.}$

Line (b) $A = \pi r^2$

Line (c) $A = 3.14 \times (12 \text{ ft.})^2$

Line (d) $A = 3.14 \times 144 \text{ sq. ft.} = 452.16 \text{ sq. ft.}$

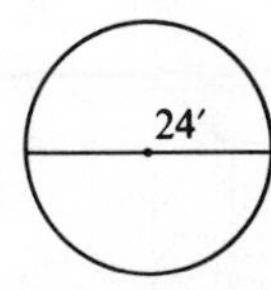

Study Exercise Six

Find the areas of circles having radii of:

1. 10 in. **2.** 7.2 mi. **3.** 4 yd.

Find the areas of circles having diameters of:

4. 4.6 yd. **5.** 42 ft.

6. How many more square inches of pizza do you get in a 14-inch diameter pan as compared to a 12-inch diameter pan?

Surface Area of a Rectangular Solid

Formula: $A = 2lw + 2lh + 2wh$

Line (a) The *surface area* of the outside surface of a rectangular solid is the area of its six faces.

Line (b) Bottom face $= lw =$ Top face

Line (c) Front face $= lh =$ Back face

Line (d) Side face $= wh =$ other side face

Line (e) $A = 2lw + 2lh + 2wh$

Line (f) $A = 2(lw + lh + wh)$

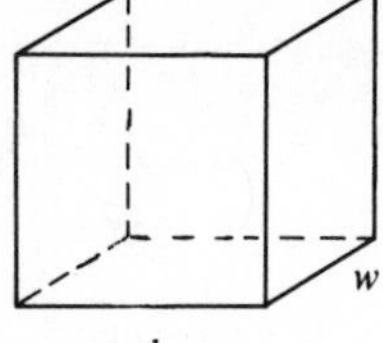

Using the Formula $A = 2lw + 2lh + 2wh$

Example 1: Find the surface area of a rectangular solid 9 ft. long by 6 ft. wide by 7 ft. high.

Solution:

Line (a) $A = 2\,lw + 2\,lh + 2\,wh$

Line (b) $A = 2 \times 9 \text{ ft.} \times 6 \text{ ft.} + 2 \times 9 \text{ ft.} \times 7 \text{ ft.} + 2 \times 6 \text{ ft.} \times 7 \text{ ft.}$

Line (c) $A = 2 \times 54 \text{ sq. ft.} + 2 \times 63 \text{ sq. ft.} + 2 \times 42 \text{ sq. ft.}$

Line (d) $A = 108 \text{ sq. ft.} + 126 \text{ sq. ft.} + 84 \text{ sq. ft.}$

Line (e) $A = 318 \text{ sq. ft.}$

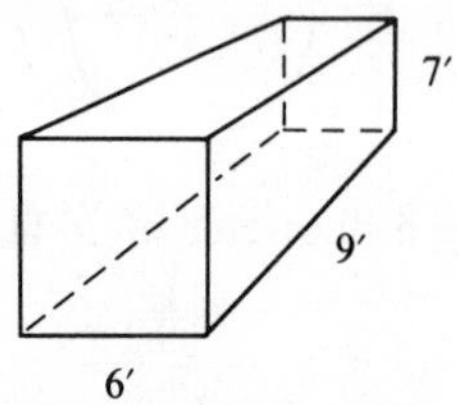

22

Surface Area of a Cube

Formula: $A = 6s^2$

Line (a) Each face has an area of s^2

Line (b) There are six faces

Line (c) $A = 6s^2$

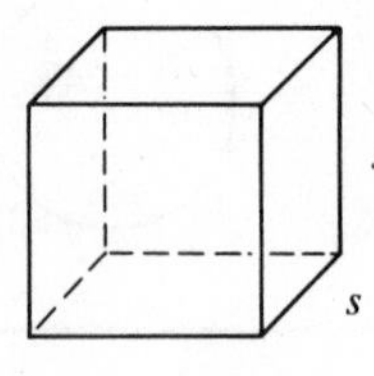

Example 1: Find the surface area of a cube whose sides measure 8 in.

Solution:

Line (a) $A = 6s^2$

Line (b) $A = 6 \times (8 \text{ in.})^2$

Line (c) $A = 6 \times 64 \text{ sq. in.}$

Line (d) $A = 384 \text{ sq. in.}$

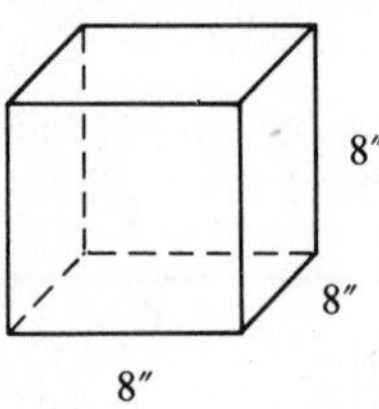

Study Exercise Seven

Find the surface areas of rectangular solids having the following dimensions:

1. Length 3 ft.
   Width 2 ft.
   Height 5 ft.
2. Length 1.6 ft.
   Width 2.4 ft.
   Height 3.1 ft.
3. Find the amount of cardboard needed to make a cereal box with dimensions 2 in. by 7 in. by 10 in.
4. Find the surface area of a cube whose sides measure 4.7 inches.

Lateral Area of a Right Circular Cylinder

Formula: $A = 2\pi rh$

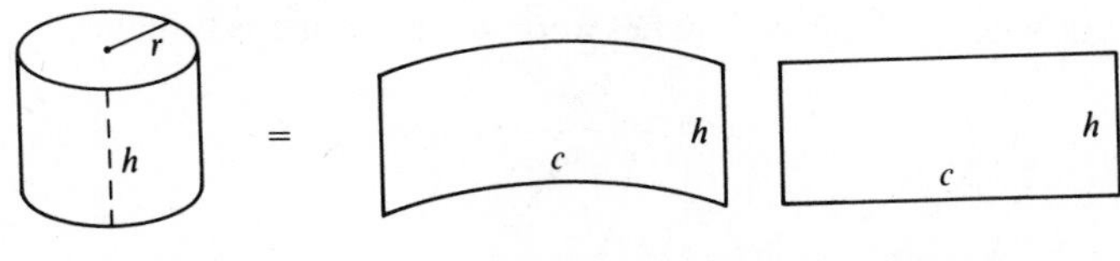

Fig. 1 *Fig. 2* *Fig. 3*

Line (a) $A = ch$
Line (b) $c = \pi d$ or $2\pi r$
Line (c) $A = 2\pi rh$

Using the Formula $A = 2\pi rh$

Example 1: Find the lateral area of a cylinder whose height is 8 in. and whose radius is 3 in.

Solution:

Line (a) $A = 2\ \pi\ r\ h$

Line (b) $A = 2 \times 3.14 \times 3\text{ in.} \times 8\text{ in.}$
Line (c) $A = 150.72\text{ sq. in.}$

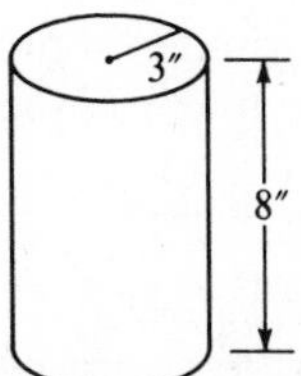

The Total Area of a Right Circular Cylinder

Line (a) Lateral Area $= 2\pi rh$
Line (b) Lower and upper bases are circles
Line (c) $A = \pi r^2$
Line (d) $A = 2\pi rh + 2\pi r^2$

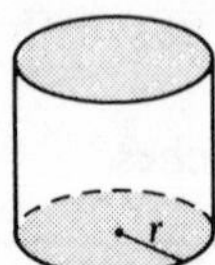

27

Using the Formula $A = 2\pi rh + 2\pi r^2$

Find the total area of a can whose height is 6 inches and whose radius is 4 inches.

Solution:

Line (a) $A = 2\ \pi\ r\ h \quad + 2\ \pi\ r^2$

Line (b) $A = 2 \times 3.14 \times 4 \times 6 + 2 \times 3.14 \times (4)^2$
Line (c) $A =$ 150.72 sq. in. + 100.48 sq. in.
Line (d) $A =$ 251.2 sq. in.

28

Study Exercise Eight

Find the lateral areas of cylinders with the following dimensions:

1. radius 4 in.
height 10 in.

2. radius 5 in.
height 3 in.

3. Find the total area of the cylinder in problem one.
4. Find the total area of the cylinder in problem two.

REVIEW EXERCISES

1. Find the cost of cementing a sidewalk 6 ft. by 30 ft. at $.55 per sq. ft.
2. What is the picture area on a television screen 21.5 in. by 16.5 in.?
3. Find the cost of covering a floor 12 ft. by 12 ft. with vinyl at $4.50 per sq. yd.
4. At a coverage of 36 sq. ft. per roll how many whole rolls of wallpaper are needed to do a room 10 ft. by 14 ft. with 8 ft. ceilings if one extra roll is bought to cover waste?
5. At $6.00 per roll what will the paper in problem 4 cost?
6. If the room in problem 4 has three windows measuring 3 ft. by 3 ft., 4 ft. by 4 ft., 7 ft. by 2 ft. and a door opening 3 ft. by 7 ft., how many whole rolls of paper will it require still keeping one extra roll for waste?
7. Find the total area of a parking lot which contains 40 parking spaces in the form of parallelograms with bases of 8 ft. and altitudes of 16 ft. and two triangular tree planters with bases 8 ft. and altitudes 16 ft.
8. Find the area of both sides of a tapered model airplane wing which has the shape of a trapezoid if the parallel sides measure 3 ft. 9 in. and 4 ft. 3 in. and these sides are 6 in. apart.
9. A television station can televise programs for a distance of 60 linear miles in all directions. What area in square miles does it cover?

SOLUTIONS TO REVIEW EXERCISES

1. 6 ft. × 30 ft. = 180 sq. ft.; 180 × $.55 = $99
2. 21.5 × 16.5 = 354.75 sq. in.
3. 12 ft. × 12 ft. = 144 sq. ft.; 144 sq. ft. ÷ 9 = 16 sq. yd.; 16 × $4.50 = $72
4. Total Area = 2 × 14 × 8 + 2 × 10 × 8 = 224 + 160
   = 384 sq. ft.

$$\begin{array}{r} 10 + 24 \text{ sq. ft. remainder} \\ 36\overline{)384} \\ 36 \\ \hline 24 \end{array} \quad = \quad \begin{array}{r} 11 \text{ rolls} \\ +\ 1 \text{ extra} \\ \hline 12 \text{ rolls} \end{array}$$

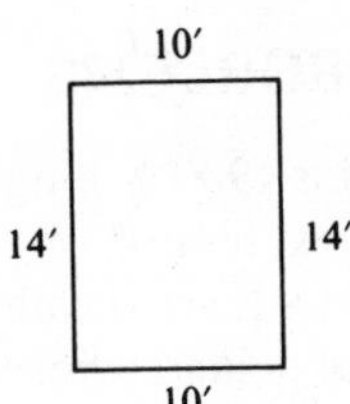

5. 12 × $6 = $72
6.

Window 1	3 ft. × 3 ft.	=	9 sq. ft.
Window 2	4 ft. × 4 ft.	=	16 sq. ft.
Window 3	7 ft. × 2 ft.	=	14 sq. ft.
Door	3 ft. × 7 ft.	=	21 sq. ft.

384 sq. ft. − 60 sq. ft. = 324 sq. ft.

$$\begin{array}{r} 9 \text{ rolls} \\ 36\overline{)324} \\ 324 \\ \hline \end{array} \qquad \begin{array}{r} 9 \text{ rolls} \\ +1 \text{ extra} \\ \hline 10 \text{ rolls} \end{array}$$

SOLUTIONS TO REVIEW EXERCISES, CONTD.

(Frame 31, contd.)

7. *Solution One*

$8 \times 16 = 128$ sq. ft.; $128 \times 40 = 5120$ sq. ft. Parking Spaces

$\frac{1}{2} \times 8 \times 16 = 64$ sq. ft.; $64 \times 2 = 128$ sq. ft. Planters

5248 sq. ft. Total area

Solution Two

Two triangles with the same bases and altitudes are equivalent to a single parallelogram

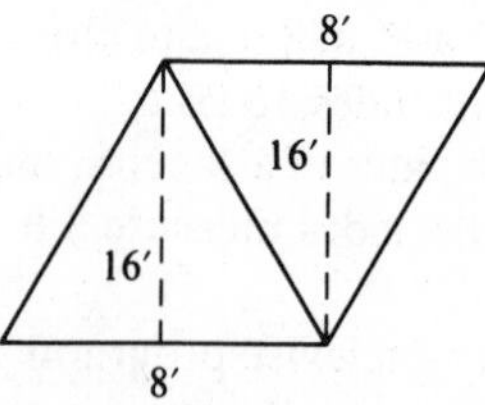

$8 \times 16 = 128$ sq. ft.; $128 \times 41 = 5248$ sq. ft. total area

8. $3 \text{ ft. } 9 \text{ in.} = 3\frac{9}{12} \text{ ft.} = 3\frac{3}{4} \text{ ft.} = \frac{15}{4} \text{ ft.}$

$4 \text{ ft. } 3 \text{ in.} = 4\frac{3}{12} \text{ ft.} = 4\frac{1}{4} \text{ ft.} = \frac{17}{4} \text{ ft.}$

$6 \text{ in.} = \frac{6}{12} \text{ ft.} = \frac{1}{2} \text{ ft.}$

$A = \frac{1}{2} \times \frac{1}{2}\left(\frac{15}{4} + \frac{17}{4}\right) = \frac{1}{2} \times \frac{1}{2} \times \frac{\overset{16}{\cancel{32}}}{\underset{2}{\cancel{4}}} = \frac{16}{8} = 2$ sq. ft.; 4 sq. ft. for both sides

9. $A = \pi r^2 = 3.14 \times (60)^2 = 3.14 \times 3600 = 11{,}304$ sq. mi. (31)

SUPPLEMENTARY PROBLEMS

A. 1. A rectangular field is 1900 ft. long and 1,700 ft. wide. How many acres are there in the field (to nearest tenth)?

2. A piece of land 1 mi. square is called a section. How many sections are contained in a triangular piece of land whose base is 10 miles and whose altitude is 8 miles?

3. If 15 sq. ft. is allowed for each student, what is the maximum number of students that should be permitted in a room 33 ft. by 27 ft.?

4. Three circular plates, each with 8 in. radius, are cut from a rectangular piece of metal 50 in. by 18 in. To the nearest whole square inch, how much material is wasted?

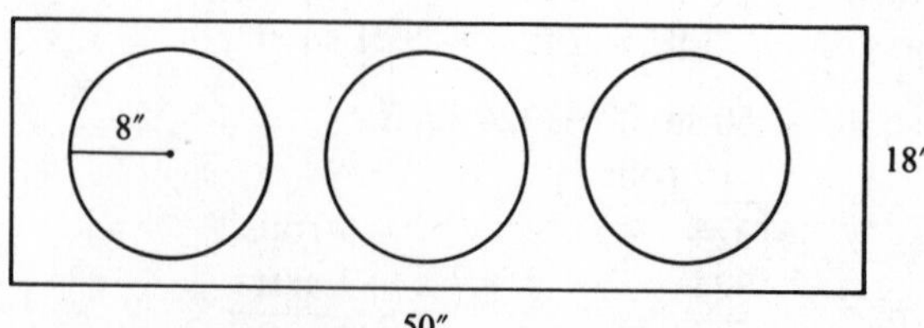

B. Find the areas of the following:

5. a square whose side is 32 in.

6. a parallelogram whose altitude is 6 ft. and whose base is 20 ft.

SUPPLEMENTARY PROBLEMS, CONTD.

7. a tin can with one lid missing whose height is 6 in. if the diameter of the missing lid is 4 in.
8. a trapezoid with bases 8 ft. and 9 ft. and altitude 14 ft.
9. the walls of a room 14 ft. by 16 ft. with 8 ft. ceilings
10. a cube with sides 6.5 ft.

C. 11. How much paper is needed to make a label for a can of tomatoes if the can is 9 inches high and has lids with 4 inch diameters?

12. If a gallon of paint covers 450 sq. ft., how many gallons are needed to do a room 40 ft. by 80 ft. with 20 ft. ceilings if all four walls and the ceiling are to be painted? (Answer to the nearest whole gallon.)

13. A city lot 100 ft. by 80 ft. is to have a lawn covering all the area except for the house and driveway. The house measures 40 ft. by 60 ft. and the drive is 20 ft. by 30 ft. If the seed is spread at the rate of 1 lb. per 400 sq. ft., how much seed will be needed?

SOLUTIONS TO STUDY EXERCISES

Study Exercise One (Frame 5)

1. 48 ft. × 22 ft. = 1,056 sq. ft.
2. 4 ft. 6 in. = $4\frac{6}{12}$ ft. = $4\frac{1}{2}$ ft. = $\frac{9}{2}$ ft.

   2 ft. 9 in. = $2\frac{9}{12}$ ft. = $2\frac{3}{4}$ ft. = $\frac{11}{4}$ ft.

   $\frac{9}{2} \times \frac{11}{4} = \frac{99}{8} = 12\frac{3}{8}$ sq. ft.
3. 9.6 yd. × 2.3 yd. = 22.08 sq. yd.
4. $7\frac{1}{4} \times 3\frac{3}{8} = \frac{29}{4} \times \frac{27}{8} = \frac{783}{32} = 24\frac{15}{32}$ sq. in.
5. 4.2 mi. × 3.6 mi. = 15.12 sq. mi.
6. 16 ft. × 4 ft. = 64 sq. ft.

   64 × $.45 = $28.80
7. 9 ft. × 12 ft. = 108 sq. ft.

   11 ft. × 14 ft. = 154 sq. ft.

   154 − 108 = 46 sq. ft. uncovered

5A

Study Exercise Two (Frame 8)

1. $(9\text{ in.})^2 = 81$ sq. in.
2. $(8.7\text{ ft.})^2 = 75.69$ sq. ft.
3. $\left(5\frac{1}{2}\text{ ft.}\right)^2 = (5.5\text{ ft.})^2 = 30.25$ sq. ft.
4. 1 ft. 5 in. = $1\frac{5}{12}$ ft. = $\frac{17}{12}$ ft.; $\left(\frac{17}{12}\text{ ft.}\right)^2 = \frac{289}{144}$ sq. ft.

   $= 2\frac{1}{144}$ sq. ft. $\approx$ 2 sq. ft.
5. $(440\text{ yd.})^2 = 193600$ sq. yd.
6. $\left(\frac{11}{16}\text{ in.}\right)^2 = \frac{121}{256}$ sq. in. $\approx$ .47 sq. in.

SOLUTIONS TO STUDY EXERCISES, CONTD.

Study Exercise Two (Frame 8, contd.)

7. $(.65 \text{ in.})^2 = .4225$ sq. in.
8. 6 ft. 8 in. $= 6\frac{8}{12}$ ft. $= 6\frac{2}{3}$ ft. $= \frac{20}{3}$ ft.;

   $\left(\frac{20}{3}\text{ ft.}\right)^2 = \frac{400}{9}$ sq. ft. $= 44\frac{4}{9}$ sq. ft.

8A

Study Exercise Three (Frame 11)

1. 17 in. × 22 in. = 374 sq. in.
2. 9.2 yd. × 3.6 yd. = 33.12 sq. yd.
3. 1 ft. 3 in. $= 1\frac{3}{12}$ ft. $= 1\frac{1}{4}$ ft. $= 1.25$ ft.;

   1.25 ft. × 3 ft. = 3.75 sq. ft.
4. $\frac{3}{8}$ in. $\times \frac{5}{16}$ in. $= \frac{15}{128}$ sq. in. $\approx .12$ sq. in.
5. 6.7 mi. × 1.3 mi. = 8.71 sq. mi.
6. 62 × 8 ft. × 16 ft. = 7,936 sq. ft.

11A

Study Exercise Four (Frame 14)

1. $\frac{1}{\underset{1}{\cancel{2}}} \times \overset{10}{\cancel{20}} \times 22 = 220$ sq. ft.
2. 1 ft. 2 in. $= 1\frac{2}{12}$ ft. $= 1\frac{1}{6}$ ft. $= \frac{7}{6}$ ft.;

   1 ft. 8 in. $= 1\frac{8}{12}$ ft. $= 1\frac{2}{3}$ ft. $= \frac{5}{3}$ ft.

   $\frac{1}{2} \times \frac{7}{6} \times \frac{5}{3} = \frac{35}{36}$ sq. ft.
3. $\frac{1}{2} \times 8.2$ yd. $\times 3.4$ yd. $= 13.94$ sq. yd.
4. $\frac{1}{\underset{1}{\cancel{2}}} \times \overset{6}{\cancel{12}} \times 18 = 108$ sq. ft.

14A

Study Exercise Five (Frame 17)

1. $\frac{8}{2}(10 + 14) = \frac{8}{2}(24) = 4 \times 24 = 96$ sq. in.
2. $\frac{1}{2} \times \frac{27}{4}\left(\frac{9}{2} + \frac{19}{2}\right) = \frac{1}{\underset{1}{\cancel{2}}} \times \frac{27}{4} \times \frac{\overset{\overset{7}{\cancel{14}}}{\cancel{28}}}{\underset{1}{\cancel{2}}} = \frac{189}{4} = 47\frac{1}{4}$ sq. ft.
3. $\frac{1}{2} \times 31(20 + 14) = \frac{1}{\underset{1}{\cancel{2}}} \times 31 \times \overset{17}{\cancel{34}} = 527$ sq. yd.

SOLUTIONS TO STUDY EXERCISES, CONTD.

Study Exercise Five (Frame 17, contd.)

4. $\frac{1}{2} \times 9(14 + 24) = \frac{1}{\cancel{2}_{1}} \times 9 \times \cancel{38}^{19} = 171$ sq. ft.

1 coat must cover 171 sq. ft.
2 coats must cover $171 \times 2 = 342$ sq. ft.
We don't have enought paint.

17A

Study Exercise Six (Frame 20)

1. $3.14(10)^2 = 3.14 \times 100 = 314$ sq. in.
2. $3.14(7.2)^2 = 3.14 \times 51.84 = 162.7776$ sq. mi.
3. $3.14(4)^2 = 3.14 \times 16 = 50.24$ sq. yd.
4. 4.6 yd. $\div 2 = 2.3$ yd.
$3.14(2.3)^2 = 3.14 \times 5.29 = 16.6106$ sq. yd.
5. 42 ft. $\div 2 = 21$ ft.
$3.14(21)^2 = 3.14 \times 441 = 1384.74$ sq. ft.
6. 14 inch pan; $14 \div 2 = 7$ in. radius
$3.14(7)^2 = 3.14 \times 49 = 153.86$ sq. in.
12 in. pan, $12 \div 2 = 6$ in. radius
$3.14(6)^2 = 3.14 \times 36 = 113.04$ sq. in.
$153.86 - 113.04 = 40.82$ sq. in.

Study Exercise Seven (Frame 24)

1. $A = 2lw + 2lh + 2wh$ or $2(lw + lh + wh)$
$A = 2(3 \times 2 + 3 \times 5 + 2 \times 5) = 2(6 + 15 + 10)$
$= 2 \times 31 = 62$ sq. ft.
2. $A = 2(1.6 \times 2.4 + 1.6 \times 3.1 + 2.4 \times 3.1)$
$= 2(3.84 + 4.96 + 7.44) = 2 \times 16.24 = 32.48$ sq. ft.
3. $A = 2(2 \times 7 + 2 \times 10 + 7 \times 10) = 2(14 + 20 + 70)$
$= 2 \times 104 = 208$ sq. in.
4. $A = 6s^2 = 6(4.7)^2 = 6(22.09) = 132.54$ sq. in.

24A

Study Exercise Eight (Frame 29)

Lateral Area
1. $A = 2\pi rh = 2 \times 3.14 \times 4 \times 10 = 251.2$ sq. in.
2. $A = 2\pi rh = 2 \times 3.14 \times 5 \times 3 = 94.2$ sq. in.
Total Area
3. $A = 2\pi rh + 2\pi r^2$
$2\pi r^2 = 2 \times 3.14 \times (4)^2 = 2 \times 3.14 \times 16 = 100.48$ sq. in.
$A = 2\pi rh + 2\pi r^2 = 251.2 + 100.48 = 351.68$ sq. in.
4. $A = 2\pi rh + 2\pi r^2$
$2\pi r^2 = 2 \times 3.14 \times (5)^2 = 2 \times 3.14 \times 25 = 157$ sq. in.
$A = 2\pi rh + 2\pi r^2 = 94.2 + 157 = 251.2$ sq. in.

unit

38

Volume

Objectives:

By the end of this unit you should:

1. be able to find the volumes of:
   a. rectangular solids
   b. cubes
   c. right circular cylinders
   d. spheres
   e. right circular cones

2. know the volume formulas for the above-named figures. (1)

Volume

Volume is the number of cubic units contained in a given space.

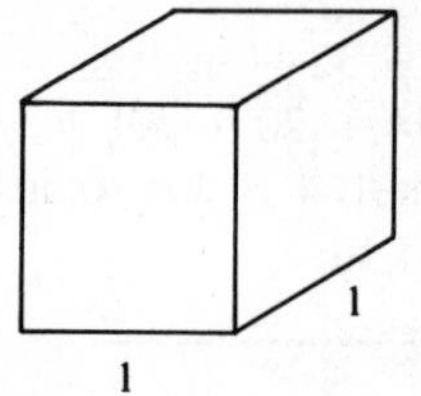

Fig. 1—1 cubic unit

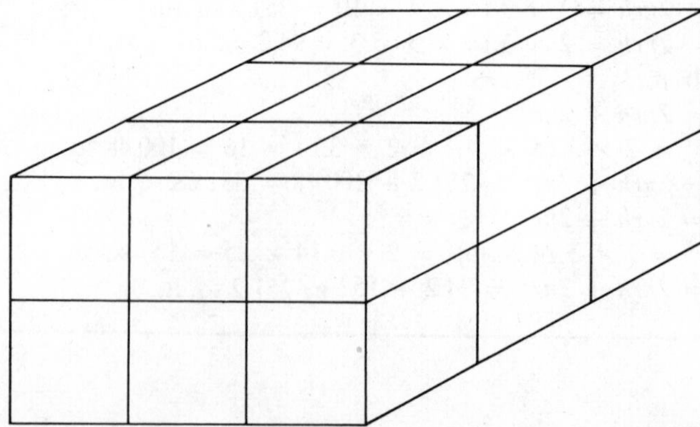

Fig. 2—9 cubic units on top + 9 cubic units on bottom = 18 cubic units total 

Volume of a Rectangular Solid

Formula: $V = lwh$

Example 1: Find the volume of a rectangular solid 3 ft. long, 2 ft. wide, 4 ft. high.

Line (a) 2 ft. × 3 ft. = 6 sq. ft. at the bottom
Line (b) 4 ft. × 6 sq. ft. = 24 cu. ft. total
Line (c) $V = lwh$
Line (d) $V = 3 \times 2 \times 4 = 24$ cu. ft.

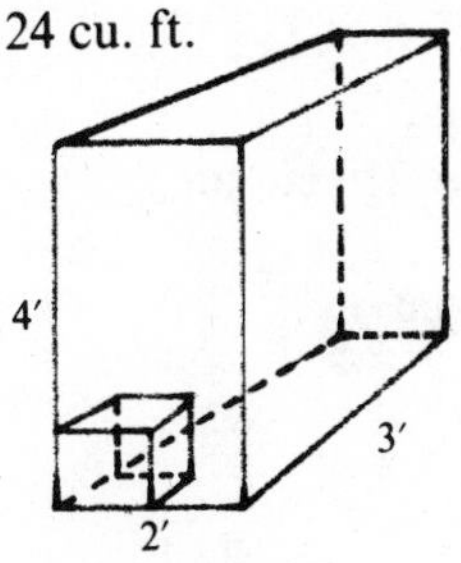

4′ 3′ 2′

Fig. 1

3

Study Exercise One

Find the volume of rectangular solids with the following dimensions:

	1.	2.	3.	4.
Length	14 in.	4 in.	2 ft. 8 in.	8.2 ft.
Width	7 in.	2 in.	3 ft. 6 in.	1.6 ft.
Height	8 in.	8 in.	4 ft. 2 in.	1.2 ft.

5. How many cubic yards of dirt are needed to fill a hole 15 ft. long, 10 ft. wide, 6 ft. deep?
6. A swimming pool 32 ft. long and 20 ft. wide is filled to an average depth of 5 ft.
   a. What is the weight of the water? (1 cu. ft. water weighs about $62\frac{1}{2}$ lb.)
   b. How many gallons of water are in the pool? (1 cu. ft. holds $7\frac{1}{2}$ gallons)

Volume of a Cube

Formula: $V = e^3$

Example 1: Find the volume of a cube whose edge is 6 in.

Solution:

Line (a) $l = w = h = e$

Line (b) $V = l\ w\ h$

Line (c) $V = e \times e \times e = e^3$

Line (d) $V = e^3$

Line (e) $V = (6)^3 = 6 \times 6 \times 6 = 36 \times 6 = 216$ cu. in.

or

Line (f) $V = (6)^3$ Using Table III, $V = 216$ cu. in.

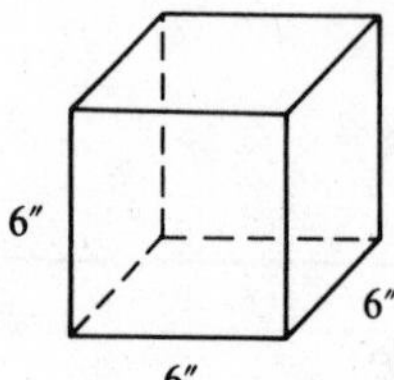

5

Study Exercise Two

Find the volumes of cubes whose sides measure:

1. 7 in.
2. 3.2 in.
3. $\frac{1}{2}$ ft.
4. 4 ft. 4 in.
5. $6\frac{1}{2}$ yd.
6. If ice weighs 57 lb. per cu. ft., what is the weight of a block in the form of a cube with edges 1.8 ft. (to nearest pound)?

6

Volume of a Right Circular Cylinder

Formula: $V = \pi r^2 h$

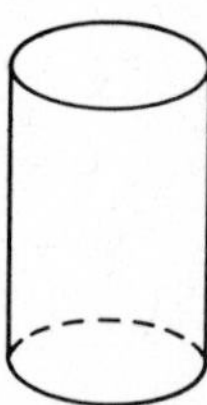

Fig. 1

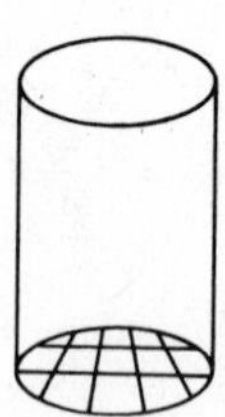

Fig. 2
Number of squares in bottom $= \pi r^2$

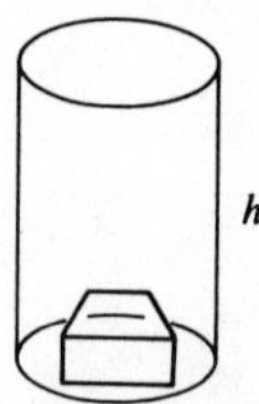

Fig. 3
Number of cubes $\pi r^2 \times h$

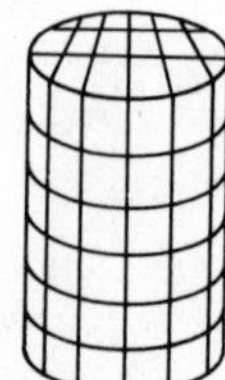

Fig. 4
$V = \pi r^2 h$

7

Example 1: Find the volume of a cylinder 30 ft. high having a radius of 10 ft.

Solution:

Line (a) $V = \pi r^2 h$

Line (b) $V = 3.14 \times (10)^2 \times 30$

Line (c) $V = 3.14 \times 100 \times 30$

Line (d) $V = 9420$ cu. ft.

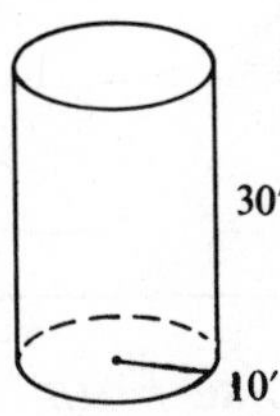

Example 2: Find the volume of a cylinder 30 ft. high having a diameter of 20 ft.

Solution:

Line (a) 20 ft. ÷ 2 = 10 ft.

Line (b) This is now the same as Example 1.

(8)

Study Exercise Three

Find the volumes of cylinders with the following dimensions:

	1.	**2.**	**3.**	**4.**
Radius	8 in.	6 ft.	$4\frac{1}{2}$ ft.	3.1 yd.
Height	10 in.	30 ft.	10 ft.	21 ft.

5. How many cubic yards of dirt must be removed to dig a well 4 ft. in diameter and 100 ft. deep? (To nearest whole cubic yard)
6. At $20 per cu. yd., how much will the excavation in problem 5 cost?

Volume of a Sphere

Formula: $V = \frac{4}{3}\pi r^3$

Example 1: Find the volume of a sphere with radius of 4 in. to nearest cubic inch.

Solution:

Line (a) $V = \frac{4}{3}\pi r^3$

Line (b) $V = \frac{4}{3} \times 3.14 \times (4)^3$

(Frame 10, contd.)

Line (c) $V = \frac{4}{3} \times 3.14 \times 64 = \frac{803.84}{3} \approx 268$

Line (d) $V = 268$ cu. in.

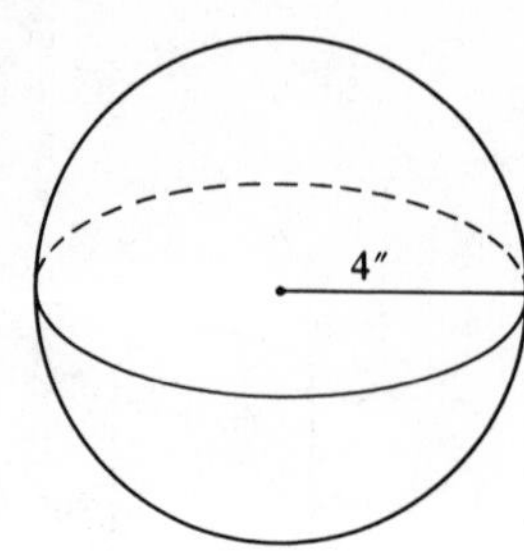

(10)

Study Exercise Four

1. Find the volume of a sphere with radius 6 ft. to the nearest cu. ft.
2. Find the volume of a sphere with a diameter of 36 in. to the nearest cu. in.

(11)

Volume of a Right Circular Cone

Formula: $V = \frac{1}{3}\pi r^2 h$

Example 1: Find the volume of a right circular cone whose radius is 3 ft. and whose height is 10 ft.

Solution:

Line (a) $V = \frac{1}{3}\pi r^2 h$

Line (b) $V = \frac{1}{3} \times 3.14 \times (3)^2 \times 10$

Line (c) $V = \frac{1}{\cancel{3}_{1}} \times 3.14 \times \overset{3}{\cancel{9}} \times 10 = 94.2$ cu. ft.

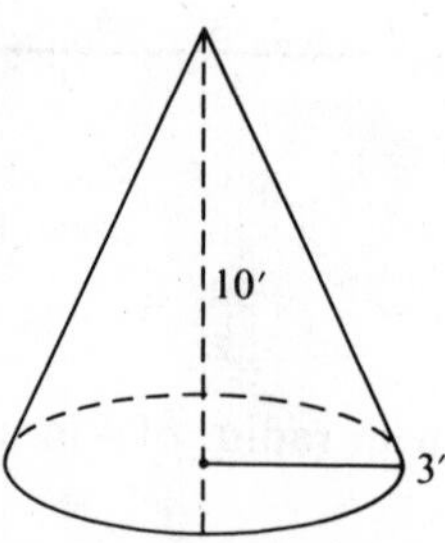

(12)

Study Exercise Five

1. Find the volume of a right circular cone whose radius is 5 in. and whose height is 6 in. to the nearest cu. in.
2. How many cu. yd. of sand are in a conical pile of sand 21 ft. in diameter and 9 ft. high? (To nearest tenth of a cubic yd.)
3. How many tons of coal are in a conical pile of coal 22 ft. in diameter and 10 ft. high if one ton of coal occupies 35 cu. ft.? (To nearest tenth of a ton)

REVIEW EXERCISES

A. Find the volume of:

1. a rectangular solid 7 ft. by 8 ft. by 10 ft.
2. a cube whose edge is 2 ft. 3 in.
3. a right circular cylinder with radius 10 in. and height 20 in.
4. a sphere with radius 1 ft. (nearest hundredth of a cu. ft.)
5. a right circular cone with radius 5 in. and height 30 in.

B. Use Tables I and III whenever possible.

6. A circular swimming pool has a diameter of 15 ft. and a depth of 42 in. $\left(\text{Use } \pi = \frac{22}{7}\right)$
   a. How many cubic feet of water does it hold? (to nearest cubic foot).
   b. How many gallons does it hold?
   c. What is the weight of the water in it?
   d. If the filter on the pool pumps 1620 gallons per hour, how long will it take to completely recirculate the entire amount of water in the pool? (To nearest whole hour.)
7. A classroom is 30 ft. long, 20 ft. wide, and 16 ft. high.
   a. Using 15 sq. ft. of floor space per pupil, how many pupils should this room hold?
   b. How many cubic feet of air does this allow for each pupil?
8. How much greater is the volume of a 10-inch cube than a 9-inch cube?
9. How many gallons of water can be stored in a cylindrical drum with a diameter of 4 ft. and height of 6 ft.?
10. How much rubber is used in making a solid toy rubber ball with a diameter of 2 in.? (To nearest whole cubic inch.)
11. An ice cream cone is filled with ice cream. If the cone has a diameter of 2 in. and a height of 6 in., how much ice cream does it contain? (14)

SOLUTIONS TO REVIEW EXERCISES

A.

1. $V = 7 \times 8 \times 10 = 560$ cu. ft.

2. $V = (2 \text{ ft. } 3 \text{ in.})^3 = \left(2\frac{3}{12} \text{ ft.}\right)^3 = \left(2\frac{1}{4} \text{ ft.}\right)^3 = \left(\frac{9}{4} \text{ ft.}\right)^3$

   $= \frac{729}{64}$ cu. ft. $= 11\frac{25}{64}$ cu. ft.

3. $V = 3.14 \times (10)^2 \times 20 = 3.14 \times 100 \times 20 = 6{,}280$ cu. in.

4. $V = \frac{4}{3} \times 3.14 \times (1)^3 = \frac{4}{3} \times 3.14 \times 1 = \frac{12.56}{3} \approx 4.19$ cu. ft.

5. $V = \frac{1}{3} \times 3.14 \times (5)^2 \times 30 = \frac{1}{\cancel{3}_{1}} \times 3.14 \times 25 \times \cancel{30}^{10} = 785$ cu. in.

B.

6. a. 42 in. $= 3$ ft. 6 in. $= 3\frac{1}{2}$ ft. $= \frac{7}{2}$ ft. height

   15 ft. $\div 2 = 7\frac{1}{2}$ ft. $= \frac{15}{2}$ ft. radius

   $$V = \frac{22}{7} \times \left(\frac{15}{2}\right)^2 \times \frac{7}{2} = \frac{\cancel{22}^{11}}{\cancel{7}_{1}} \times \frac{225}{4} \times \frac{\cancel{7}^{1}}{\cancel{2}_{1}} = \frac{2475}{4} \approx 619 \text{ cu. ft.}$$

SOLUTIONS TO REVIEW EXERCISES, CONTD.

(Frame 15, contd.)

b. From Table I, 1 cubic foot $= 7\frac{1}{2}$ gallons

$619 \times 7.5 = 4{,}642.5$ gallons

c. From Table I, 1 cubic foot of fresh water $= 62\frac{1}{2}$ pounds

$619 \times 62.5 = 38{,}687.5$ pounds

d.

```
       2.8   ≈ 3 hours
1620 | 4642.5
       3240
       1402 5
       1296 0
        106 5
```

7. a. $A = 30 \times 20 = 600$ sq. ft.;

$600 \div 15 = 40$ pupils in the room

b. $V = 30 \times 20 \times 16 = 9600$ cu. ft.

```
     240        240 cu. ft. of air per pupil
40 | 9600
     86
      160
      160
        0
```

8. $V = (10)^3 = 1000$ cu. in.
$V = (9)^3 = 729$ cu. in.
1000 cu. in. − 729 cu. in. = 271 cu. in. more

9. 4 ft. ÷ 2 = 2 ft. radius
$V = 3.14 \times (2)^2 \times 6 = 3.14 \times 4 \times 6 = 75.36$ cu. ft.

From Table I, 1 cu. ft. $= 7\frac{1}{2}$ gal.

$75.36 \times 7.5 = 565.2$ gal.

10. 2 in. ÷ 2 = 1 in. radius

$V = \frac{4}{3} \times 3.14 \times (1)^3 = \frac{4}{3} \times 3.14 \times 1 \approx 4$ cu. in.

11. 2 in. ÷ 2 = 1 in. radius

$V = \frac{1}{3} \times 3.14 \times (1)^2 \times 6 = \frac{1}{\cancel{3}_1} \times 3.14 \times 1 \times \cancel{6}^{2} = 3.14 \times 2 = 6.28$ cu. in.

(15)

SUPPLEMENTARY PROBLEMS

1. To the nearest tenth of a yard, how many cubic yards of dirt are needed to fill a hole 14 ft. long, 12 ft. wide and 6 ft. deep?
2. A swimming pool 60 ft. long and 30 ft. wide is filled to an average depth of 4 ft.
   a. What is the weight of the water?
   b. How many gallons are in the pool?
3. How many cubic feet of sand are in a conical pile of sand 15 ft. in diameter and 9 ft. high?
4. How many cubic yards of dirt must be removed to dig a well 6 ft. in diameter and 75 ft. deep?
5. At $22 per cubic yard, how much will the excavation in problem 4 cost?

SUPPLEMENTARY PROBLEMS, CONTD.

6. To the nearest whole cubic yard, how many cubic yards of concrete will be needed to pave an area 21 ft. × 36 ft. to an average depth of 4 inches?
7. Mr. Jones wishes to place a sand bed 4 inches high under a circular pool with a diameter of 18 ft. To the nearest whole cubic yard, how much sand must he buy?
8. Mr. Smith builds children's sand boxes 4 ft. by 4 ft. by 1 ft. high. He wishes to give each customer enough sand to fill the box 4 inches. If he plans to sell 50 boxes, how much sand should he have? (To nearest whole cubic yard.)
9. An aquarium measures 3 ft. by 6 ft. by 2 ft. How many gallons of water does it contain?
10. If the aquarium in problem 9 is used for ocean fish, what will be the weight of the water?
11. A tank is to be designed to hold 100 gal. of water. How many cubic feet must it contain?
12. If the tank in problem eleven is to be a rectangular solid with the bottom a 2 ft. square, how high should it be? (To the nearest tenth of a foot)
13. If the tank in problem eleven is to be a cylinder with a circular bottom of 2 ft. in diameter, how high should it be? (To the nearest tenth of a foot)
14. If the tank in problem eleven is to be a cone with a circular bottom of 2 ft. in diameter, how high should it be? (To the nearest tenth of a foot)

SOLUTIONS TO STUDY EXERCISES

Study Exercise One (Frame 4)

1. $V = 14 \times 7 \times 8 = 784$ cu. in.
2. $V = 4 \times 2 \times 8 = 64$ cu. in.
3. 2 ft. 8 in. $= 2\frac{8}{12}$ ft. $= 2\frac{2}{3}$ ft. $= \frac{8}{3}$ ft.;

   3 ft. 6 in. $= 3\frac{1}{2}$ ft. $= \frac{7}{2}$ ft.;

   4 ft. 2 in. $= 4\frac{2}{12}$ ft. $= 4\frac{1}{6}$ ft. $= \frac{25}{6}$ ft.

   $V = \frac{\overset{2}{\cancel{\overset{4}{\cancel{8}}}}}{3} \times \frac{7}{\underset{1}{\cancel{2}}} \times \frac{25}{\underset{3}{\cancel{6}}} = \frac{350}{9} = 38\frac{8}{9}$ cu. ft.
4. $V = 8.2 \times 1.6 \times 1.2 = 15.744$ cu. ft.
5. $V = 15 \times 10 \times 6 = 900$ cu. ft.; $900 \div 27 = 33\frac{1}{3}$ cu. yd.
6. $V = 32 \times 20 \times 5 = 3200$ cu. ft.
   a. $3200 \times 62.5 = 200{,}000$ pounds
   b. $3200 \times 7.5 = 24{,}000$ gallons

(4A)

Study Exercise Two (Frame 6)

1. $V = (7 \text{ in.})^3 = 343$ cu. in. (Use Table III)
2. $V = (3.2 \text{ in.})^3 = 32.768$ cu. in. (Use Table III)
3. $V = \left(\frac{1}{2}\right)^3 = \frac{1}{2} \times \frac{1}{2} \times \frac{1}{2} = \frac{1}{8}$ cu. ft.

SOLUTIONS TO STUDY EXERCISES, CONTD.

Study Exercise Two (Frame 6, contd.)

4. $4 \text{ ft. } 4 \text{ in.} = 4\frac{4}{12} \text{ ft.} = 4\frac{1}{3} \text{ ft.} = \frac{13}{3} \text{ ft.}$

$V = \left(\frac{13}{3}\right)^3 = \frac{2197}{27} = 81\frac{10}{27} \text{ cu. ft.}$

5. $6\frac{1}{2}$ yd. = 6.5 yd.

$V = (6.5 \text{ yd.})^3 = 274.625$ cu. yd. (Use Table III)

6. $V = (1.8 \text{ ft.})^3 = 5.832$ cu. ft. (Use Table III)

$5.832 \times 57 = 332.424 \approx 332$ lb.

Study Exercise Three (Frame 9)

1. $V = 3.14 \times (8)^2 \times 10 = 3.14 \times 64 \times 10 = 2009.6$ cu. in.

2. $V = 3.14 \times (6)^2 \times 30 = 3.14 \times 36 \times 30 = 3391.20$ cu. ft.

3. $V = 3.14 \times (4.5)^2 \times 10 = 3.14 \times 20.25 \times 10 = 635.85$ cu. ft.

4. 21 ft. ÷ 3 = 7 yd.

$V = 3.14 \times (3.1)^2 \times 7 = 3.14 \times 9.61 \times 7 = 211.2278$ cu. yd.

5. 4 ft. ÷ 2 = 2 ft.

$V = 3.14 \times (2)^2 \times 100 = 3.14 \times 4 \times 100 = 1256$ cu. ft.

$1256 \div 27 \approx 47$ cu. yd.

6. $47 \times \$20 = \$940$

9A

Study Exercise Four (Frame 11)

1. $V = \frac{4}{3}\pi r^3$

$V = \frac{4}{3} \times 3.14 \times (6)^3 = \frac{4}{\cancel{3}_1} \times 3.14 \times \overset{72}{\cancel{216}} = 904.32 \approx 904$ cu. ft.

2. $V = \frac{4}{3} \times 3.14 \times (18)^3 = \frac{4}{\cancel{3}_1} \times 3.14 \times \overset{1944}{\cancel{5832}} = 24416.64$

$\approx 24{,}417$ cu. in.

11A

Study Exercise Five (Frame 13)

1. $V = \frac{1}{3}\pi r^2 h$

$V = \frac{1}{\cancel{3}_1} \times 3.14 \times 25 \times \overset{2}{\cancel{6}} = 3.14 \times 50 = 157$ cu. in.

2. 21 ft. ÷ 3 = 7 yd.; 9 ft. ÷ 3 = 3 yd.

$V = \frac{1}{3} \times 3.14 \times \left(\frac{7}{2}\right)^2 \times 3 = \frac{1}{\cancel{3}_1} \times \overset{1.57}{\cancel{3.14}} \times \frac{49}{\cancel{4}_2} \times \frac{\overset{1}{\cancel{3}}}{1} = \frac{76.93}{2} = 38.465$

$V \approx 38.5$ cu. yd.

3. $V = \frac{1}{3} \times 3.14 \times 121 \times 10 = 1266.47$ cu. ft.

$1266.47 \div 35 \approx 36.2$ tons

13A

Practice Test—Measurement—Units 34–38

1. Change 6 days to hours.
2. Change 18 gallons to quarts.
3. Change 31 feet to the nearest tenth of a yard.
4. How many pounds is 12 kilograms?
5. Add and simplify:
   5 hr. 20 min. 6 sec.
   3 hr. 42 min. 38 sec.
   2 hr. 15 min. 22 sec.
6. Subtract:
   5 lb. 7 oz.
   2 lb. 12 oz.
7. Multiply and simplify:
   3 yd. 2 ft. 2 in.
   8
8. Divide:
   6)39 lb. 3 oz.
9. Find the perimeter of a rectangle whose width is 9 ft. and whose length is 4 yd. Express your answer in yards.
10. A polygon with 5 sides is called a ___________.
11. An octagon has ___________ sides.
12. Find the perimeter of a square with a side of 3 ft. 4 in.
13. How much molding would be needed to go around the edge of a circular table if the diameter of the table is 4 ft.? (To nearest whole foot)
14. $\pi = \frac{22}{7}$ (True or False)
15. What distance do you travel on one turn of a Ferris wheel if you sit 10 ft. from the center.
16. The area of the interior of any plane figure is given by the number of ___________ units it contains.
17. Find the area of a rectangle whose width is 9 ft. and whose length is 12 ft., expressing your answer in square yards.
18. Find the cost of cementing a sidewalk 3 ft. by 60 ft. at $.50 per square foot.
19. Use Table III to find the area of a square with side 97 ft.
20. State the formula for the area of a parallelogram.
21. Find the area of a trapezoid with bases 8 ft. and 9 ft. and altitude of 16 ft.
22. To the nearest tenth, how many square yards of material are contained in a circular tablecloth with a radius of 4 ft.?
23. a. Find the surface area of a cereal box 8 in. by 6 in. by 10 in.
    b. What volume does this box contain?
24. A vegetable can is to be made with a height of 10 in. and a radius of 3 in.
    a. What area will the label of the can cover?
    b. What total area of metal will be necessary?
    c. What will the volume of the can be?
25. A solid rubber ball with a 3 in. radius will contain what volume of rubber?

Answers—Practice Test—Measurement

1. 144 hr.
2. 72 qt.
3. 10.3 yd.
4. 26.4 lbs.
5. 11 hr. 18 min. 6 sec.
6. 2 lb. 11 oz.
7. 29 yd. 2 ft. 4 in.
8. 6 lb. $8\frac{1}{2}$ oz.
9. 14 yd.
10. pentagon
11. 8
12. 13 ft. 4 in.
13. 13 ft.
14. false
15. 62.8 ft.
16. square
17. 12 sq. yd.
18. $90
19. 9,409 sq. ft.
20. $A = bh$
21. 136 sq. ft.
22. 5.6 sq. yd.
23. a. 376 sq. in.
    b. 480 cu. in.
24. a. 188.4 sq. in.
    b. 244.92 sq. in.
    c. 282.6 cu. in.
25. 113.04 cu. in.

Appendix

Table I
TABLES OF MEASURE

Measure of Length

1 foot (ft.)	=	12 inches (in.)
1 yard (yd.)	=	3 ft. (ft.)
	=	36 inches (in.)
1 rod (rd.)	=	$16\frac{1}{2}$ feet (ft.)
	=	$5\frac{1}{2}$ yards (yd.)
1 statute mile (stat. mi.)	=	5,280 feet (ft.)
	=	1,760 yards (yd.)
	=	320 rods (rd.)
	=	0.87 nautical mile
1 nautical mile (naut. mi.)	=	6,080 feet
	=	1.15 statute miles
1 fathom (fath.)	=	6 feet (ft.)

Measure of Area

1 square foot (sq. ft.)	=	144 square inches (sq. in.)
1 square yard (sq. yd.)	=	9 square feet (sq. ft.)
1 square rod (sq. rd.)	=	30.25 square yards (sq. yd.)
1 acre	=	160 square rods (sq. rd.)
	=	4,840 square yards (sq. yd.)
	=	43,560 square feet (sq. ft.)
1 square mile (sq. mi.)	=	640 acres

Measure of Volume

1 cubic foot (cu. ft.)	=	1,728 cubic inches (cu. in.)
1 cubic yard (cu. yd.)	=	27 cubic feet (cu. ft.)

Dry Measure

1 quart (qt.)	=	2 pints (pt.)
1 peck (pk.)	=	8 quarts (qt.)
1 bushel (bu.)	=	4 pecks (pk.)

Measure of Time

1 minute (min.)	=	60 seconds (sec.)
1 hour (hr.)	=	60 minutes (min.)
1 day (da.)	=	24 hours (hr.)
1 week (wk.)	=	7 days (da.)
1 year (yr.)	=	12 months (mo.)
	=	52 weeks (wk.)
	=	365 days (da.)

Metric Measures of Length

10 millimeters (mm.)	=	1 centimeter (cm.)
10 centimeters (cm.)	=	1 decimeter (dm.)
10 decimeters (dm.)	=	1 meter (m.)
10 meters (m.)	=	1 dekameter (dkm.)
10 dekameters (dkm.)	=	1 hectometer (hm.)
10 hectometers (hm.)	=	1 kilometer (km.)

Metric—English Equivalents

1 meter	=	39.37 inches
	=	3.28 feet
	=	1.09 yards
1 centimeter	=	.39 inches
1 millimeter	=	.04 inches
1 kilometer	=	.62 miles
1 liter	=	1.06 liquid quarts
1 liter	=	.91 dry quart
1 gram	=	.04 ounce
1 kilogram	=	2.2 pounds
1 metric ton	=	2,204.62 pounds
1 inch	=	25.4 millimeters
1 foot	=	.3 meter
1 yard	=	.91 meter
1 mile	=	1.61 kilometers
1 liquid quart	=	.95 liter
1 dry quart	=	1.1 liters
1 ounce	=	28.35 grams
1 pound	=	.45 kilogram
1 short ton	=	.91 metric ton

Liquid Measure

1 tablespoon (tbsp.)	=	3 teaspoons (tsp.)
	=	$\frac{1}{2}$ ounce (oz.)
1 cup (c.)	=	16 tablespoons (tbsp.)
	=	8 ounces (oz.)
1 pint (pt.)	=	2 cups (c.)
	=	16 ounces (oz.)
1 quart (qt.)	=	2 pints (pt.)
1 gallon (gal.)	=	4 quarts (qt.)

Measure of Weight—Avoirdupois

1 pound (lb.)	=	16 ounces (oz.)
1 short ton (sh. tn. or T.)	=	2,000 pounds (lb.)
1 long ton (1. ton)	=	2,240 pounds (lb.)

Volume, Capacity, and Weight Equivalents

1 gallon (gal.)	=	231 cubic inches (cu. in.)
1 cubic foot (cu. ft.)	=	$7\frac{1}{2}$ gallons (gal.)

1 cu. ft. of fresh water weighs $62\frac{1}{2}$ pounds (lb.)

1 cu. ft. of sea water weighs 64 pounds (lb.)

Angles and Arcs

1 circle	=	360 degrees (°)
1 degree (°)	=	60 minutes (′)
1 minute (′)	=	60 seconds (″)

Measure of Speed

1 knot	=	1 nautical m.p.h.

TABLE II—COMPOUND INTEREST TABLE SHOWING HOW MUCH $1 WILL AMOUNT TO AT VARIOUS RATES

Yrs/Periods	$\frac{1}{2}$%	1%	$1\frac{1}{2}$%	2%	3%	4%	5%	6%
1	1.0050	1.0100	1.0150	1.0200	1.0300	1.0400	1.0500	1.0600
2	1.0100	1.0201	1.0302	1.0404	1.0609	1.0816	1.1025	1.1236
3	1.0151	1.0303	1.0457	1.0612	1.0927	1.1249	1.1576	1.1910
4	1.0202	1.0406	1.0614	1.0824	1.1255	1.1699	1.2155	1.2625
5	1.0253	1.0510	1.0773	1.1041	1.1593	1.2167	1.2763	1.3382
6	1.0304	1.0615	1.0934	1.1262	1.1941	1.2653	1.3401	1.4185
7	1.0355	1.0721	1.1098	1.1487	1.2299	1.3159	1.4071	1.5036
8	1.0407	1.0829	1.1265	1.1717	1.2668	1.3686	1.4775	1.5939
9	1.0459	1.0937	1.1434	1.1951	1.3048	1.4233	1.5513	1.6895
10	1.0511	1.1046	1.1605	1.2190	1.3439	1.4802	1.6289	1.7909
11	1.0564	1.1157	1.1779	1.2434	1.3842	1.5395	1.7103	1.8983
12	1.0617	1.1268	1.1956	1.2682	1.4258	1.6010	1.7959	2.0122
13	1.0670	1.1381	1.2136	1.2936	1.4685	1.6651	1.8856	2.1329
14	1.0723	1.1495	1.2318	1,3195	1.5126	1.7317	1.9799	2.2609
15	1.0777	1.1610	1.2502	1.3459	1.5580	1.8009	2.0789	2.3966
16	1.0831	1.1726	1.2690	1.3728	1.6047	1.8730	2.1829	2.5404
17	1.0885	1.1843	1.2880	1.4002	1.6528	1.9479	2.2920	2.6928
18	1.0939	1.1961	1.3073	1.4282	1.7024	2.0258	2.4066	2.8543
19	1.0994	1.2081	1.3270	1.4568	1.7535	2.1068	2.5270	3.0256
20	1.1049	1.2202	1.3469	1.4859	1.8061	2.1911	2.6533	3.2071
25	1.1328	1.2824	1.4509	1.6406	2.0938	2.6658	3.3864	4.2919
50	1.2832	1.6446	2.1052	2.6916	4.3839	7.1067	11.4674	18.4202

TABLE III—POWERS AND ROOTS

n	n^2	n^3	$\sqrt{n}$	$\sqrt[3]{n}$
0	0	0	0.000	0.000
1	1	1	1.000	1.000
2	4	8	1.414	1.260
3	9	27	1.732	1.442
4	16	64	2.000	1.587
5	25	125	2.236	1.710
6	36	216	2.449	1.817
7	49	343	2.646	1.913
8	64	512	2.828	2.000
9	81	729	3.000	2.080
10	100	1 000	3.162	2.154
11	121	1 331	3.317	2.224
12	144	1 728	3.464	2.289
13	169	2 197	3.606	2.351
14	196	2 744	3.742	2.410
15	225	3 375	3.873	2.466
16	256	4 096	4.000	2.520
17	289	4 913	4.123	2.571
18	324	5 832	4.243	2.621
19	361	6 859	4.359	2.668
20	400	8 000	4.472	2.714
21	441	9 261	4.583	2.759
22	484	10 648	4.690	2.802
23	529	12 167	4.796	2.844
24	576	13 824	4.899	2.884
25	625	15 625	5.000	2.924
26	676	17 576	5.099	2.962
27	729	19 683	5.196	3.000
28	784	21 952	5.292	3.037
29	841	24 389	5.385	3.072
30	900	27 000	5.477	3.107
31	961	29 791	5.568	3.141
32	1 024	32 768	5.657	3.175
33	1 089	35 937	5.745	3.208
34	1 156	39 304	5.831	3.240
35	1 225	42 875	5.916	3.271
36	1 296	46 656	6.000	3.302
37	1 369	50 653	6.083	3.332
38	1 444	54 872	6.164	3.362
39	1 521	59 319	6.245	3.391
40	1 600	64 000	6.325	3.420
41	1 681	68 921	6.403	3.448
42	1 764	74 088	6.481	3.476
43	1 849	79 507	6.557	3.503
44	1 936	85 184	6.633	3.530
45	2 025	91 125	6.708	3.557
46	2 116	97 336	6.782	3.583
47	2 209	103 823	6.856	3.609
48	2 304	110 592	6.928	3.634
49	2 401	117 649	7.000	3.659

n	n^2	n^3	$\sqrt{n}$	$\sqrt[3]{n}$
50	2 500	125 000	7.071	3.684
51	2 601	132 651	7.141	3.708
52	2 704	140 608	7.211	3.733
53	2 809	148 877	7.280	3.756
54	2 916	157 464	7.348	3.780
55	3 025	166 375	7.416	3.803
56	3 136	175 616	7.483	3.826
57	3 249	185 193	7.550	3.849
58	3 364	195 112	7.616	3.871
59	3 481	205 379	7.681	3.893
60	3 600	216 000	7.746	3.915
61	3 721	226 981	7.810	3.936
62	3 844	238 328	7.874	3.958
63	3 969	250 047	7.937	3.979
64	4 096	262 144	8.000	4.000
65	4 225	274 625	8.062	4.021
66	4 356	287 496	8.124	4.041
67	4 489	300 763	8.185	4 062
68	4 624	314 432	8.246	4.082
69	4 761	328 509	8.307	4.102
70	4 900	343 000	8.367	4.121
71	5 041	357 911	8.426	4.141
72	5 184	373 248	8.485	4.160
73	5 329	389 017	8.544	4.179
74	5 476	405 224	8.602	4.198
75	5 625	421 875	8.660	4.217
76	5 776	438 976	8.718	4.236
77	5 929	456 533	8.775	4.254
78	6 084	474 552	8.832	4.273
79	6 241	493 039	8.888	4.291
80	6 400	512 000	8.944	4.309
81	6 561	531 441	9.000	4.327
82	6 724	551 368	9.055	4.344
83	6 889	571 787	9.110	4.362
84	7 056	592 704	9.165	4.380
85	7 225	614 125	9.220	4.397
86	7 396	636 056	9.274	4.414
87	7 569	658 503	9.327	4.431
88	7 744	681 472	9.381	4.448
89	7 921	704 969	9.434	4.465
90	8 100	729 000	9.487	4.481
91	8 281	753 571	9.539	4.498
92	8 464	778 688	9.592	4.514
93	8 649	804 357	9.644	4.531
94	8 836	830 584	9.695	4.547
95	9 025	857 375	9.747	4.563
96	9 216	884 736	9.798	4.579
97	9 409	912 673	9.849	4.595
98	9 604	941 192	9.899	4.610
99	9 801	970 299	9.950	4.626
100	10 000	1 000 000	10.000	4.642

Index